A PEARSON ORIGINAL

FOOLPROOF GUIDE TO STATISTICS USING IBM SPSS

2nd Edition

Dr Adelma Hills
University of Western Sydney

Pearson Originals is an imprint of Pearson Australia, a division of Pearson Australia Group Pty Ltd. It has been established to provide academics throughout Australia and New Zealand with fast and efficient access to the printing, warehousing and distribution services of Australia's leading educational publisher, ensuring a smooth supply channel to your campus bookseller.

For more information about the **Pearson Originals** service, contact the Editorial Department, Pearson Australia, Unit 4, Level 3, 14 Aquatic Drive, Frenchs Forest, New South Wales, 2086. Telephone: 02 9454 2200.

Managing Editor: Jill Gillies
Senior Consultant: Danielle Woods
Project Manager: Melissa Faulkner
Production Controller: Barbara Honor

Printed and bound in Australia by Pegasus Media & Logistics

ISBN 978 1 442 549821

Pearson Australia
Unit 4, Level 3,
14 Aquatic Drive
Frenchs Forest NSW 2086
www.pearson.com.au

PEARSON ORIGINALS
An imprint of Pearson Australia

What is PASW 18?

PASW 18 is Version 18 of what used to be known as SPSS (Statistical Package for the Social Sciences) until the SPSS company was taken over by IBM in 2009. Version 19 of the software, released late in 2010, returns to the SPSS brand name as IBM SPSS 19.

About this book

The name SPSS will be used throughout this book, which is intended to provide a foundation for understanding statistics and the use of SPSS. It has been developed primarily for students of psychology, but has application to other disciplines where quantitative research is used (provided the reader can stand having examples drawn from psychology).

New in this edition:

- *How to* boxes for SPSS procedures, updated for PASW 18
- A List of *How To* Boxes following the Table of Contents
- The new style for nonparametric tests shown in Appendix 2
- Sample Results sections updated consistent with the new edition of the *APA Manual*[1]
- A new section on overlapping confidence intervals and statistical significance (p. 141)

Learning by doing

Each chapter contains a worked example, and students are advised to treat these as tutorial exercises, working through each one, exploring the other options available in each of the SPSS procedures, examining the output that is produced, and comparing it to the extracts provided in the chapter.

The shaded *How to* boxes show the steps needed to perform various operations and analyses, and pages illustrating output from analyses are identified by shaded bars down the outside margin.

Occasionally website addresses are provided. While all were active at the time of publication, some may become inactive over time.

Online datasets

Several chapters use quite large datasets. These can be downloaded from a link on the Pearson website. Go to www.pearson.com.au/9781442549821. Scroll down the page. You will see the link in blue letters, 'Student downloads'. Click on the link to download the datasets.

---

[1] American Psychological Association (2010). *Publication manual of the American Psychological Association* (6th ed.). Washington, DC: American Psychological Association.
(The abbreviation *APA Manual* is used throughout this book.)

Table of Contents

List of How to Boxes

1 INTRODUCTION

About this book

The intention with this guide is to provide a foundation from which readers can develop their understanding of statistics, and their skill in using SPSS. It is not intended as an exhaustive text on either. The domain of statistics, in its full complexity, is vast. Many modern text books aim to be relatively comprehensive in their coverage, but in doing so, they can overwhelm the limited information processing capabilities of the poor human user—particularly the new and/or infrequent user. SPSS, like so many software programs, is similarly vast and can do pretty much "anything" statistical, but to make full use of the program you need an extensive knowledge of its capabilities and operations. Many of these capabilities go beyond the user-friendly, point-and-click procedures and require the use of syntax (instructions written in the code language of SPSS). A basic introduction to syntax is provided at various points in this guide.

The aim then, is to provide the essential information students and researchers need in order to have a fundamental grasp of a range of statistical techniques, and essential practical skills in using SPSS for Windows. This guide should always be accompanied by more comprehensive texts that can be consulted for more detailed or specialised information. For example, texts[1] by Field (2005), Green and Salkind (2005), Howell (2002), Pallant (2007), and Tabachnick and Fidell (2007) are all useful as alternative sources of information, or more comprehensive texts.

Important . . .

Throughout the guide procedures are described for getting things done in SPSS, but very often there are alternative ways of achieving the same outcome. You might find the alternatives easier or preferable to the one provided. Therefore, as you acquire confidence, it is essential that you explore all the menus and option buttons; and consult the extensive Help (F1 key) that is provided in SPSS, so you can develop a high level of expertise, solve any problems you encounter, and learn about advanced topics. SPSS for Windows is actually quite easy to use, but real skill only comes with practice. One learns by doing! Once you can "think" like SPSS you can wean yourself away from guidebooks and work out how to do things yourself.

It is also important to be aware that computers and their associated software programs are marvellous tools when they serve us; but they have a sinister potential when we serve them. Increasingly, software programs automate many operations, but this increases the helpless dependence of the human user. SPSS has innumerable default options, and this book relies on many of them, but as one becomes more experienced default options should always be investigated and a decision should be made as to whether or not they are appropriate. Users of SPSS can easily generate misleading results—or even utter garbage—if they mindlessly point-and-click without really knowing what they are doing.

Changing technology

Constant technological change is now a fact of life, and one of the most important skills to develop is the ability to adapt to change. Once you understand how computers and software programs work it is usually relatively easy to figure out changes and adapt to them yourself. SPSS upgrades frequently so be prepared for changes that may not yet have been updated in textbooks or guidebooks. Similarly, be

---

[1] Text books are updating all the time, so always check to see if a later edition is available.

prepared for differences if you are using an older version of SPSS, or even if you are using a student version. If you encounter any such changes from the material in this text, try to work them out yourself, rather than depending on an explanation from someone else.

Computers and computing

To use SPSS for Windows effectively you need to make sure you are familiar with the computer system you are using. Make this your first task. This includes knowing where to find and start programs such as SPSS, knowing where to find files on the hard disk or network, and knowing how to log on to networks if you are using one (e.g., in university computer labs).

One issue in the use of computers is document location. Windows programs control the storing of files (documents) in various folders, but the user may have no idea where they are on the hard disk. Some users, however, prefer to control this process themselves, keeping all their data in a personal folder separate from programs—usually the **My Documents** folder. The **Windows Explorer** program allows you to navigate through the folder structure, create your own folders, and exchange files among them, as well as easily copy to disks, CDs or DVDs, or USB flash drives. (Note that **Windows Explorer** is a Windows accessory program for managing files; it is not to be confused with the internet browser **Internet Explorer**.)

Users of Macintosh computers need to have a similar understanding of their system.

VERY important: Backing up your work

Always be sure to back up any work you want to keep, as a safeguard against system failure. There are automated ways of doing this, or in the interests of maintaining some sense of control users like me prefer to do this manually—it's easy once you develop the habit. At the end of each session (or during the session if you are particularly obsessive) you can use Windows Explorer to quickly copy files to other storage devices (e.g., USB flash drives).

Basic statistical concepts

The remainder of this book provides the fundamentals you need in order to use the various statistical techniques, and SPSS to perform the analyses. Early chapters in each section deal with the basic techniques and include manual calculations to enhance your understanding of what the analysis is actually doing. Later chapters deal with advanced techniques.

Before proceeding, make sure you are familiar with the following statistical concepts. These must be part of the general knowledge of any graduate in disciplines that involve quantitative research.

Variables

A **variable** is any attribute that can vary (e.g., age, gender, religion, self-esteem, air temperature, circle diameter, etc.), as opposed to a **constant** that always has the same value (e.g., pi, the ratio of the circumference of a circle to its diameter). Constants are relatively rare in the behavioural sciences, although a variable can be **held constant** in a research study by only considering one of its values (e.g., women in the case of gender). **Measures** of variables are the **data** of quantitative research.

Population

In research, interest is in understanding the nature of variables in a large group of people—the **population of interest** (e.g., the ages of first year students at a particular university, or in a particular state, or in the whole country). A summary measure of some population variable (e.g., average age) is known as a **parameter**.

Samples and random samples

Usually it is not practicable to measure every member of the population in order to determine a population parameter. Instead we take a **sample** or subset of people from the population. A summary measure from a sample is known as a **statistic**, and it is used as an **estimator** of the population parameter.

For the sample statistic to be a good estimator of the population parameter the sample must be **representative** of the population, and not biased in any way. Would a sample of students from an evening class be likely to give an unbiased estimate of the average age of first year psychology students at a university? I hope you can see that the answer is "no"—why?—because it is likely to be biased toward older people who are working during the day.

The best way to achieve a representative sample is via random sampling. A **random sample** is one in which every member of the population has an **equal chance** of being selected in the sample.

Descriptive and inferential statistics

There are two goals of analysis. The first is simply to **describe** the sample (or sometimes the population), using techniques that organise and summarise the data. This is the province of **descriptive** statistics.

A second goal is to use sample statistics to make **inferences** about population parameters, or to use relationships found in a sample to make **inferences** about the relationships that exist in the population. This is the province of **inferential** statistics.

Hypothesis

The basis of research is a clear and concise **research question**. Reference to extant theory then leads wherever possible to expression of the research question in terms of a **hypothesis**. This is a tentative statement about the relationship between two or more variables. The aim of the research is then to **test** the research hypothesis, by finding evidence that either **supports** or **refutes** it. (Note that a hypothesis is never "proved"[1]; it can only be supported.) Variables can be positively related (as one increases the other increases), negatively related (as one increases the other decreases), or unrelated (changes in one are not associated with any predictable change in the other). In **experimental** research **causal** relationships are investigated by looking for differences between groups treated differently. For example, if a negative causal relationship is hypothesised between test difficulty and performance, this can be tested by giving one group of research participants a difficult test and another group an easy test. If the group with the difficult test performs worse the hypothesis is supported.

The main types of variables in research

In research we hypothesise that one variable (e.g., X) will affect or be related to another (Y), that is, we hypothesise that X will **cause** changes in Y, or we hypothesise that X will simply be **related** to Y. To test this hypothesis we **manipulate** (in experimental research) or **select** (in correlational research) the levels of X and measure participants' responses on Y as a function of the different levels of X.

The variable that is manipulated or selected in the research design is known as the **independent variable**. It is usually abbreviated as the **IV**.

---

[1] We are very tiny creatures inhabiting a small planet, orbiting a nondescript star, one of billions and billions and billions in the known universe. We only have access to a part of reality and can't be absolutely certain of anything. It helps to remember this, and to continually question what we *think* we know.

The variable that is observed and measured in response to the independent variable is known as the **dependent variable**; that is, its values **depend** on the levels of the independent variable. **DV** is the abbreviation.

Note that in correlational research, which tests relationship not causation, it is actually more correct to refer to the IV as the **predictor**, and the DV as the **criterion**.

Control variables are those that are held constant in a study.

Internal and external validity

Be very careful of the term **validity**; it is a very important term that has somewhat different meanings in different contexts—be sure you always fully understand it. In the context of research, there are two main kinds of validity.

Internal validity refers to the accuracy of any conclusions we draw about the **causal** relationship between the IV and DV. It is threatened to the extent that the observed relationship can be attributed to other things. Consider the test difficulty example used previously. If all the participants working on the difficult test did so in a hot, confined room, while those working on the easy test were in a comfortable room, then performance differences might have been **caused** by the environmental conditions and not by the difficulty of the test.

External validity refers to the extent to which research conclusions can be generalised beyond the specific research context, that is, to different people, places, and times. For example, can research findings in Australia be generalised to Inuit people living in Alaska; or research findings with 20-year-olds be generalised to 80-year-olds? The answers depend on the circumstances of the particular research study.

Quantification of variables: Levels of measurement

In order to conduct **quantitative** research we need to be able to quantify or measure variables. There is, of course, another type of research, namely **qualitative** research that uses thematic analysis of qualitative data (e.g., interviews, texts, videos of behaviour), not statistical analysis of quantitative data; however, it is not the subject of this book. In quantifying variables there are four types of measurement scale that you must understand:

Nominal or categorical scales involve using numbers simply as **codes** for some attribute. For instance, we might code different religions as:

1= Protestant 2= Catholic 3=Baptist 4=Other

In nominal scales there is no mathematical relationship between the numbers (i.e., 1 is in no way bigger, smaller, better, more than, or less than 2). Nominal variables are often referred to as **categorical** variables, or even as qualitative variables.

Nominal variables that can have only **two** values (e.g., gender) are known as **dichotomous** variables.

Ordinal scales involve numbers that **do** indicate some mathematical **rank order**, but the intervals between ranks are not necessarily equal. For example, when asked to list six life goals in order of their importance to her, a participant produces this list:

1 Material Wealth
2 Pleasure
3 Security
4 Freedom
5 A World of Peace
6 Salvation

This is a rank order. It tells us, for instance, that material wealth is more important to the person than pleasure, but we cannot say that the intervals between goals are the same. The first four could be of near equal importance, while the fifth and sixth ones may be much further removed.

Interval scales also assign numbers to a characteristic, but in this case there is a strong mathematical relationship between the numbers, as each interval is equal. The classic example of an interval scale is the temperature scale (Fahrenheit or Centigrade).
The thing to note about an interval scale is that it does not have a **true zero**; 0°C or 0°F does not indicate a complete absence of heat. In the absence of a true zero it is NOT the case, for example, that 4 can be regarded as twice as much as 2. Forty degrees centigrade is **not** twice as hot as 20°C, although the **difference** between 40°C and 50°C is the same as the **difference** between 20°C and 30°C.

Ratio scales involve an even stronger mathematical relationship; not only are the intervals between the numbers or scale values equal, but there **is** a **true zero** so that 4 **is** twice as much as 2. Distance is a ratio scale as 0 indicates no distance, and 10 km is twice as far as 5 km.

Discrete and continuous variables

Nominal variables are sometimes referred to as **discrete** variables, because they have **fixed values**, and it is not possible to have smaller values between them (e.g., you cannot *really* be .5 of a man).

Interval or ratio variables, on the other hand, are often **continuous** variables, because they can be broken up into any number of finer divisions. Variables such as distance, for example, are continuous. Depending on how finely we measure the distance between two points there can be anything up to an infinite number of measures (e.g., 15 km, 14.91 km, 14.907 km, 14.9068 km etc.). However, interval or ratio variables can also be discrete (e.g., number of children in the family, where it is not possible for a family to have 3.65 or 3.642 children).

Levels of measurement in psychology

In psychology we often use rating scales similar to the following, where participants tick a box to indicate where they stand:

Political advertising should not be permitted on television in the last week before an election.

☐	☐	☐	☐	☐	☐	☐
Strongly Disagree	Disagree	Slightly Disagree	Neutral	Slightly Agree	Agree	Strongly Agree

We then code these scales numerically (e.g., 1 to 7, or -3 to 3), and strictly speaking they are **ordinal** scales. Nonetheless, it is common in psychology to regard them **as if** they are **interval** scales. It is assumed that the psychological intervals are about the same.

Be aware this is quite a controversial issue, and different researchers and textbooks can take different points of view.

Introduction to research design and statistics

Despite the dread induced in students by research and, even worse, statistics, quantitative research design and analysis (statistics) are really based on very simple ideas. If you understand the simple foundations you will not be so overwhelmed by the details, **although you must be sure to retain concepts**, as later ones build on earlier ones, and you **will** soon be overwhelmed if you forget what is learned at each step.

In essence, quantitative research is about commonsense **pattern recognition**. One of the simplest forms of pattern recognition involves identifying what things occur together, so you can predict one on the basis of the other. This in essence is what **correlation** is about, but you need to be very careful with the interpretation. Just because things occur together does **not** necessarily mean that one **causes** the other. Perhaps the most fundamental issue of all is that of **causation**.

Usually, what we most want to identify are **cause** and **effect** relationships. The commonsense way to do this is to **experiment**. If you think X might **cause** Y, then **manipulate** X and see if Y changes accordingly, and do this repeatedly so you can rule out chance as an explanation.

Research designs

These commonsense approaches to understanding how the world works are the basis of the two main types of research design, namely nonexperimental and experimental.

Nonexperimental or correlational designs assess **relationships** among variables.

Experimental designs attempt to assess **cause** and **effect**. We measure group differences on the **effect** variable (**the dependent variable, DV**) for groups of research participants treated differently on the hypothesised **causal** variable (**the independent variable, IV**). The defining feature of experimental designs is that the researcher actively **manipulates** the IV. **True experiments** use **randomly** formed groups of participants; **quasi experiments** use **intact** (i.e., preexisting) groups.

Two types of experimental designs are **between-groups** (or **between-subjects**) where different groups of participants receive the different manipulations of the IV, and **repeated measures** (or **within-subjects**) where only one group of participants receives the different manipulations of the IV on different occasions.

When conducting and writing up research, **never get confused between experimental and correlational research designs**. In particular, never use causal terminology such as "effect of" or "influence of" or "impact of" etc. when you only have a correlational design. Instead, be sure to only talk about **relationships**. You may wish to argue on logical grounds for a causal relationship, but you must not assume it from the design. **Do not ever forget this!**

Analysis techniques

Even though research designs and analysis techniques are intimately related, they are not one-in-the-same. Analysis of Variance (abbreviated as ANOVA), for instance, is the analysis technique for analysing group differences, and is the main analysis for **experimental** research designs when the groups receive different levels of a **manipulated** variable (e.g., different amounts of a drug, different room temperatures). However, it can be also used for **nonexperimental correlational** research designs when the groups comprise a **naturally occurring variable** such as gender.

An Area of Confusion

Always be alert to the **natural groups** nonexperimental design depicted on the next page. Arguably, the term natural groups (Shaughnessy, Zechmeister, & Zechmeister, 2000, pp. 235-238) best describes this design, but different texts often use different terms. Very often too, the design is included under **quasi experiments**, which is **not** appropriate—it is **not** an experiment because there is no manipulation. Furthermore, the design is fundamentally correlational in nature; it tests relationship, not causation.

The different types of research design and analysis are illustrated on the next page.

Types of quantitative research design and statistical analysis

Experimental Design		Nonexperimental (Correlational) Design	
True experiment	**Quasi experiment**	**Natural groups**	**Correlational**
Manipulate IV: Administer different levels of IV to different groups of participants. Groups formed by **random assignment** (in between-groups designs). **Control** other *relevant* variables (i.e., hold them constant).	**Manipulate IV:** Administer different levels of IV to different groups of participants. **Intact groups** used, often in natural settings (e.g., school classes, shifts of workers). **Control** of other variables is problematic.	**"IV" occurs naturally and is not manipulated:** Levels of IV are *selected* and represented by discrete groups (e.g., gender, age-group, high versus low self-esteem groups, religion).	**No manipulation or selection into groups:** Simply measure all variables of interest and assess relationships among these naturally occurring variables.
Causal inference can be inferred. Can infer *from the design* that IV *causes* the DV.	**Causal inference is problematic.** Depends on explicitly ruling out threats to causal inference (i.e., to internal validity).	**Cannot infer causation from design.** Control of other variables is problematic, and other variables may be the cause of observed relationships; hence causal inference is much more difficult and *cannot be made from the design itself.* Researcher may be able to argue causation on logical grounds alone.	

Significance of group differences **Analysis of Variance (ANOVA) "family" of statistics (and their nonparametric alternatives)**	Relationship among variables **Correlational "family" of statistics (and their nonparametric alternatives)**
Between-groups designs: 2 groups: Independent *t* test (Mann-Whitney) 3+ groups: One-way ANOVA (Kruskal-Wallis) 2+ IVs: Factorial ANOVA **Repeated measures designs:** 2 levels: Dependent *t* test (Wilcoxon) 3+ levels: One-way repeated ANOVA (Friedman) 2+ IVs: Factorial repeated ANOVA **Other ANOVA designs:** Mixed ANOVA (Between- and repeated IVs) ANCOVA Analysis of Covariance MANOVA Multivariate ANOVA (multiple DVs) **Frequency (Nominal) data:** One-way and two-way chi-square (nonparametric) Multiway frequency analysis (nonparametric)	**Two-variables:** Pearson bivariate correlation and regression Spearman nonparametric alternative **Relationship between several "IVs" (predictors) and one "DV" (outcome or criterion):** Multiple regression Logistic regression (nonparametric) **Structure (relationship pattern) in set of variables:** Principal components or factor analysis Cluster analysis (nonparametric) Multidimensional scaling (nonparametric) **Frequency (Nominal) data:** One-way and two-way chi-square (nonparametric) Multiway frequency analysis

2 Entering And Saving Data in Spss

Getting things done in SPSS is usually a matter of pointing and clicking (or double-clicking) on the appropriate icon, then pointing and clicking instructions for what you want to do. Many operations can also be performed by selecting from pull-down menus. You are free to choose which technique you prefer to use. Also, there are often several different ways of doing the same thing: which technique you use is a matter of personal preference. You may find and prefer alternative ways of doing many of the things covered in this text.

The main SPSS window appears as follows. Data are entered in this **Data View** window, but not before you have defined your variables by clicking on the **Variable View** tab at the bottom to go to the **Variable View** window. Note that the colour scheme and style may vary, depending on the Windows software version you are using and the style preferences that have been selected.

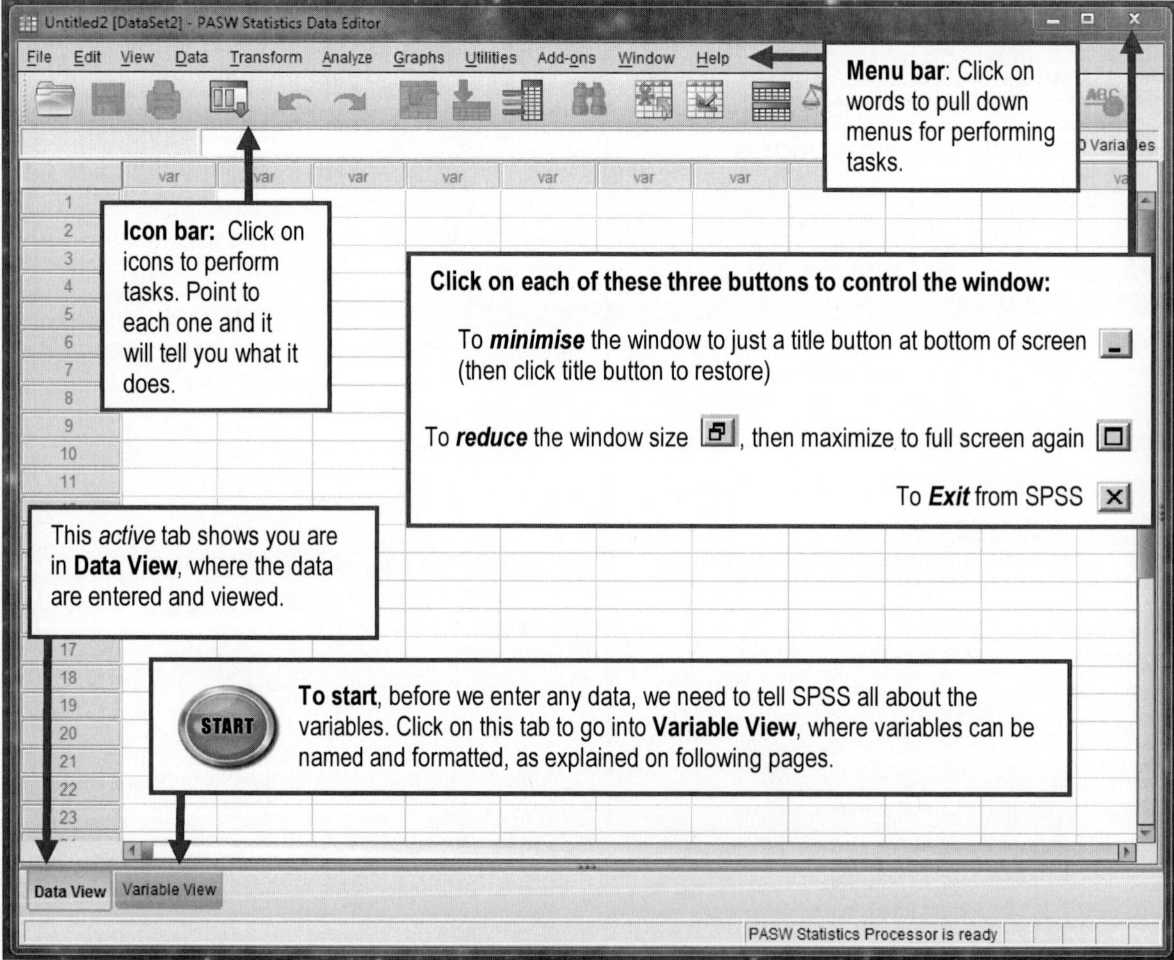

Saving the contents of SPSS windows

The contents of any window can be saved to a file on the hard disk, floppy disk, CD or DVD, USB flash drive, or network etc. SPSS works with three types of windows (and files):

Data windows are where variables are defined and data are entered (as explained on the following pages). Data window contents are saved in data documents or files, to which SPSS gives the extension **sav**.

Output windows are where SPSS places the output or results of statistical analyses. The contents of output windows are saved in SPSS viewer documents, to which it gives the extension **spo**.

Syntax windows are where SPSS commands or control lines can be typed, then run to perform any statistical analysis. We only tend to use syntax for more complex analyses. The contents of syntax windows are saved in text documents, to which SPSS gives the extension **sps**.

Whenever you are working with any of these windows and you want to keep the contents, remember to ask SPSS to **save** the document. The contents of any window can be saved as follows: .

1 How to: Save window contents to documents (files)

[For example, to save the contents of a Data window, select Menu options by clicking as follows:]

File

Save As...

This calls up the **Save Data As** dialog box, as demonstrated below.

When you have specified the location and file name click on **Save** button.

Important: If successfully saved, the window heading will change to the new file name. If this does not happen you have done something wrong, and the file has not been saved!

Saving subsequent versions of a file:
Once a file has been named, subsequent versions can be saved by choosing the **File** option from the **Menu bar**, then simply click on the **Save** option, **OR** just click on the **save icon** (second from left on the icon bar). Note that subsequent versions **replace** the earlier version.

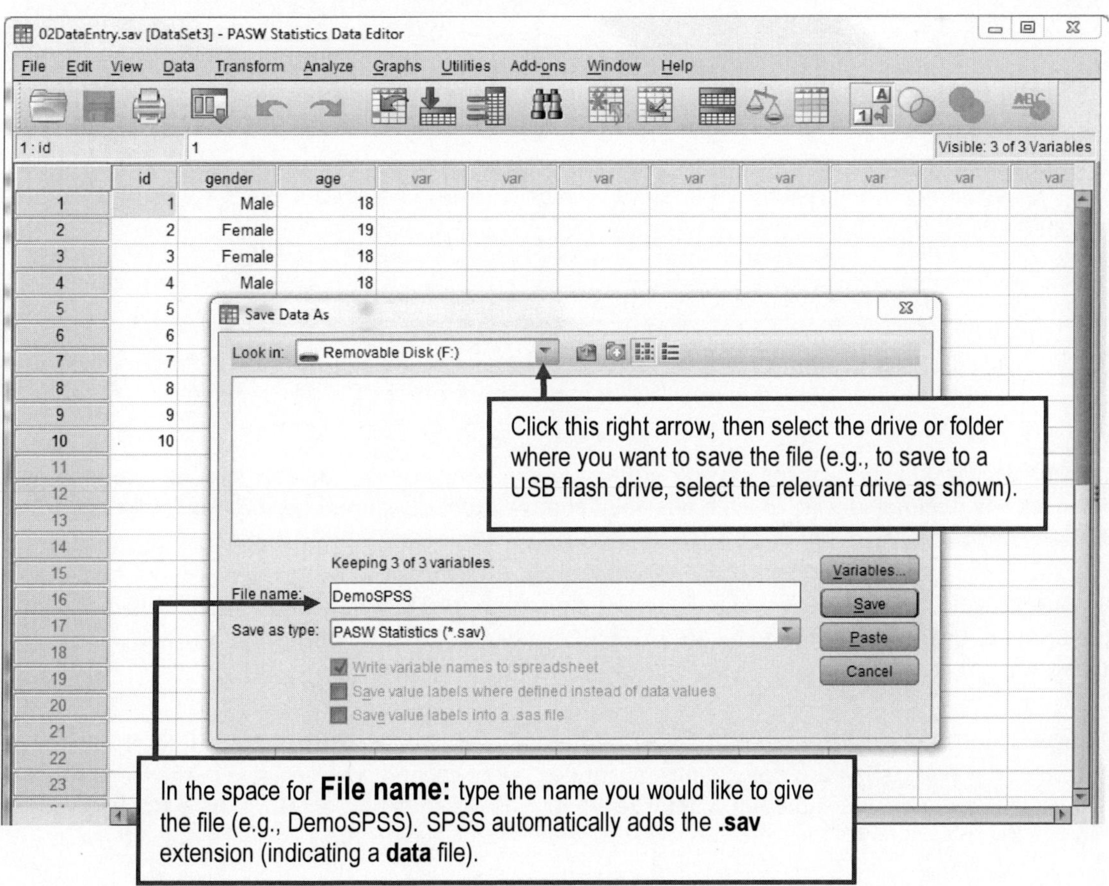

Defining variables

Let us move on now to learn about how to enter data into SPSS. Suppose that data on age and gender are to be recorded for a class of 10 students. We must first define these variables in **Variable View**.

The Variable View window is shown here, with column reference numbers added for the important columns to cross reference to the explanations below.

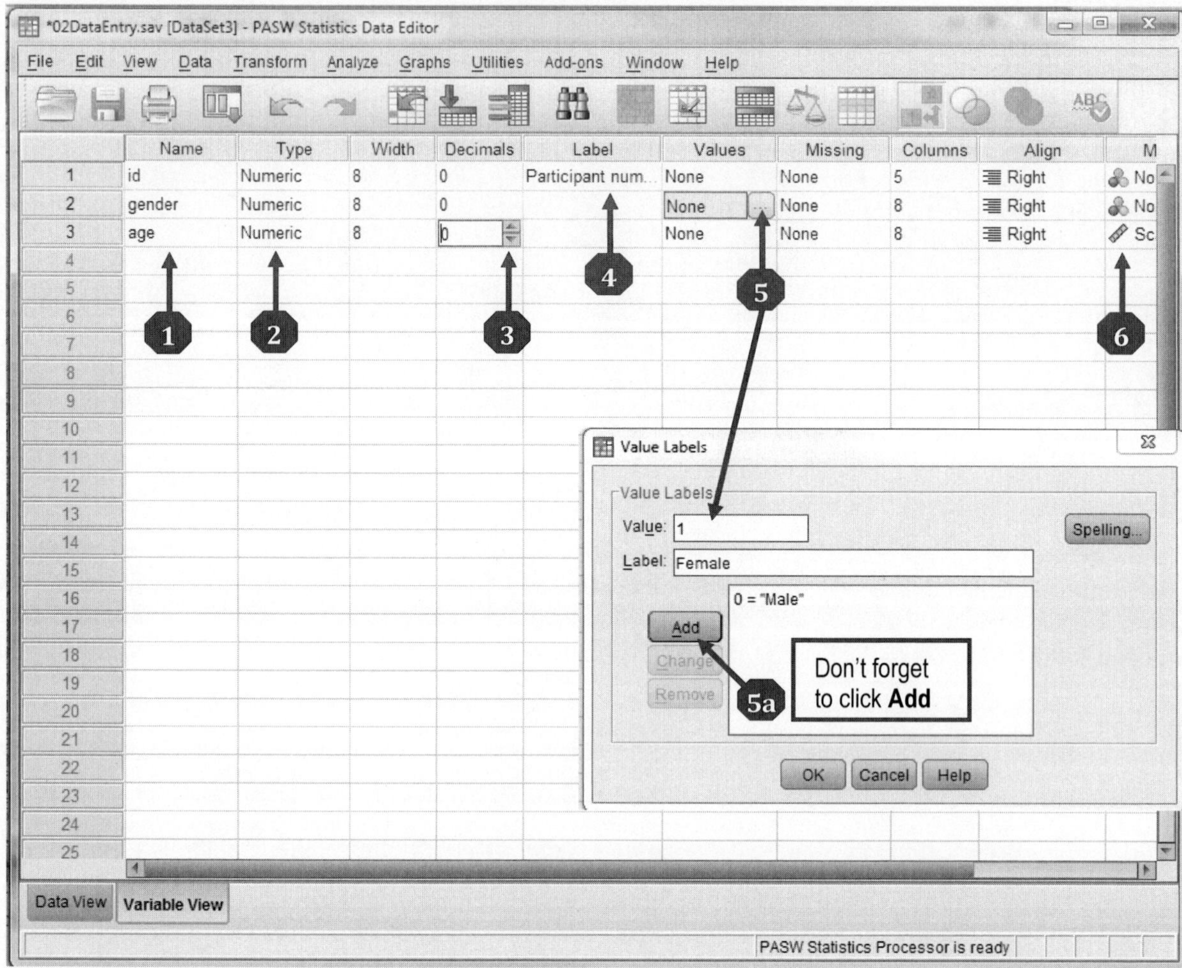

If you click on a cell in any column except **Name** and **Label** an arrow or arrows selector (as shown in column 3 for **Decimals**) or option indicator (as shown in column 5 for **Values**) appears to the right of the cell.

Arrow selectors: If you click on the arrows you can change the settings. In this example, none of the variables uses any decimal places, so each has been changed to 0, as shown here for age. (To save time you can then copy and paste from this one to the others.)

Option indicators: If you click on cells in any of the columns **Type**, **Values**, or **Missing** an option indicator appears to the right of the cell, as shown here in the **Values** column for gender (this is explained below).

1. **Name column:** Here, you type the short name of each variable. For this example, ID has been typed in the first row of the Name column, gender in the second row, and age in the third row. Once you name the variables in **Variable View**, SPSS heads the columns in **Data View** with these names.

Although SPSS already has case numbers down the left hand side of the data window, it is a very good idea to include an ID variable as a **unique** identifier for each participant, because SPSS often rearranges the data during analysis. When this happens the numbers down the side no longer correspond to the correct participant number.

2. **Type column:** This can be left as the default **Numeric** for variables that are numbers. If, however, you have alphabetic variables (e.g., if you wish to literally type in "male" and "female" for gender) you would need to activate the **Numeric** dialogue box, and select the ⊙ **String** option, which means alphabetic characters.

3. **Decimals column:** This is where you specify the number of decimal places you want for each variable, by clicking on the arrow indicators accordingly.

4. **Label column (optional):** Where a longer, more explanatory, variable name is desired in the printed output, type such names in the **Label** column, as shown here for ID.

5. **Values column:** Here is an example of an option indicator. When it is clicked upon, a dialog box appears for **Value Labels** (as shown). In order to perform statistical analyses on a nominal (or categorical) variable such as gender it must be assigned numerical codes (e.g., 0=Male, 1=Female). This column enables you to specify how such variables are coded.

   To tell SPSS how gender is coded in this example, type in the **Value Labels** dialogue box as follows (Note: these steps have already been completed in the example for the 0 value):

   [In the space for] **Value** [type] **0** [note this is zero, not letter O]
   [Hit the Tab key, then in the space for] **Value Label** [type] **Male**
   [Click on the button] **Add**
   [Cursor moves to the space for] **Value** [type] **1**
   [Hit the Tab key, then in the space for] **Value Label** [type] **Female**
   [Click on the button] **Add**
   [Click on the button] **OK**

6. **Measure column:** There are three options here for specifying the level of measurement of the variables, namely, **Scale** (interval or continuous), **Ordinal**, and **Nominal** (categorical). In this example ID and gender have been changed to nominal.

Of the remaining columns, **Width** only needs to be changed if your data are likely to need more than the default of 8 characters (as may be the case if you are entering text for a string variable). **Missing** is used if you want to assign codes for different types of missing data, while **Columns** and **Align** control the physical width and alignment respectively of the data columns in the **Data View** window.

Entering data

Now that the variables have been defined in **Variable View** you can click on the **Data View** tab at the bottom left to move back to the data window where the actual data (listed here) can be entered. Below are versions of the Data View window after these data have been entered.

ID	Gender	Age
1	0	18
2	1	19
3	1	18
4	0	18
5	1	32
6	1	19
7	0	40
8	1	18
9	1	19
10	0	20

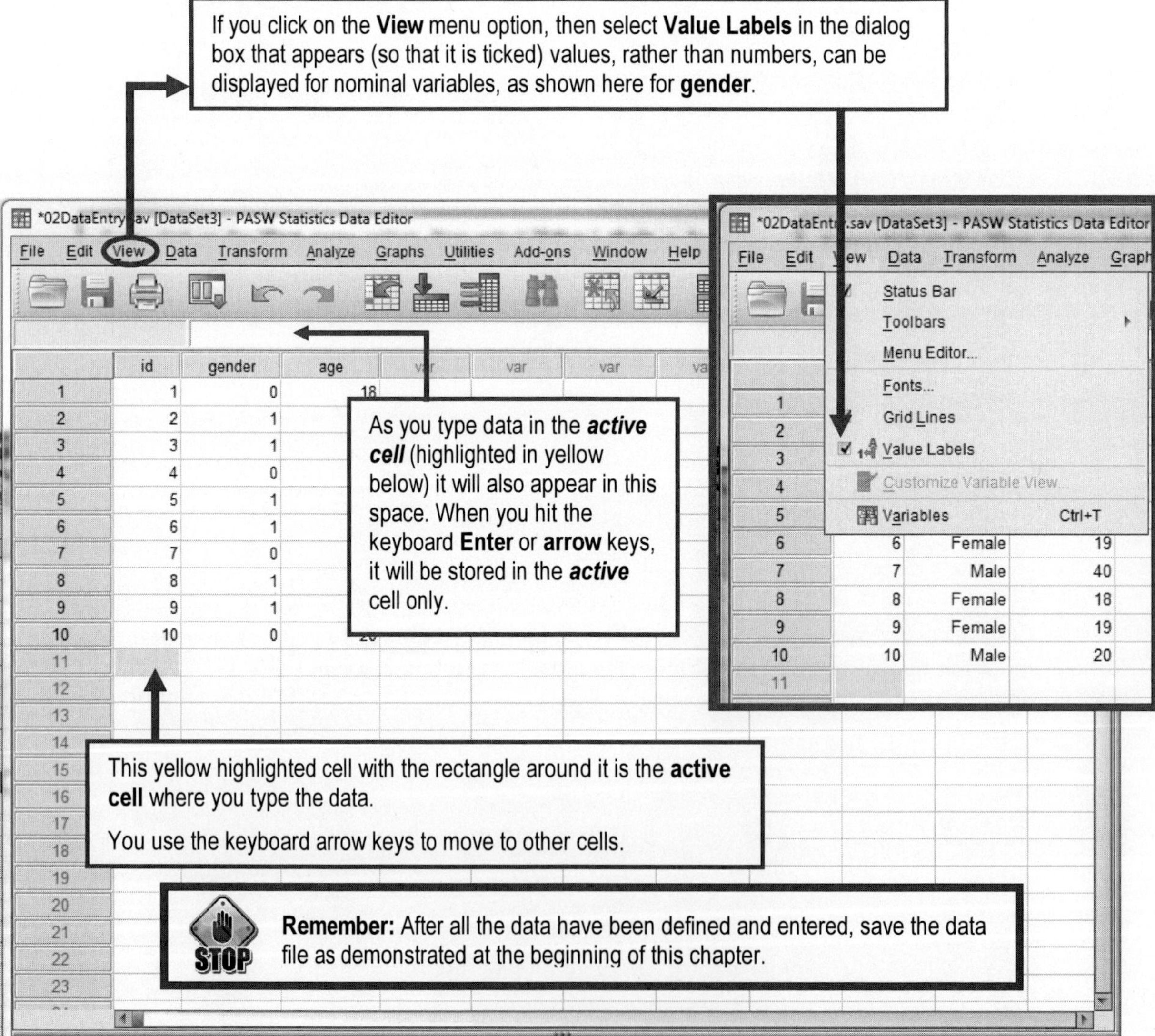

Using SPSS to perform a simple frequency analysis

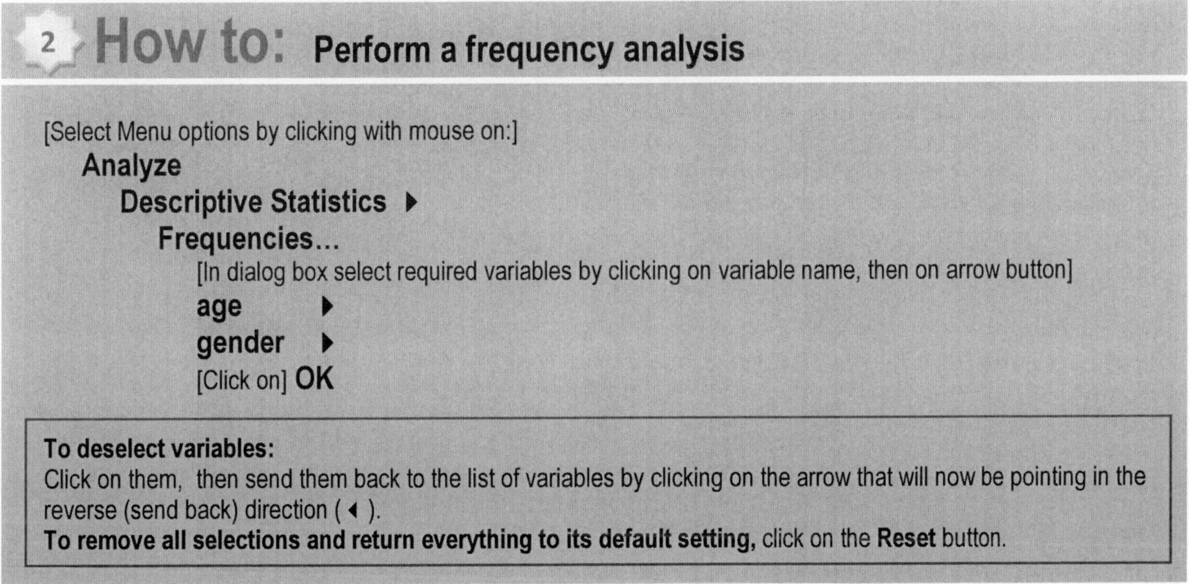
2 How to: **Perform a frequency analysis**

[Select Menu options by clicking with mouse on:]
Analyze
Descriptive Statistics ▸
Frequencies...
[In dialog box select required variables by clicking on variable name, then on arrow button]
age ▸
gender ▸
[Click on] **OK**

To deselect variables:
Click on them, then send them back to the list of variables by clicking on the arrow that will now be pointing in the reverse (send back) direction (◂).
To remove all selections and return everything to its default setting, click on the **Reset** button.

Examining the output

SPSS moves to the Output window where the results of this analysis (the output) appear. .

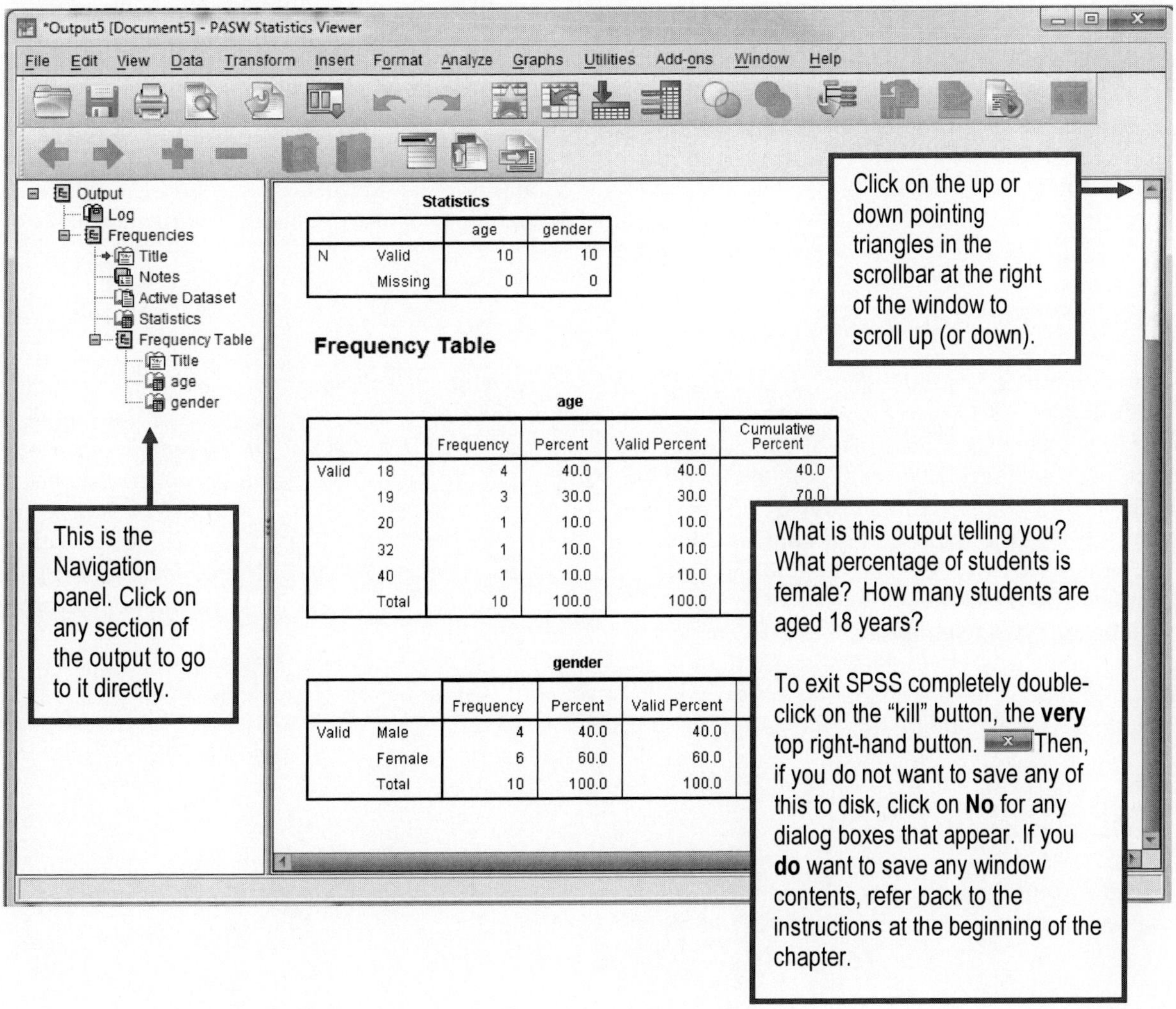

*Output5 [Document5] - PASW Statistics Viewer

File Edit View Data Transform Insert Format Analyze Graphs Utilities Add-ons Window Help

Output
Log
Frequencies
Title
Notes
Active Dataset
Statistics
Frequency Table
Title
age
gender

Statistics

		age	gender
N	Valid	10	10
	Missing	0	0

Frequency Table

age

		Frequency	Percent	Valid Percent	Cumulative Percent
Valid	18	4	40.0	40.0	40.0
	19	3	30.0	30.0	70.0
	20	1	10.0	10.0	
	32	1	10.0	10.0	
	40	1	10.0	10.0	
	Total	10	100.0	100.0	

gender

		Frequency	Percent	Valid Percent
Valid	Male	4	40.0	40.0
	Female	6	60.0	60.0
	Total	10	100.0	100.0

Click on the up or down pointing triangles in the scrollbar at the right of the window to scroll up (or down).

This is the Navigation panel. Click on any section of the output to go to it directly.

What is this output telling you? What percentage of students is female? How many students are aged 18 years?

To exit SPSS completely double-click on the “kill” button, the **very** top right-hand button. Then, if you do not want to save any of this to disk, click on **No** for any dialog boxes that appear. If you **do** want to save any window contents, refer back to the instructions at the beginning of the chapter.

3 The Problem of Variability: Descriptive Statistics For Central Tendency and Variance

If human beings and other living organisms were like pieces of metal, copper for instance, behavioural science research would be relatively easy. Any piece of copper behaves like any other piece of copper of equivalent purity under the same conditions. Therefore, if you want to establish the effect on copper of changes in temperature any individual sample will suffice.

This is not the case in the biological and behavioural sciences where there exists the problem of variability! Living organisms are complex entities whose behaviour (both physical and behavioural) is determined by innumerable variables. Thus, for example, the effect of a given drug on one individual may be quite different to the effect of the same drug on another individual; it may have a large effect on one individual and little or no effect on another. Moreover, if it is a drug hypothesised to affect blood pressure, for instance, it will be difficult to compare effects across individuals because those individuals are likely to have a range of different blood pressure levels to begin with for a range of different reasons.

The way this problem has traditionally been dealt with in disciplines such as psychology is to average across individuals and determine how much of the entire variance in a sample of individual scores can be explained or accounted for by a particular variable (the drug in this example). Inferential statistics are then used to generalise from the sample to the population. Of course, what this means is that we have an estimate of the average effect in the whole population, but we are not able to specify the effect for any given individual, especially when the proportion of variance explained is quite small—as it very often is in the social and behavioural sciences.

The problem of individual variability underpins the whole of behavioural research and the whole of statistics. In fact, statistics is nothing more than an elaborate device for trying to deal with variability.

Summary statistics

In averaging across individuals, two summary descriptive statistics are needed to summarise or describe a sample. The first is a measure of the **central tendency** (i.e., the average score in the sample), and the second is a measure of the average variability or **variance** about that average score. Both measures are necessary to adequately describe the sample. To know, for example, that the average income in a country is $150,000 is not enough; you would not rush over there until you knew the variability. You might be less enthusiastic if you discovered that there is a large variance and that incomes actually range from $5,000 (for most of the population) to $900,000 (for the ruling elite). If, on the other hand, the variance was small, with incomes ranging from $145,000 to $155,000, you may well be applying to emigrate.

The remainder of this chapter reviews measures of central tendency and variance, but first a word about statistical notation.

Statistical notation

In order to understand statistics you to need to be familiar with statistical notation, that is, the set of abbreviations used to represent statistical concepts. Among the main symbols are:

N refers to the total number of scores (participants) in a sample *n* refers to the number of scores in a subset of a sample

X refers to any score

Σ means **the sum of** (it is the Greek capital letter, sigma)

$\overline{X}$ is the symbol used in **calculations** for the mean ("X-bar"), however, when writing research reports use *M* for the mean and *SD* for the standard deviation.

Note that most symbols are in italics.

Measures of central tendency and variability

The mode

The simplest measure of central tendency is the **mode**, which is just the **most frequently occurring score** or scores.

To find the mode, list the scores in order, then locate the score or scores with the highest frequency.

The mode can be found for **nominal** level data and above.

There is no specific measure of variability to accompany the mode, although with ordinal, interval or ratio level data, the **range** should be indicated (see below).

2 3 4 4 5 [6 6 6] 7 8 **Mode = 6**

The median, the range, and the semi-interquartile range

The **median** is the measure of central tendency that corresponds to the 50th percentile, that is, the score that separates the bottom 50% of scores from the top 50% of scores.

To find the median, first list the scores in order, then locate the **middle** score.

The median is used with **ordinal** level data and above.

The abbreviation used for the median in written reports is *Mdn*.

2 3 4 4 5 [ ] 6 6 6 7 8 ***Mdn*** **= 5.5**

1 1 2 2 3 [4] 6 6 6 7 8 ***Mdn*** **= 4**

3 4 5 [ ] 8 8 9 ***Mdn*** = 5 + ((8 - 5) / 2)
= 5 + 3 / 2
= 5 + 1.5
= **6.5**

The **range** is the measure of variability that accompanies the median. It is the distance between the highest and lowest scores. Scores must be listed in order to determine the range.

2 3 4 4 5 6 6 6 7 8 **Range = 8 - 2 = 6**

The **semi-interquartile range** is half the distance between the 25th and 75th percentiles (i.e., half the distance between the bottom 25% of scores and the top 25% of scores; or the 50% of scores that fall either side of the median).

The range and semi-interquartile range are used with **ordinal** level data and above.

The **median** is the most appropriate measure of central tendency for **ordinal** data. The **range** can be used as the measure of variability, although the presence of extreme scores or **outliers** can make it misleading. Hence, some texts recommend the use of the **semi-interquartile range** as the best indicator of variability in ordinal level data (accompanying the median).

The mean, the variance, and the standard deviation

The mean is the most frequently used measure of central tendency. It is the arithmetic "average".

Note that the statistical symbol for the mean used in calculations is $\overline{X}$.

The symbol for the mean used in written reports is *M*.

The mean is used with **interval** level data and above.

To demonstrate how the mean is calculated, let us use the same set of figures as for the mode and median, but note there is no need to arrange them order.

Participant Number (ID)	*X*
1	6
2	4
3	3
4	2
5	6
6	8
7	7
8	5
9	6
10	4
$N = 10$	$\Sigma X = 51$

Using the formula for the mean, we calculate is as follows:

$$\overline{X} = \frac{\Sigma X}{N}$$

$$= \frac{51}{10}$$

$$= \mathbf{5.1}$$

For this set of data note that the mean is **5.1**, the median is **5.5**, and the mode is **6**.

This discrepancy in the three measures of central tendency is due to the **negative skew**[1] in the scores. In a perfectly symmetrical distribution the mode, median, and mean will be the same, but in skewed distributions they are spread apart.

The standard deviation approximates the average amount by which scores deviate from the mean, and the **variance** is the standard deviation squared.

For data that are **interval** level or above, the **mean** is the most appropriate measure of central tendency, and the **standard deviation** is the most appropriate measure of variability.

The statistical symbol for standard deviation used in most calculations is *s*.

The symbol for standard deviation used in written reports is *SD*.

The statistical symbol for variance used in most calculations is: s^2.

There are two ways to calculate the variance and standard deviation. The most useful way for **conceptual understanding** is to use the **deviation score formula**, as follows (note it requires the mean to be calculated first):

ID	X	$X - \bar{X}$ x (Deviation score)	$(X - \bar{X})^2$ x^2 (Squared deviation score)
1	7	1.2	1.44
2	3	-2.8	7.84
3	9	3.2	10.24
4	4	-1.8	3.24
5	6	0.2	0.04
$N = 5$	$\sum X = 29$	$\sum x = 0.0$ Sum of deviation scores Always equals 0	$\sum x^2 = 22.80$ Sum of **squared** deviation scores **SUM OF SQUARES (*SS*)**

$$\bar{X} = \frac{\Sigma X}{N} = \frac{29}{5} = \mathbf{5.8}$$

$$s^2(variance) = \frac{\Sigma x^2}{N - 1} = \frac{22.80}{4} = \mathbf{5.7}$$

$$s(SD) = \sqrt{s^2} = \sqrt{5.7} = 2.3875 = \mathbf{2.39}$$

---

[1] Skewness is explained in the next chapter.

The most **efficient** formula for **calculation** purposes is the **raw score formula**, as follows:

ID	X	X^2
1	7	49
2	3	9
3	9	81
4	4	16
5	6	36
$N = 5$	$\sum X = 29$	$\sum X^2 = 191$

Variance calculation:

$$s^2 = \frac{\Sigma X^2 - \frac{(\Sigma X)^2}{N}}{N - 1}$$

$$= \frac{191 - \frac{(29)^2}{5}}{5 - 1}$$

$$= \frac{191 - \frac{841}{5}}{4}$$

$$= \frac{191 - 168.2}{4}$$

$$= \frac{22.8}{4}$$

$$= \mathbf{5.7}$$

Standard deviation calculation (square root of variance):

$$s(SD) = \sqrt{s^2}$$

$$= \sqrt{5.7}$$

$$= 2.3875$$

$$= \mathbf{2.39}$$

Obtaining central tendency and variability with SPSS

Hypothetical data set

Imagine you are a school psychologist concerned about indications of low self-esteem among senior students at your school. To investigate this you select a random sample of 30 Year 12 students and give them a self-esteem questionnaire. Imagine this consists of 20 items relating to self-esteem (e.g., *I am confident I can succeed in life*). Participants respond on a 5-point, agree-disagree scale, coded from 1 to 5 so that a high number always corresponds to high self-esteem. The full questionnaire has a possible range of 20 (indicating lowest possible self-esteem) to 100 (indicating highest possible self-esteem).

The data from your 30 participants are as listed below.

Refer to the previous chapter to enter these data in SPSS and save them in a file that you might call ***Describe.sav*** (alternatively, you can give this file a name of your own choosing, perhaps a longer one that better indicates its contents).

It used to be the rule that variable names cannot be longer than eight characters in SPSS. Although this no longer applies, it can still be useful to give variables short names, expanding on them if necessary through the **Label column** in **Variable View**. Here the variable code names are ID (for participant number) and SELFEST (for self-esteem scores). Give SELFEST the **Variable Label** Self-esteem.

Note also that your data are entered in the form of a **data matrix**, with **rows** corresponding to the **participants** (or cases), and **columns** corresponding to the **variables**.

ID	SELFEST	ID	SELFEST	ID	SELFEST
1	55	11	39	21	64
2	58	12	53	22	38
3	67	13	47	23	61
4	63	14	43	24	57
5	63	15	78	25	51
6	41	16	35	26	49
7	72	17	31	27	45
8	57	18	57	28	53
9	59	19	45	29	54
10	52	20	58	30	52

3 How to: Obtain descriptive statistics

Analyze
Descriptive Statistics ▸
Frequencies...
[Select variables(s) you want]
Self-esteem [selfest] ▸

Statistics...
[Click on following boxes to request descriptive statistics]
Percentile Values
☑ **Quartiles**
Central Tendency
☑ **Mean**
☑ **Median**
☑ **Mode**
☑ **Sum**
Dispersion
☑ **Std Deviation**
☑ **Variance**
☑ **Range**
☑ **Minimum**
☑ **Maximum**

Continue
[Click off box for ☐ **Display Frequency Tables** if you do NOT want a frequency distribution]
OK

> ***Note:*** You can also obtain some descriptive statistics automatically using:
> **Analyze**
> **Descriptive Statistics ▸**
> **Descriptives...**

4 How to: Save the output window to a file

Make sure you are in the Output window.
File
Save as...
[Type in **File Name**: space] **F:\Describe**[1] [SPSS automatically adds the extension .SPO]
OK

---

[1] This assumes you want to save to a USB stick or other device that is designated as the F:\ drive. You need to understand the particular computer system you are using and where and how to save files.

SPSS output for descriptive statistics

Frequencies

Statistics

Self-esteem

N	Valid	30
	Missing	0
Mean		53.23
Median		53.50
Mode		57
Std. Deviation		10.769
Variance		115.978
Range		47
Minimum		31
Maximum		78
Sum		1597
Percentiles	25	45.00
	50	53.50
	75	59.50

To calculate the semi-interquartile range:

Semi-interquartile range = 75th percentile - 25th precentile / 2
= 59.5 - 45.0 / 2
= 14.5 / 2
= **7.25**

4 The Problem of Probability: Frequency Distributions, The Normal Distribution, and Standard Scores

It has been argued that the problem of variability underpins the whole of behavioural science research and statistics. There is, however, a second fundamental problem, and that is the difficult problem of **probability**. It is difficult because we humans seem instinctively to want to *misunderstand* probability.

Scores on psychological variables vary—by definition—but the various scores that are possible for a variable rarely, if ever, have the same **frequency** of occurrence within a sample or within the population. That is, they occur with different degrees of **probability**. Common scores have a high probability of occurrence (e.g., heights of around 6 foot or 150 cm); extreme scores have a low probability of occurrence (e.g., heights of around 8 foot or 240 cm).

Frequency distributions

Rather than just relying on measures of central tendency and variability to describe a sample, it is always important to examine the **distribution** of the scores, to ascertain which scores occur in the sample and how frequently they occur.

This is achieved by constructing a **frequency distribution**.

The normal distribution

A curious thing happens with many variables in nature, including variables that measure human physical characteristics (e.g., height, weight) or psychological characteristics (e.g., IQ). If one takes a large sample of scores (i.e., many hundreds or thousands), constructs a frequency distribution, then graphs this frequency distribution, the result is a consistent pattern known as the **normal distribution** or **normal curve**. Although the fit is not exact, the **normal curve** is a remarkably good model of the population distributions of many variables.

An understanding of the normal curve is of critical importance, for it underpins the main **inferential statistics** that are used in psychology, namely, **parametric statistics** that are based on the parameters of the normal curve.

Normal curves have the following properties:

1. They are unimodal (only one peak) and symmetrical.
2. The mean, median, and mode have the same value in normal distributions.
3. They are asymptotic—they never touch the abscissa or *x*-axis (i.e., the horizontal axis).
4. They can have different means and variances (and *SD*s) that determine whether they are "tall and skinny" or "short and fat".

The normal curve, standard scores, and percentiles

If scores in a normal distribution are converted to units of standard deviation (e.g., a score may be ½ *SD* above the mean, or 2 *SD* below the mean etc.), one finds that the area under the curve between units of standard deviation corresponds to the approximate **proportion** of scores that are found between those units.

Some other characteristics of the normal curve are:

1. The area **below** the mean contains 50% of scores, as does the area **above** the mean (because the curve is **symmetrical** and the mean and the median are the same).
2. Approximately 34% of scores occur within the area 1*SD* **below** the mean, and 34% of scores occur within 1 *SD* **above** the mean (i.e., approximately 68% of scores are ± 1*SD* from the mean).
3. The other proportions of the normal curve are as illustrated below, with the mean always having a value of 0 in *SD* units.

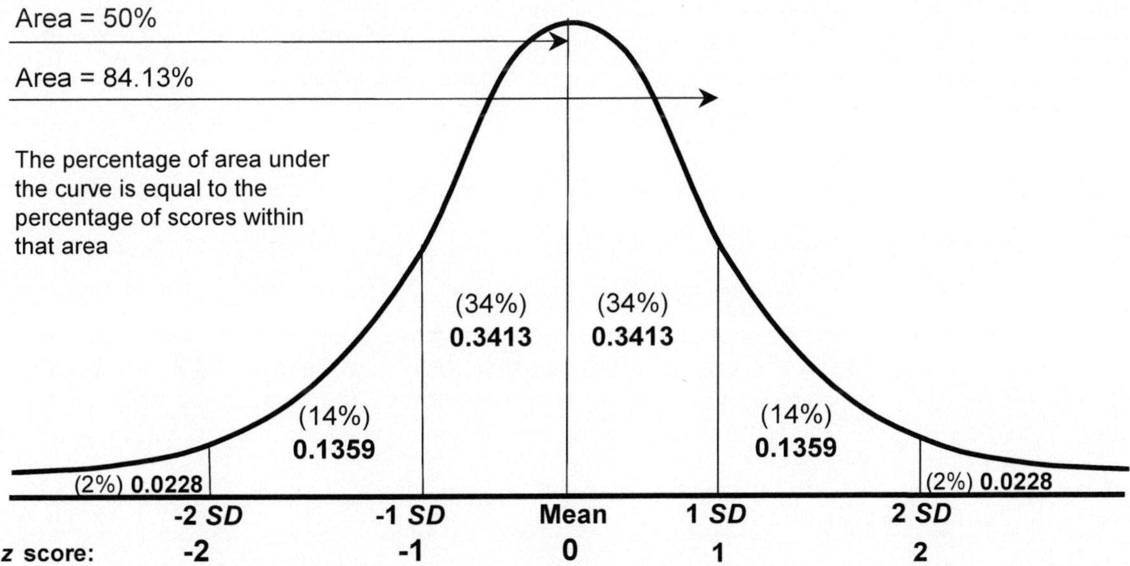

This correspondence between areas under the curve and the proportion of scores means that if we have a raw score from a normally distributed variable, we can work out that score's **percentile rank** (i.e., the percentage of scores that are lower than that score) by converting the raw score into units of standard deviation, which are known as **standard scores** or ***z* scores**.

The **percentile rank** of a score is the percentage of lower scores.

A **percentile** is the **raw score point** below which a given percentage of scores in the distribution fall; that is, it is the raw score at a given percentile rank.

For example, if in a given population a height of 153 cm had a percentile rank of 47, it would mean that 47% of people in the population have a height less than 153 cm; 153 cm would be the 47th percentile.

From the proportions of the normal distribution illustrated above, it can be seen that one standard score above the mean (i.e., +1 *z* score) is approximately the 84th percentile. It has a percentile rank of 84, meaning that 84% of scores are lower. In contrast, one standard score below the mean (i.e., -1 z score) is approximately the 16th percentile. It has a percentile rank of 16, meaning that only 16% of scores are lower.

Parametric statistics and the normal curve

One of the reasons why the normal curve is so important is that most of the major inferential statistics are based on the assumption that the variable being measured has a **normal distribution** in the population. These are known as **parametric statistics**.

Before using parametric statistics, this assumption needs to be checked by looking at the distribution of a variable to see if it resembles a normal curve.

Where the assumptions of parametric statistics are seriously violated, alternative **nonparametric statistics** can sometimes be used.

Grouped frequency distributions

With large amounts of data it can still be difficult to make sense of frequency distributions, so it may be necessary to group the data into a smaller number of **class intervals**—a **grouped frequency distribution**, as demonstrated on the next page.

This results in a loss of detail, but a gain in understanding the overall pattern.

Histograms

Graphical representations of grouped frequency distributions are usually in the form of **histograms**, which will be demonstrated later.

These can help to establish if the distribution resembles, or is quite different to, a normal distribution. In particular, one looks to see if there is evidence of positive or negative **skewness**, or **bimodality**. A good way to remember which is which for skewness is to draw from the heraldic notion of the left as the "sinister" side, as illustrated below.

Positive skew: The distribution tails off at the high end. For positive skew, make a fist with the "good", **positive**, **right** hand facing toward you, then extend the thumb (tail).

Negative skew: The distribution tails off at the low end. For negative skew, make a fist with the "sinister", **negative**, **left** hand facing toward you, then extend the thumb (tail).

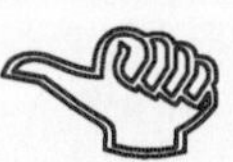

Bimodal distribution: A bimodal distribution has two peaks not one.

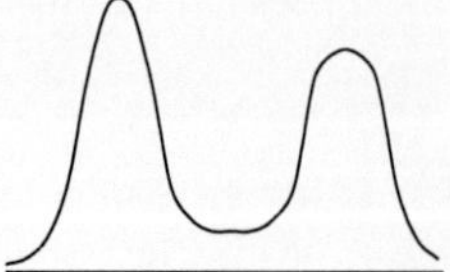

In addition to skewness, distributions can be assessed for **kurtosis**. Distributions that appear flatter than the normal curve are termed **platykurtic**, while those that are thinner and more peaked are termed **leptokurtic**.

Manually constructing a grouped frequency table

Using established statistical conventions, the following procedures would be used to produce a grouped frequency table for the self-esteem data from 30 participants in the previous chapter.

First, calculate the **class interval** or **group interval** by hand, as follows:

1. Normally use around 10-20 intervals; 10 will be suitable for this analysis.
2. Range = highest score - lowest score = 78 - 31 = **47**.
3. **Class interval size** = Range / Number of intervals = 47/10 = 4.7 = **5**.
4. Begin with closest lower score divisible by interval size (i.e., 30), or zero if that is the closest.
5. Construct the grouped frequency table in **descending** order.

Lower Real Limit Lower class interval - ½ unit	Class Interval	Upper Real Limit Upper class interval + ½ unit	Midpoint	Freq. (*f*)	Cumul. *f*	Relative *f* *f* / *N* *N* = Total number of scores	Cumul. %
74.5	**75-79**	79.5	77	**1**	30	.033	100.00
69.5	**70-74**	74.5	72	**1**	29	.033	96.67
64.5	**65-69**	69.5	67	**1**	28	.033	93.33
59.5	**60-64**	64.5	62	**4**	27	.133	90.00
54.5	**55-59**	59.5	57	**7**	23	.233	76.67
49.5	**50-54**	54.5	52	**6**	16	.200	53.33
44.5	**45-49**	49.5	47	**4**	10	.133	33.33
39.5	**40-44**	44.5	42	**2**	6	.067	20.00
34.5	**35-39**	39.5	37	**3**	4	.100	13.33
29.5	**30-34**	34.5	32	**1**	1	.033	3.33

The **relative frequency distribution** (**Relative *f***) shows the proportion or percentage of the total number of scores (*N*) that occur in each interval.

A **cumulative frequency distribution** (**Cumul. *f***) shows the number of scores falling below the upper real limit of each interval—the number of scores added to that point.

A **cumulative proportion distribution** or **cumulative percentage distribution** (**Cumul. %**) shows the cumulative frequency as a proportion or percentage of the total number of scores, that is, the proportion or percentage of scores that fall below the *upper real limit* of each interval.

Using SPSS to produce frequency tables, grouped frequency tables, and histograms

SPSS easily produces frequency tables, and histograms of grouped frequency distributions.

However, it will not automatically produce a grouped frequency table. Should you ever want one, you can force SPSS to do it by first working out the class intervals manually, then creating a new variable in SPSS that specifies these class intervals. An example of how to do this is provided at the end of the chapter.

To produce a **frequency table** and **histogram** of the self-esteem data in the previous chapter, first retrieve that data into SPSS.

Assuming you saved the data on a USB flash drive, the following steps would apply (if saved elsewhere, such as on a floppy disk in A: drive, you would need to alter accordingly).).

5 How to: Open a file in SPSS

Enter SPSS and make sure you are in the **Data View** window. Then either click on the **Open File** icon (🗁) first from left on icon bar, OR choose from the menu as follows:

File
Open
▸ **Data...**

Select the appropriate drive by clicking on the arrow ▾ to the right of the **Look in**: box

Then, click on the drive you want, for example, **Removable disk (F:)**

Then double click on the file you want (i.e., ***Describe***).

6 How to: Request a frequency table and histogram

Analyze
Descriptive Statistics ▸
Frequencies...
[Select variables you want]
Self-esteem [selfest] ▸

[Click on button for **Charts...** at bottom of dialogue box]
[From **Chart Type** select]
⊙ **Histograms**
☑ **Show normal curve on histogram**
Continue
OK

7 How to: Modify graphs in the output

When the histogram appears in the **Output Window** double-click anywhere on the histogram to open a **Chart Editor** window.

Select Menu options as follows to format the chart according to the class intervals from the Grouped Frequency Table. *(Note that the SPSS Chart Editor prior to version 12 worked differently, and you may need to use some initiative to work out how to obtain the desired effects if using an early version.)*

Chart
Edit...
[Select] **X Select X Axis** [or choose **X** from icon bar]
[Click on **Scale** tab in the separate **Properties** window that opens, unless already in **Scale** window]
[Click on ☑ in **Auto** column to remove tick for **Major Increment**]
[Type] **5** [in box in **Custom** column]
[Click on **Apply** button at bottom]
[Click on **Close** button at bottom]

To close the **Chart Editor** window and return to the Output window, click on the close window button (☒) at the top right of the **Chart Editor window**.

The histogram produced by SPSS is as follows. It has a normal curve superimposed to give an idea of how closely the distribution fits a normal curve. Since version 12, SPSS also labels the bars according to the beginning of the class interval, not the midpoint.

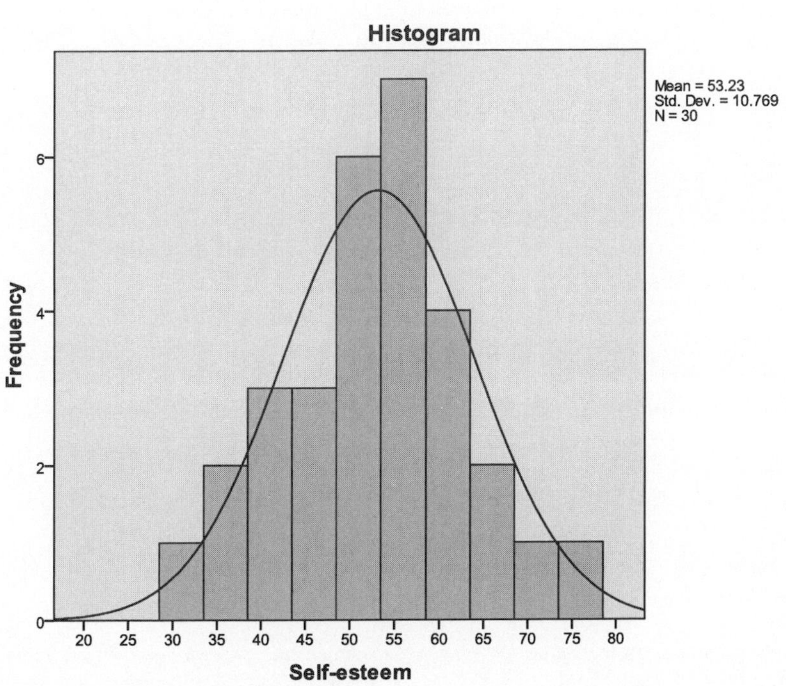

Some things to do

What would you say about the shape of the distribution in the histogram? Is it roughly symmetric? How well does it fit a normal curve? Check the footnote after you've thought about it yourself. [1]

Examine the frequency table and histogram produced by SPSS. What do they tell you about the self-esteem among your sample of Year 12 students? Does it appear to be relatively high or relatively low? [2]

Go back to the ***Chart Editor*** *in SPSS and explore* ***all*** *the menu options. All sorts of modifications can be made to charts, including changing the colours, titles, and labels; removing the line border; altering the axes etc. Look especially at the* ***Options*** *menu, then* ***Bin Element****, then the* ***Binning*** *tab, where you can change the number of intervals used for your histogram.*

Using standard scores (or *z* scores)

To convert raw scores to *z* scores, the following formula is used:

$$z = \frac{X - \overline{X}}{s}$$

Note that *z* scores always have a mean of 0 and a standard deviation of 1.

By transposing (rearranging) the formula, we obtain the formula for converting *z* scores back to raw scores:

$$X = zs + \overline{X}$$

---

[1] *With small samples histograms nearly always look irregular. With experience you are more able to judge them. This one is "roughly" symmetric and normal.*

[2] *To interpret the self-esteem scores you* ***must*** *refer back to how the self-esteem questionnaire was scored. It had a possible range of 20 to 100. We discovered in the last chapter that the median score was 53.50, but from the histogram we can see that 80% of students have scores below 62 (since the cumulative percent is 80 for a score of 61, and the score above that is 63). This all suggests the self-esteem is tending to be relatively low overall.*

To find the percentile rank of a score by converting it to a z score

Imagine I have a measure of social competence in children and this variable is normally distributed in the population (the higher the score, the greater the social competence). Imagine that studies with large samples have shown it has a mean of 50 and a standard deviation of 5.

Suppose you have a daughter whose raw score is 60 on this measure. What is her percentile rank? First calculate her *z* score:

$$z = \frac{X - \overline{X}}{s}$$

$$= (60\text{-}50)/\ 5$$
$$= 10/5$$
$$= \mathbf{2}$$

Then, refer back to the graph of the normal curve just after the beginning of this chapter, and add up the proportions lying below a *z* score of +2:

50% (for the half below the mean) + 34.13% + 13.59% = **97.72%**

Therefore, your daughter's percentile rank is 97.72. This tells you that 97.72% of children have lower social competence scores than she has, and only 2.28% have higher social competence scores.

Now, suppose you also have a son, and his raw score is only 43 (he's a boy, what do you expect?). What is his percentile rank?

$$z = \frac{X - \overline{X}}{s}$$

$$= (43\text{-}50)/\ 5$$
$$= \text{-}7/5$$
$$= \mathbf{-1.4}$$

The normal curve that has been illustrated cannot help us this time, because it does not show a *z* score of -1.4. But it so happens that mathematicians have worked out the proportions for a great number of *z* scores, and these are shown in tables of the normal distribution that are found in the Appendix of most statistics textbooks. Increasingly, too, there are internet sites that will directly calculate proportions under the normal curve for any *z* score you specify. Here is one you can use:
http://www.danielsoper.com/statcalc/calc02.aspx

These tables typically have columns that show the following for **positive** *z* scores (their meaning is illustrated on the next page):

The area (proportion of scores) between the mean and *z*
The area (proportion of scores) **below** *z* (larger portion).
The area (proportion of scores) **above** or **beyond** *z* (smaller portion)

First, look up your daughter's *z* score of 2. We want to know your daughter's percentile rank (the proportion of scores lower than hers). This is given in the **below** *z* column. Surprise, surprise, you will find it is .9772, just the same as the percentage of 97.72 obtained from the curve.

Also, from the **above** *z* column it can be seen that only .0228 (2.28%) of scores are above your daughter's.

Your son's case is slightly more difficult as he has a **negative** *z* score (his raw score is **below** the mean), and most tables only give **positive** *z* scores.

But we know the normal curve is **symmetric**; it is just the same below the mean as it is above. So, we look up a *z* score of + 1.4. The **above** *z* column tells us that .0808 (8.08%) of scores are **above** this *z* score.

Because the normal curve is symmetric, the same proportions will apply to a negative *z* score of -1.4, except that what is **above** will now be **below** and vice versa. You just swap these terms around for negative *z* scores. So, for your son's negative *z* score of -1.4, 8.08% are **below** (i.e., his percentile rank is 8.08).

Only 8.08% of children have lower social competence scores than your son, while 91.92% have higher social competence scores. It is not looking too good for this kid!

Understanding tables of the normal distribution

Note that normal distribution tables can be presented differently in different text books (and on the internet), and you may need to use your initiative to work out what the various columns represent, in relation to the examples given below.

For positive z scores (shown in most tables):
1 = The area between the mean and z
2 = The area **below** z (larger portion)
3 = The area **above** or **beyond** z (smaller portion)

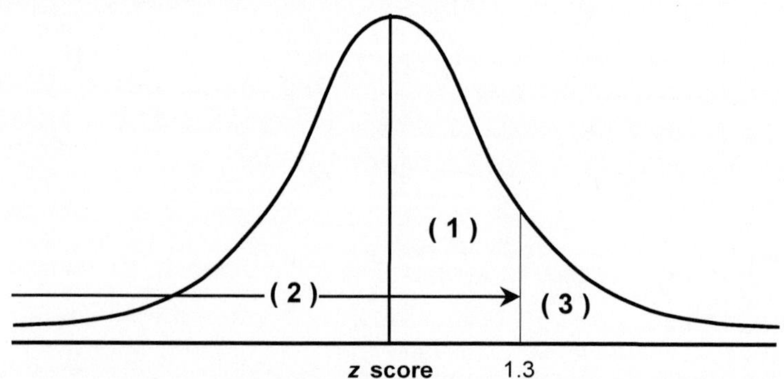

For negative z scores:
1 = The area between the mean and z
2 = The area **above** z (larger portion)
3 = The area **below** or **beyond** z (smaller portion)

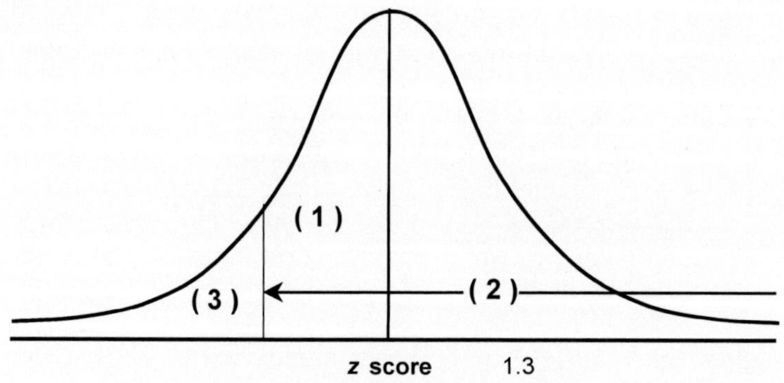

To find what proportion of scores fall between two raw scores

At this stage you might want to know what proportion of children fall between your son's raw score of 43 and your daughter's raw score of 60. This easy:

1. Convert the raw scores to *z* scores, as previously (-1.4 and 2).
2. Ascertain the percentile ranks, as previously (8.08% and 97.72%).
3. Subtract the smaller percentile rank from the larger: 97.72 - 8.08 = **89.64%**

Thus, 89.64% of children fall between your son's low social competence score and your daughter's high social competence score.

To find the proportion of scores that fall between a raw score and the mean

This is also easy. You just look at the column for the area between the mean and z.

1. Convert raw score to *z* score (i.e., -1.4 for your son, 2 for your daughter).
2. For the appropriate *z* score (ignoring its sign) read off the column for the area between the mean and *z* (i.e., .4192 or 41.92% for a *z* score of 1.4, and .4772 or 47.72% for a *z* score of 2).

Alternatively, since we know 50% of scores are in each half of the distribution, just subtract percentile ranks from fifty (for negative *z*), or vice versa (for positive *z*): 50 -8.08 = 41.92; 97.72-50 = 47.72.

What raw score falls above (or below) a given percentage of scores

This is the final permutation for *z* scores. Suppose you intend to work on your son's social skills to get him to the point where around 75% of children have a lower score than he has.

You need to know what raw score to aim for. That is, you need to ascertain the raw score corresponding to the 75th percentile.

1. Go to a normal distribution table and find the *z* score corresponding as closely as possible to a value of .7500 in the below *z* column. The closest value turns out to be .7486 and corresponds to a *z* score of 0.67. (Or use can use this internet URL: http://www.danielsoper.com/statcalc/calc19.aspx)

2. Use the formula for converting a *z* score to a raw score:

$$\begin{aligned} X &= zs + \overline{X} \\ &= (0.67\text{x}5) + 50 \\ &= 3.35 + 50 \\ &= \mathbf{53.35} \end{aligned}$$

In order for around 75% of children to have lower social competence scores than your son, you need to work on his social skills until he manages a raw score of 53.35.

Standard scores and normality

Standard scores (or *z* scores) are not automatically normally distributed. If the raw scores are normally distributed, the *z* scores will be normally distributed, but if the raw scores are skewed the *z* scores will be skewed.

The normal distribution or *z* score tables can only be used to work out percentile ranks when the variable's raw scores are normally distributed.

Using SPSS to convert raw score to *z* scores

For example, to obtain *z* scores (i.e., standard scores) for the self-esteem data.

8 How to: Obtain standard scores (z scores)

Analyze
 Descriptive Statistics ▸
  Descriptives...
   [Select variables you want]
   Self-esteem [selfest] ▸

   [Click on the following box to request *z* scores]
   ☑ **Save standardized values as variables**
 OK

This will add to the data file a variable called **Zselfest** that consists of the standardised self-esteem scores (i.e., **selfest** scores converted to ***z* scores**).

Examine the data window and note the new variable Zselfest.
Remember to save this updated data file by choosing menu options **File**, then **Save**.

Using SPSS to calculate exact percentile scores

9 How to: Obtain percentile scores

Analyze
 Descriptive Statistics ▸
  Frequencies...
   [Select variables you want]
   Self-esteem [selfest] ▸
   Statistics...
    [From **Percentile Values** select]
    ☑ **Percentile(s)** [then type] **95**
    Add
  Continue
  OK

Examine the output for this analysis in the **Output** window. The 95th percentile for this set of self-esteem scores is 74.70.

Using the transform menu in SPSS to create new variables or modify existing ones

The **Transform** menu in SPSS allows you to create new variables. For example, the **Compute** option can be used to compute a new variable that is the sum of other variables, as when scores on items in a scale need to be added to give a total score.

Alternatively, new variables can be created by modifying existing ones. For example, suppose you really wanted SPSS to produce the grouped frequency table that was previously calculated manually. To do this you could create a new variable that groups the self-esteem scores according to your calculated class intervals. Self-esteem scores of 30-34 will have a group score of 1, scores of 35-39 will have a group score of 2, and so on. A name for this variable might be ESTGRP (self-esteem **group**).

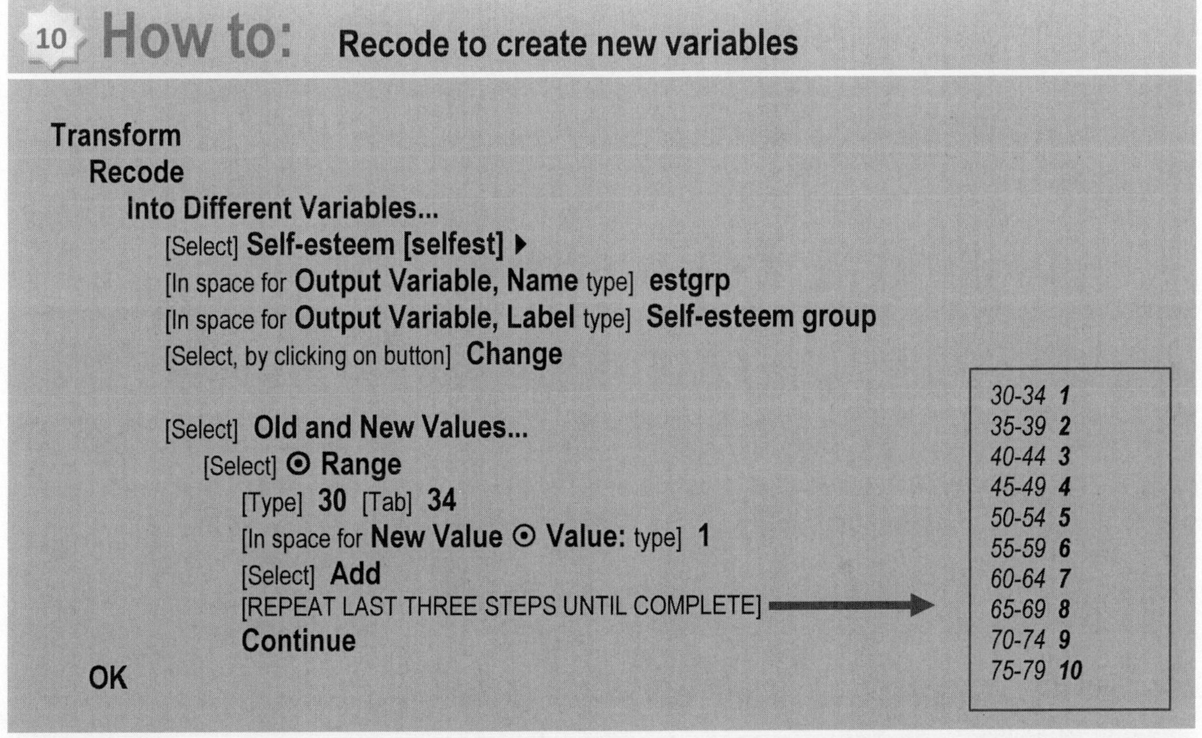

10 **How to:** **Recode to create new variables**

Transform
Recode
Into Different Variables...
[Select] **Self-esteem [selfest]** ▸
[In space for **Output Variable, Name** type] **estgrp**
[In space for **Output Variable, Label** type] **Self-esteem group**
[Select, by clicking on button] **Change**

[Select] **Old and New Values...**
[Select] ⊙ **Range**
[Type] **30** [Tab] **34**
[In space for **New Value** ⊙ **Value:** type] **1**
[Select] **Add**
[REPEAT LAST THREE STEPS UNTIL COMPLETE] →
Continue
OK

30-34 ***1***
35-39 ***2***
40-44 ***3***
45-49 ***4***
50-54 ***5***
55-59 ***6***
60-64 ***7***
65-69 ***8***
70-74 ***9***
75-79 ***10***

Specify value labels for the new variable, estgrp

[Double-click on the variable **estgrp** in the data window]
[In the **Values** column, click on word **None**, then click on [...] to right of this word]
[Click on the button] **Labels...**
[In the space for] **Value** [type] **1**
[Hit Tab key, then in the space for] **Value label** [type] **30-34**
[Click on the button] **Add**
[REPEAT LAST THREE STEPS UNTIL COMPLETE]
OK

Save the updated data file (File then Save, or click on the save icon).
Then request a frequency table from the Analyze menu, for the variable Self-esteem group [estgrp].

5 THE LOGIC OF SIGNIFICANCE TESTING: ONE- AND TWO- SAMPLE *Z* TESTS

Inferential statistics draw **inferences** about **populations** from information available in **samples**. The logic behind statistical inference is not difficult, but very confusing. Thus, it is generally the most difficult aspect of statistics to grasp. Furthermore, just when you think you have it, it tends to revert to being hopelessly confusing again. This is quite normal, so do not despair. However, do not ignore it either. You need to work hard to get your mind around it, as it is critical to understanding the use of statistics in research. Keep working at it, and eventually, over time, it will become clear. Be prepared to read a number of text books on the subject until you find one that explains it clearly for you.

The material in this text is essentially a summary of what you need to understand; however, you will need to read widely on the subject to develop that understanding.

One-sample research designs

One-sample research designs use the **sampling distribution of means**. If one takes a large number of random samples of a given size from a population and constructs a frequency polygon of their **means**, this frequency polygon approximates a normal curve. It tends to be normal with sample sizes of 30+ even if raw scores are not normally distributed in the population.

The mean of this sampling distribution of means equals the population mean, which in statistics books has the symbol for the Greek letter mu (μ). Its standard deviation is known as the **standard error of the mean,** with the Greek letter sigma as its symbol (σ).

Now, *z* scores that correspond to units of standard error can be calculated, just as *z* scores that correspond to units of standard deviation can be calculated with the normal curve for raw scores.

Hence, one can calculate a *z* score for a sample mean, then use the *z* score tables (area beyond *z*) to ascertain the **probability** (*p*) of this *z* score **occurring by chance** in a population with a given mean.

If this *z* turns out to be **very unlikely**, that is, to have a **low probability** of occurring in a population with a given mean, then we conclude that it has **not** come from that population, but rather from some different population.

Null and alternative hypotheses

These concepts are expressed formally in the language of **null** and **alternative hypotheses**:

The null hypothesis in this case would be that the sample comes from a population with a given mean. (Null hypothesis symbol: H_0)

The alternative or research hypothesis would be that the sample does **not** come from a population with the given mean, but rather from a different population with a different mean. (Alternative hypothesis symbol: H_A or H_1)

Statistical significance

Inferential statistics test the **null** hypothesis, and draw their conclusions in terms of **statistical significance**.

Not significant (retain null hypothesis)

We conclude that a sample mean **has** come from a population with a given mean if:

- it has a relatively **small** absolute z score;
- that has a **high probability** of occurring **by chance** in such a population.

Under these circumstances we **retain** the **null hypothesis**; and decide that the z test is **not statistically significant** (i.e., shows no evidence of a **real** difference in population mean).

Significant (reject null hypothesis)

We conclude a sample mean has **not** come from a population with a given mean if:

- it has a relatively **large** absolute z score;
- with a **low probability** of occurring **by chance** in such a population.

In this case, we **reject** the **null hypothesis** in favour of the **alternative hypothesis**; and decide the z test **is statistically significant** (i.e., shows evidence of a **real** difference in population mean, not just one due to chance).

One- and two-tailed tests

Usually the **alternative hypothesis** is **nondirectional** (i.e., the sample comes from a population with either a smaller or larger mean—along one tail or the other of the normal distribution for the given mean). A nondirectional hypothesis uses a **two-tailed test**.

Sometimes prior research or compelling logic allows us to specify a **directional hypothesis**, for example, that the sample comes from a population with a **larger** mean; a mean that lies in only one tail of a normal distribution for the given mean. A directional hypothesis makes use of a **one-tailed test**.

Decision rules, and alpha

In testing the null hypothesis there can be no certainty. Our decision is made on the basis of probability, which means we have to decide **how unlikely a result has to be** before we deem it to be **statistically significant** (i.e., indicative of a **real** effect or **real** difference, not just one due to chance).

The cutoff probability on which our decision is based is known as **alpha** (α). Alpha is the **rejection level** or the **significance level**.

By convention:

$\alpha = .05$ and corresponds to a probability (p) of $\leq .05$ (i.e., the test result would only occur 5% of the time or less by chance).

A p of .05 corresponds to the following **critical values** of z:

1.96 in a nondirectional **or two-tailed test** ($p = .025$ in **each** of two tails);
1.65 (some texts use 1.64) in a directional or **one-tailed test** ($p = .05$ in **one-tail only**).

To help get your mind around this, you need to realise that:

A **two-tailed** *p* of .05 equates to a *p* of .025 in **each** tail.

A **one-tailed** *p* of .05 equates to a *p* of .05 in only **one** tail, but it must only be the tail specified by the directional hypothesis in order to count (it equates to a two-tailed *p* of .10).

For example, a two-tailed *p* of .08 (and not significant) would equate to a one-tailed *p* of .04 (which would be significant).

If we want to be even more certain that a test result is significant, we can choose a more conservative cutoff criterion (a more extreme probability) as our significance level. By convention, we use α **= .01**. This corresponds to a p of $\leq .01$ (i.e., the test result would only occur 1% of the time or less by chance).

Decision Rules for Alpha = .05

With $\alpha = .05$ these are the decision rules—**LEARN and REMEMBER THEM**:

Note:

< means "less than"
> means "greater than"
≤ means "less than or equal to"
≥ means "greater than or equal to"
Absolute means to disregard the sign (+ or -) and only consider the number

Two-tailed test :

If $z < 1.96$ **absolute**, p must be $>.05$, thus z is **not significant**, **retain** H_0

If $z \geq 1.96$ **absolute**, p must be $\leq .05$, thus z is **significant**, **reject** H_0 (and accept H_A)

One-tailed test (there are two steps to the decision rule for a one-tailed test):

If the difference is **not** in the predicted direction, **automatically retain** H_0

If the difference **is** in the predicted direction:

If $z < 1.65$ absolute, p must be $>.05$, thus z is not significant, **retain** H_0

If $z \geq 1.65$ absolute, p must be $\leq .05$, thus z is significant, **reject** H_0 (and accept H_A)

Important note

When we decide to **retain** the null hypothesis this does **not mean that it is true**. Rather, it means that we do not have sufficient evidence to reject it.

Example of one-sample *z* test

Note: All examples are purely hypothetical, and are not intended necessarily to be good examples of research methodology. They merely serve to illustrate use of the statistical tests.

Let us work through a hypothetical example to reinforce these concepts. Suppose you have made it as a psychologist, and you are called in to conduct research in a community that is concerned about lead levels in the environment. Lead affects intellectual development, so you are asked to establish if the mean IQ of local children is the same as or different to that in the general population. Suppose the mean IQ in the general population is 100, and the standard deviation is 15. You draw a random sample of 25 local children and find that the mean IQ in this sample is 95. You use inferential statistics to answer the question: Has this sample come from a population with a mean of 100, or from a different population with a different mean?

Step 1: Specify hypotheses and decision criteria

H_0: $\mu = 100$

H_1: $\mu \neq 100$

$\alpha = .05$, two-tailed

Retain H_0 if absolute value of observed $z <$ critical z of 1.96

Reject H_0 if absolute value of observed $z \geq$ critical z of 1.96

Step 2: Set down appropriate formula, substitute values, and calculate observed *z*

The statistical test is based on the *z* score formula for raw scores, that is, *z* = the score minus the mean, divided by the standard deviation

$$z = \frac{X - \overline{X}}{s}$$

By converting it to the appropriate formula for the sampling distribution of means, we obtain:

$$z_{\overline{X}} = \frac{\overline{X} - \mu}{\sigma_{\overline{X}}}$$

$$= \frac{95 - 100}{3}$$

$$= -5/3$$

$$= -1.6667$$

$$= -1.67$$

$$= \mathbf{1.67}\ (\text{absolute})$$

The standard error of the mean in the denominator is the population standard deviation divided by the square root of *N*:

$$\sigma_{\overline{X}} = \frac{\sigma}{\sqrt{N}}$$

$$= \frac{15}{\sqrt{25}}$$

$$= \frac{15}{5}$$

$$= \mathbf{3}$$

Step 3: Draw statistical conclusion

The absolute value of the observed z (1.67) is $<$ critical z (1.96), therefore retain H_0: $\mu = 100$.

Step 4: Specify the research conclusion

There is no statistically significant difference in mean IQ between children in the lead-affected community and children in the general population.

Confidence interval of the mean

Knowledge of a sample mean can be used to work out the range of values within which the population mean is likely to fall.

For example, the **95% confidence interval** of the population mean (μ) is given by formula:

$$\overline{X} - 1.96\sigma_{\overline{X}} \leq \mu \leq \overline{X} + 1.96\sigma_{\overline{X}}$$

What this formula says is that the population mean (μ) lies between the **95% confidence limits** of:

the sample mean - 1.96 $SE_{\text{of the mean}}$ and the sample mean + 1.96 $SE_{\text{of the mean}}$

The 95% confidence interval for the lead affected population would be:

$$95 - (1.96 \times 3) \leq \mu \leq 95 + (1.96 \times 3)$$
$$95 - 5.88 \leq \mu \leq 95 + 5.88$$
$$89.12 \leq \mu \leq 100.88$$

Note that this confidence interval includes 100. This is consistent with our z test (α=.05) conclusion that **we cannot reject** the null hypothesis that the lead-affected population has the same mean as the normal population (i.e., a mean of 100). The 95% confidence interval from this sample shows that the lead-affected population could have a mean of 100.

Be careful about the interpretation of confidence limits. Strictly speaking one cannot say here that there is a 95% chance that the population mean lies between 89.12 and 100.88 (it either does or it does not). Rather, the interpretation is that in 95% of the samples that one draws from this population the confidence interval will contain the population mean. Hence, there is a good chance that this sample **is** one of those 95 out of 100 whose confidence limits contain the population mean.

Important Note

Increasingly, it is argued that confidence intervals should be used instead of significance testing in the interpretation of inferential statistics. Significance testing is misleading in that it implies an absolute "right" or "wrong" decision.

Confidence intervals make clear that one is making a **probability judgement**. They also provide extra information by indicating the range of values within which the true population mean is likely to fall. In this case, 100 **is** within the range, but only just. While we cannot rule out the possibility that the lead-affected population has the same mean as the normal population, the confidence interval based on this sample favours the probability of a lower mean. This should prompt us to conduct more research with larger samples that can give more accurate estimates of population means, and that have greater *power* (see pages 42-43 for an explanation of power).

The "problem" with one-tailed tests

Suppose in this example that all the theoretical and empirical literature was unequivocal that lead has a **detrimental** affect on intellectual development. In such circumstances a **directional alternative hypothesis** could be justified (i.e., the population mean would be hypothesised to be **lower** for local children).

The steps (and the conclusion in this case) would change as follows:

Step 1: Specify hypotheses and decision criteria

H_0: $\mu = 100$

H_1: $\mu < 100$

$\alpha = .05$, one-tailed

If difference is **not** in predicted direction, retain H_0

If difference **is** in predicted direction:

Retain H_0 if absolute value of observed $z <$ critical z of 1.65

Reject H_0 if absolute value of observed $z \geq$ critical z of 1.65

Step 3: Draw statistical conclusion

The difference **is** in the predicted direction,
and the absolute value of the observed z (1.67) is > critical z (1.65),
therefore reject H_0: $\mu = 100$, and accept H_1: $\mu < 100$.

Step 4: Specify your research conclusion

Children in the lead-affected community **do** have a significantly lower IQ than the general population IQ of 100 (i.e., the mean IQ of children from the lead-affected population is lower than the mean IQ of children from the general population).

Confidence interval

A one-tailed test leads to rejection of the null hypothesis, as the 95% upper confidence interval does not include 100: 95 + (1.65 x 3)

95 + 4.95

99.95

So what's the "problem" ?

Well, one-tailed tests are certainly **desirable** because they are more **powerful**, in that they are more likely to produce a significant result. But the problem is this: guess what bad researchers are tempted to do **after** they fail to achieve a significant result with a two-tailed test? Duh – "invent" a "justification" for using a one-tailed test all along, of course. Journal editors know this, and sometimes there can be reluctance to accept studies that use one-tailed tests.

One-tailed tests should only be used if you play by the rules. There must be **sound theoretical** or **empirical justification** for using a one-tailed test, and the decision to use a one-tailed test is made on this basis **before**, not after, you perform the test. When there is justification it is indeed better to use a one-tailed test.

Arguably, it is not really appropriate to use a one-tailed test just because you have a directional hypothesis—not unless that directional hypothesis is **so strong** (theoretically) that you are prepared to **ignore a significant difference in the other direction** on the grounds that it must be an anomaly that is of no interest. These are the rules!

Two-sample research designs

Two-sample research designs use the **sampling distribution of *differences* between means**. If one takes a large number of **pairs** of random samples of a given size from a population and constructs a frequency polygon of **differences between their means**, this frequency polygon approximates a normal curve. It tends to be normal with sample sizes of 30+ even if raw scores are not normally distributed.

The mean of this sampling distribution of differences between means equals **zero**, and its standard deviation is known as the **standard error of the *difference* between means**. This makes sense, because, **when averaged**, pairs of samples taken from the same population will have no difference in means, because they are all from the same population and differences in means will cancel out in the long run.

Again, *z* scores that correspond to units of standard error can be calculated, just as *z* scores that correspond to units of standard deviation can be calculated with the normal curve for raw scores.

Hence, one can calculate a *z* score for the difference between a pair of sample means, then use the *z* score tables (area beyond *z*) to ascertain the **probability** (*p*) of this *z* score occurring by chance for samples taken from populations with the same mean (i.e., when the difference between the population means is 0).

Exactly the same logic of significance levels can be used to decide whether to retain or reject the null hypothesis.

Example of a two-sample *z* test

Suppose you draw two random samples (each with $n = 25$) from your lead affected community. The first sample is drawn from an area within a radius of 10 km from the centre of the lead pollution (e.g., an industrial plant) and the mean IQ in this sample is 93. The second sample is drawn from an area between 10 and 20 km from the centre of lead pollution and the mean IQ in this sample is 95.5. You use inferential statistics to answer the question: Have these samples come from populations with the same mean, or different means? *In other words are differences of less than 10 km compared to 10-20 km distance from the site of lead pollution related to IQ?* Suppose that the standard deviation in each population is 15. Also, suppose that a nondirectional hypothesis is necessary as there is no evidence to indicate the range of effect for lead.

Step 1: Specify hypotheses and decision criteria

H_0: $\mu_1 = \mu_2$
H_1: $\mu_1 \neq \mu_2$
$\alpha = .05$, two-tailed
Retain H_0 if observed $z <$ critical z of 1.96 absolute
Reject H_0 if observed $z \geq$ critical z of 1.96 absolute

Step 2: Set down appropriate formula, substitute values, and calculate observed *z*

The statistical test is again based on the *z* score formula for raw scores, that is, *z* = the score minus the mean, divided by the standard deviation.

$$z = \frac{X - \overline{X}}{s}$$

By converting it to the appropriate formula for the sampling distribution of differences between means, we obtain:

$$z_{\overline{X}_1-\overline{X}_2} = \frac{\left(\overline{X}_1-\overline{X}_2\right)-\left(\mu_1-\mu_2\right)}{\sigma_{\overline{X}_1-\overline{X}_2}}$$

$$= \frac{(95.5-93)-0}{2.4495}$$

$$= \frac{2.5}{2.4495}$$

$$= 1.0206$$

$$= \mathbf{1.02}$$

The standard error of the difference between means in the denominator is given by the formula:

$$SE_{pop_1} = SD_{pop_1} \Big/ \sqrt{n_1}$$

$$= 15/\sqrt{25}$$

$$= 15/5$$

$$= \mathbf{3}$$

$$SE_{pop_2} = SD_{pop_2} \Big/ \sqrt{n_2}$$

$$= 15/\sqrt{25}$$

$$= 15/5$$

$$= \mathbf{3}$$

$$\sigma_{\overline{X}_1-\overline{X}_2} = \sqrt{\left(SE_{pop_1} + SE_{pop_2}\right)}$$

$$= \sqrt{3+3}$$

$$= \sqrt{6}$$

$$= \mathbf{2.4495}$$

Step 3: Draw statistical conclusion

Observed z of 1.02 is < critical z of 1.96 absolute, therefore, retain H_0: $\mu_1 = \mu_2$, and reject H_1: $\mu_1 \neq \mu_2$.

Step 4: Specify your research conclusion

There is no statistically significant difference in the IQ means for the population within 10 km of the centre of lead pollution, and the population between 10 and 20 km from the centre of lead pollution.

In other words differences of less than 10 km compared to 10-20 km distance from the site of lead pollution do not appear to be related to IQ.

Note that we need to speak in terms of relationship here, rather than cause and effect, because this is a correlational design, not an experiment. We have **selected** participants for the two groups; we have **not manipulated** distance and **randomly assigned** them to the different distances.

Confidence interval of difference between means

The 95% confidence interval for two-sample tests is given by the following formula, which says that the difference between the population means lies between the confidence limits of:

$$\text{difference between sample means} \pm 1.96\, SE_{\text{of difference between means}}$$

The difference between our sample means is 95.5 - 93 = 2.5.

$$2.5 - (1.96 \times 2.4495) \leq \mu \leq 2.5 + (1.96 \times 2.4495)$$
$$2.5 - 4.8010 \leq \mu \leq 2.5 + 4.8010$$
$$-2.30 \leq \mu \leq 7.30$$

There is a 95% probability that a randomly selected confidence interval such as the one arising from these samples includes the true difference between population means.

Note that this confidence interval includes 0; therefore, it confirms that with a two-tailed test and $\alpha = .05$, we cannot reject the null hypothesis that the two populations (within 10 km and 10-20 km) have no difference in mean (i.e., have the same mean).

Decisions, error, and power

There are four possibilities in statistical inference, and you need to **LEARN and REMEMBER** them:

Your Decision:	**In reality H_0 is TRUE**	**In reality H_0 is FALSE**
Retain H_0	Your decision is RIGHT Probability = **$1 - \alpha$**	Your decision is WRONG **TYPE II ERROR** Probability = **β**
Reject H_0	Your decision is WRONG **TYPE I ERROR** Probability = **α**	Your decision is RIGHT **POWER** Probability = **$1 - \beta$**

Type I error: This occurs when you **falsely reject** the null hypothesis: that is, the null hypothesis is true, but your statistical test leads you to reject it. The Type I error rate corresponds to α, that is, our alpha determines the level to which we control Type I error. With $\alpha = .05$, we can expect to make a Type I error 5 times in every 100 tests.

Correctly retaining H_0: You make a correct decision when you retain the null hypothesis and it is true: that is, the null hypothesis is true, and your statistical test leads you to retain it. The rate for correct acceptance of the null hypothesis is $1 - \alpha$.

Type II error: This occurs when you **falsely retain** the null hypothesis: that is, the null hypothesis is false and the alternative hypothesis is true, but your statistical test leads you to retain the null hypothesis. The Type II error rate is a different value altogether and is referred to as beta (β).

Power, correctly rejecting H_0: Power represents the correct decision that is made when you reject the null hypothesis when it really is false (i.e., the null hypothesis is false and the alternative hypothesis is true, and your statistical test leads you to reject the null hypothesis). The rate for power is $1 - \beta$.

Understanding power

It is very important to clearly understand the meaning of **power**. Power is the sensitivity of a test, its ability to lead to rejection of the null hypothesis if the null hypothesis really is false.

In research, it is important to have sufficiently powerful tests; tests with high enough power to be able to detect a real (and meaningful) effect if one actually exists (e.g., a real difference between population means). Ideally, one should strive for power of .70 or .80 in research.

Power increases as:

1. **Alpha increases** (e.g., $\alpha = .05$ results in a more powerful test than $\alpha = .01$).
2. **Effect size increases**, that is, the magnitude of the treatment effect increases (e.g., the greater the difference between population means in reality, the easier it can be detected; in other words the greater the power).
3. **Variability decreases**, that is, the less variability in the population, the greater the power.
4. **Sample size increases**.

Of these factors the one that researchers have most control over is sample size, hence the inevitable question "How big a sample do I need?" Determining the appropriate sample size is a problem in research (and a failing on the part of many researchers). Formulae are provided in various texts, but they are quite complex and depend on some prior knowledge or estimates of points 2 and 3 above (i.e., effect size and variability).

Various text books may be consulted for more detailed information on power (e.g., Howell, 2002, Chapter 8). One of the most important sources is the book by Cohen (1988). There is also a useful freeware program called **G*Power** that can be used to compute post hoc power values, and sample sizes needed to achieve specified power for a given effect size. It is available at:
http://www.psycho.uni-duesseldorf.de/aap/projects/gpower/

6 PREPARING FOR ANALYSIS: DATA SCREENING FOR PARAMETRIC STATISTICS

The logic of the one-sample and two-sample *z* tests applies to the inferential statistics of the Analysis of Variance (ANOVA) "family", namely *t* tests, and *F* tests, except that instead of being based on the normal distribution of *z* scores they are based on similar sorts of *t* and *F* distributions. All one really needs to know is that the logic is exactly the same.

Moreover, the logic of significance, and the null and alternative hypotheses, is the basic logic used to interpret all inferential statistics. Understand this and you understand the essence of inferential statistics.

ANOVA is one of the **parametric** statistics, which depend on the assumption that scores on the dependent variable are normally distributed in the population. However, parametric statistics are **robust** with respect to normality for sample sizes of 30 or more. Violation of the assumption of normality means that the *p* values for the tests will be inaccurate, but the degree of inaccuracy tends to be small with larger samples.

When the assumptions of parametric statistics are not met, especially when samples are small (<30), alternative **nonparametric** tests can be performed instead. The accuracy of nonparametric tests is not dependent on the satisfaction of population assumptions; however, they tend to be less powerful than parametric tests (although this is not always the case) and, hence, they are not automatically used in place of parametric tests.

Thus, the first step in the analysis process is always to screen the data, to ensure its accuracy to begin with, and then to check assumptions. On the basis of this screening process a decision may need to be made as to which statistical test to use.

Data screening and the analysis process

Data screening

Screening data involves:

1. Checking the accuracy of data entry;
2. Examining any missing data to assess if there is evidence that they might be **systematic** (i.e., caused by something specific, as opposed to being due to random errors);
3. Looking for **outliers**;
4. Checking parametric test assumptions (e.g., normality).

Outliers

Outliers are a particular problem for parametric tests. These are extreme scores, and the problem is that they distort the results. For example, they drag the mean toward themselves and exert "more than their fair share" of influence over the mean and the results of statistical tests. They can also be the cause of non-normality.

First, it is necessary to identify any outliers. SPSS identifies extremes in some of its procedures (e.g., box plots), but it can to be too liberal in doing so. Extreme scores may also be apparent from a visual inspection of histograms.

When suspected outliers are identified, it is recommended that the raw scores be converted to *z* scores, then, according to the recommendations of Tabachnick and Fidell (2007, pp. 73-77) *z* scores in excess of ±3.29 ($p < .001$) may be regarded as outliers, although when *N* is very large a few such scores are to be expected.

Second, it is necessary to deal with outliers. Having identified outliers, a decision must then be made about how to deal with them. One strategy is to delete them from any analysis, but some people object to this as it changes the nature of the sample. It is appropriate to delete outliers, however, if there is other evidence to suggest that they are not really from the population of interest (e.g., a person with an intellectual disability, when the population of interest is people with normal intellectual functioning).

Another strategy is to change the raw score for an outlier so it is still extreme, but not so extreme as to distort the results. Tabachnick and Fidell (2007, p. 77) advise changing outliers so they are only one unit more extreme than the next most extreme score. For example, if an outlier had a score of 45, and the next most extreme score was 23, one would change the outlier to 24; if the outlier had a score of 3 and the next most extreme score was 15 one would change the outlier to 14. Although this may seem like cheating with the data, it is actually the more honest thing to do. This way, the outlier still exerts an influence, but not more than its fair share of influence.

If there is a problem with non-normality and outliers are either deleted or changed, then quite often the data will now meet the assumption of normality, so that one can proceed to conduct parametric tests. **Any action in identifying and dealing with outliers must be reported, albeit succinctly, in the Results section of a research report**.

Other options for dealing with outliers and non-normality are to transform scores (see Tabachnick & Fidell, 2007, pp. 86-88), or to use nonparametric statistics. Nonparametric statistics are based on ranks that are not affected by extreme scores, and they do not require normality.

Deciding on the analysis strategy and conducting appropriate tests

After screening the data and dealing with outliers, the **second** step in the analysis process is to decide on the most appropriate test to use, before proceeding to conduct the analysis.

The chapters that follow provide details for the most common inferential statistics used in psychology (nonparametric alternatives for parametric tests are also provided where they exist).

When to use nonparametric tests

Nonparametric tests should be used:

- when the data (scores on the DV) are at the ordinal level of measurement, rather than being at the interval or ratio level;
- when samples are **small** (<30), and initial data screening indicates that the parametric test assumptions are violated, particularly the normality assumption;
- with **very small** samples (<10), where it can be virtually impossible to conduct an accurate check of assumptions;
- when sample sizes are **small and unequal**.

Nonparametric tests can be used as an alternative strategy for dealing with outliers, because unlike parametric tests they are not distorted by outliers (however, they may be less powerful).

Possibly the best reference for nonparametric tests is Siegel and Castellan (1988).

Data screening with SPSS

Hypothetical example

Imagine you are a school psychologist who is conducting an experiment to find out if mild exercise affects mental alertness. Because you want to determine if exercise has a **causal** effect on mental alertness you conduct a **true experiment** in which you randomly assign 34 student volunteers to one of two groups: a Control group and an Exercise group. Exercise is the IV, and the DV is the number of errors on a proofreading task. Students in the Control group (no exercise) spend the morning in sedentary activities, reading their course materials and taking notes. Between 11:00am and 11:30am they perform the proofreading task. Students in the Exercise condition follow exactly the same procedure, except that they engage in a light aerobics workout between 10:00am and 10:30am.

The data (i.e., number of errors) that result from your experiment are as follows:

ID	Control Group	ID	Exercise Group
1	**4**	18	**5**
2	**3**	19	**4**
3	**5**	20	**4**
4	**2**	21	**4**
5	**5**	22	**2**
6	**4**	23	**1**
7	**6**	24	**3**
8	**4**	25	**2**
9	**3**	26	**3**
10	**6**	27	**2**
11	**18**	28	**3**
12	**3**	29	**5**
13	**2**	30	**2**
14	**4**	31	**3**
15	**5**	32	**2**
16	**6**	33	**3**
17	**5**	34	**5**

Step 1: Enter data in SPSS and save data file

First, enter the data into SPSS and save them in a file that might be called ***Screen.sav***.
There are three variables, with variable labels, and value labels as follows:

ID	Participant number	*Set **Measure** column to **Nominal***
Group	Experimental group 0 = Control 1 = Exercise	*Set **Measure** column to **Nominal***
Errors	Number of proofreading errors	*Set **Measure** column to **Scale***

Data for the first participant would be entered in three columns of the first row as follows: 1 0 4, indicating that this participant has ID number 1, is in the control group (coded 0), and had an errors score of 4. The remaining 33 participants would be entered similarly in the next 33 rows.

Before performing any statistical tests the data must be screened.

A very important point to note is that when there is more than one group (e.g., in this case there are two, Control and Exercise) screening needs to be conducted for each group separately.

Step 2: Check for data accuracy and missing data

First, print out the data and check them for accuracy. Frequency tables can help to check accuracy. For example, if in the SPSS frequency table there is a proofreading score of 60 this would suggest an error in data entry, as would a Group score of 2, because it is **out-of-range**, in that the only possible scores for Group are 0 or 1.

Examine any **missing data**. With experimental research like this, look for evidence of **differential mortality**. For example, if one participant was missing for the Control group and nine were missing for the Exercise group, this may suggest something else about the effect of exercise that could have implications for your study and its interpretation.

Step 3: Screen data, checking for outliers and normality

The SPSS **Explore** procedure is useful for this purpose.

11 How to: Use the Explore procedure to screen data

Analyze
Descriptive Statistics ▸
Explore...
[Select variables you want]
Number of proofreading errors [errors] ▸ Dependent List:
Experimental group [group] ▸ Factor List:

[Click button for **Statistics...** at bottom of dialogue box, then select]
☑ Descriptives
☑ Outliers
Continue

[Click button for **Plots...** then select]
[From **Boxplots** select]
⊙ Factor levels together [This is normally ticked by default]

[From **Descriptive** select]
☐ Stem-and leaf [It is suggested that you untick this]
☑ Histogram

[Select **☑ Normality plots with tests**]
Continue

OK

Step 4: Examine and interpret the SPSS output for Explore

The SPSS Explore procedure produces a great deal of output. Only the most important output is included here. However, you can examine all the output when you run the analysis yourself.

Note that the **Explore** procedure automatically gives separate results for groups if the grouping variable is assigned to the **Factor List** (as above). Other procedures that can be useful for data screening are **Descriptive Statistics ▸ Frequencies...** and **Descriptive Statistics ▸ Descriptives...**, but with each of these you need to use **Split File** (as described later in this chapter) if you want to screen the data separately for each groups.

SPSS output for Explore

Descriptives

	Experimental group			Statistic	Std. Error
Number of proofreading errors	Control	Mean		5.00	.870
		95% Confidence Interval for Mean	Lower Bound	3.16	
			Upper Bound	6.84	
		5% Trimmed Mean		4.44	
		Median		4.00	
		Variance		12.875	
		Std. Deviation		3.588	
		Minimum		2	
		Maximum		18	
		Range		16	
		Interquartile Range		3	
		Skewness		3.247	.550
		Kurtosis		12.153	1.063
	Exercise	Mean		3.12	.296
		95% Confidence Interval for Mean	Lower Bound	2.49	
			Upper Bound	3.74	
		5% Trimmed Mean		3.13	
		Median		3.00	
		Variance		1.485	
		Std. Deviation		1.219	
		Minimum		1	
		Maximum		5	
		Range		4	
		Interquartile Range		2	
		Skewness		.219	.550
		Kurtosis		-.875	1.063

1

First, there are **Descriptive statistics** for the two groups. They include **skewness** and **kurtosis,** which are measures of normality. Skewness, as its name suggests, indicates the extent to which the distribution is positively or negatively skewed. Kurtosis relates to how peaked, with thin or fat tails, is the distribution. The more these depart from zero, the less normal is the distribution.

Dividing the skewness and kurtosis values by their standard errors in the second column, gives the *z* scores for skewness and kurtosis. Refer to Tabachnick and Fidell (2007, pp. 79-83), who suggest we can evaluate the significance of skewness or kurtosis based on the *z* scores, but using alpha levels of .01 or .001 when samples are small to moderate.

If you requested Outliers, SPSS then lists the most extreme scores in the two groups (not included here).

Tests of Normality

		Kolmogorov-Smirnov[a]			Shapiro-Wilk		
	Experimental group	Statistic	df	Sig.	Statistic	df	Sig.
Number of proofreading errors	Control	.331	17	.000	.602	17	.000
	Exercise	.186	17	.123	.912	17	.107

a. Lilliefors Significance Correction

2

The Shapiro-Wilk is the standard test of normality. Both it and the Kolmogorov-Smirnov test the **null hypothesis** that the data are from a normal distribution. They very often lead to the same conclusion, but in the event of disagreement it is advisable to use the Shapiro-Wilk.

If the test is **not significant** (i.e., *p* or Sig. > .05), the null hypothesis is **retained** and the normality assumption is met; the data are assumed to have come from a normal population.

If the test **is significant** (i.e., *p* or Sig. ≤ .05), the null hypothesis is **rejected** in favour of the alternative hypothesis and the normality assumption fails; the test indicates that the data have not come from a normal population.

Make sure you can see that in this case the normality assumption **is met**, for the **Exercise** group because the **Shapiro-Wilk** test is **not significant**, but **NOT met** for the **Control** group because the test **is significant**.

SPSS Results Coach:

If at any time you want an explanation of SPSS results, you can try right clicking on any section of output, then click on **Results Coach** if it appears as an option in the dialog box. The various parts of the output are then described. This option is not available in PASW 18.

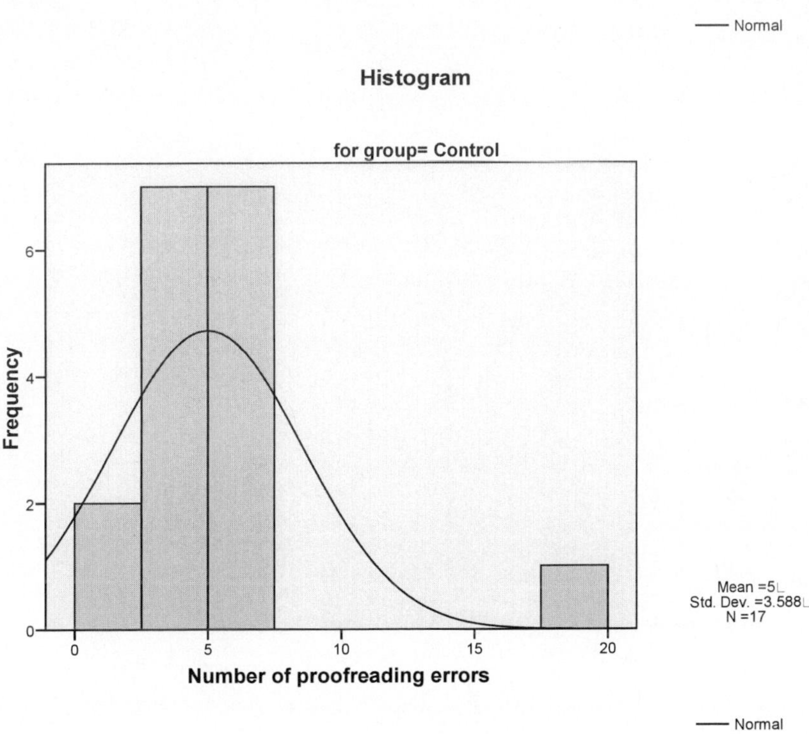

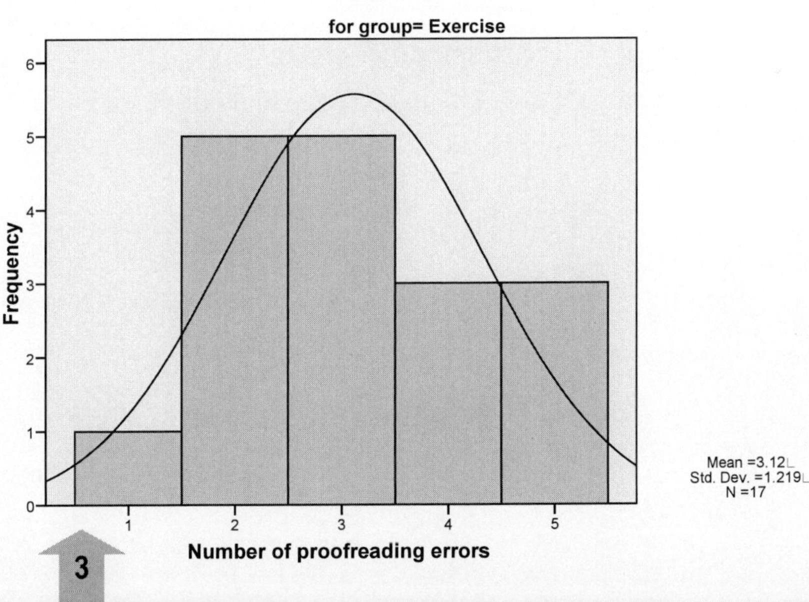

3

Histograms provide the best visual indication of normality, and also give a good indication of just how extreme potential outliers might be.

A normal curve is not automatically superimposed when you obtain a histogram using the **Explore** procedure (as it is through **Frequencies**). To obtain one, or to make other modifications, double-click on the histogram to open the **Chart Editor** window, then select the **Elements** menu; choose **Show Distribution Curve**; and select **Normal** from the dialog box that opens (then Close the dialog box and exit the Chart Editor).

Here, we can clearly see the normality problem in the Control group, and also that it may be due to an outlier.

Output: **Data screening using Explore**

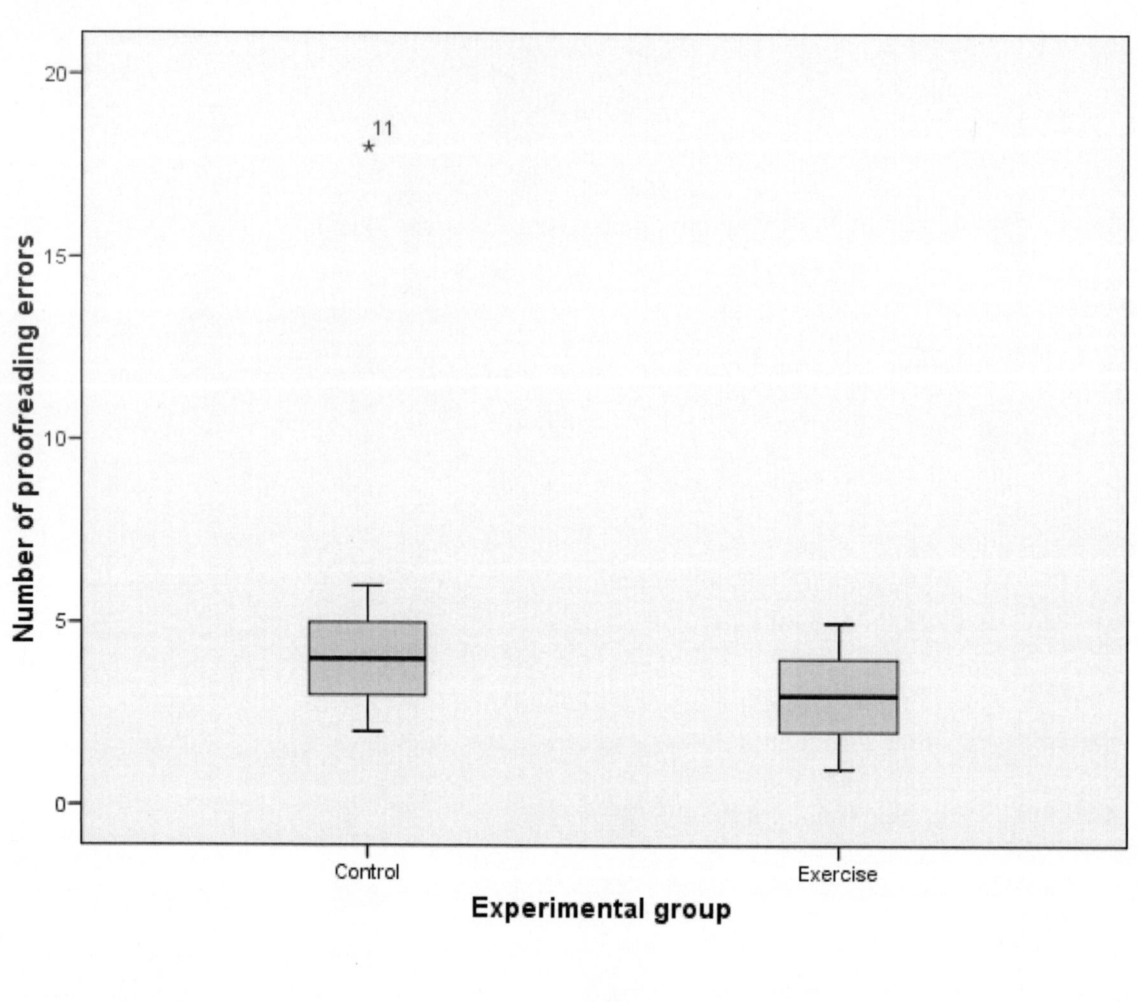

4

In addition to histograms, SPSS provides graphical output in the form of stem-and-leaf plots (if you asked for them), various normality plots, and box plots. Text books such as Tabachnick and Fidell (2007) and Howell (2002) explain how to interpret all of these.

Box plots:

Box plots (also known as box-and-whisker plots) can be useful because they also provide a good representation of distributions, and identify potential outliers.

Here is the box plot for this analysis. The dark bar in the centre of each shaded box corresponds to the **median**; the upper and lower limits of the boxes correspond to the **25th and 75th percentiles** (i.e., they mark the inter-quartile range); and the whiskers mark the smallest and largest observed values that are not potential outliers.

Potential outliers are indicated outside the whiskers. Note that case number 11 in the Control group is identified as a potential outlier here. But, be careful, this is case number 11 according to the permanent row numbers in the SPSS **Data View** window. You must check to see which ID number is located at row 11. (Because we entered the data in order, case or row number and ID number are the same, but data sometimes get re-sorted in analyses, so you always need to check.)

Step 5: Confirm any suspected outliers by obtaining *z* scores for each group

Participant number 11 in the Control Group certainly looks like an outlier.

This can be confirmed by obtaining the *z* score for this participant.

Because there are two groups, *z* scores need to be obtained for each group separately, using the Split File option in SPSS.

12 How to: Use Split File to perform analyses separately for different groups

Data
Split File...
⊙ **Compare Groups**[1]
[Select variable for which you want separate analyses]
Experimental group [group] ▸ Groups Based on:

OK

(***Note:*** A **Split File On** message appears in the bar at the bottom right of the data window.)

Turning Split File OFF

Should you want to do later analyses on **all** the data, you must remember to turn Split File **off** beforehand, as follows:

Data
Split File...
⊙ **Analyze all cases, do not create groups**
OK

[1] This displays results side-by-side. To display separately use **Organize output by groups.**

13 How to: Create a column of z scores for each group separately in the data file

Split File would be off if you wanted to create *z* scores on the whole sample.

However, in this case, where you want to create *z* scores separately for each group, **Split File must be on**.

Analyze
  Descriptive Statistics ▸
    Descriptives...
    [Select variables you want]
    Number of proofreading errors [errors] ▸ Variable(s):

    [Click on the following box to request *z* scores]
    ☑ Save standardized values as variables
  OK

The *z* scores are actually added to the data file as a variable with the same name as the raw scores variable, but preceded by a Z (i.e., **Zerrors**).

If you look down the data column created by SPSS in the data window for the new variable **Zerrors,** you will see it confirmed that participant number 11 in the Control group, with a raw score of 18 and a *z* score of 3.623 (>3.29) is an outlier.

Step 6: Repeat data screening after dealing with outliers

Suppose, on further investigation, you find out that participant number 11 is a man who normally needs to wear reading glasses, but had performed the proofreading task without them. Under such circumstances this participant's data could be deleted from the dataset on the grounds that he is **not** a member of the desired population of non-vision-impaired readers.

Alternatively, suppose this was not the case. Suppose, instead, that there is no special reason for the poor performance of participant number 11. This person is just unusually poor at proofreading. You might choose to reduce the distorting effect of this participant on the Control group according to the advice of Tabachnick and Fidell (2007). That is, you would change the raw score so it is only one unit more extreme than the next most extreme score in the Control group, as follows:

- Use SPSS to produce a frequency table for the Control group. You will see the next most extreme score is 6.
- Now change the raw score of participant number 11 from 18 to 7 (one unit larger than the next most extreme score of 6).

Now, run the data screening again with this revised score. This time you will see that there are no outliers, and in this example the normality assumption is now met for the Control group. You can proceed to use parametric statistics to analyse the data.

Important

It is strongly recommended that you always retain a file of the original data, before any changes are made.

If you decide to delete or change outliers, or make other modifications to a data file, it is recommended that you save these in a different file with a different file name. In this case ***Screen.sav*** might be the original data file, and the modified version, where the score for participant number 11 is changed, might be saved in a file called ***Screen1.sav***.

The file names used throughout this book are just suggestions. You can give your files whatever names you like. However, it is a good idea to name files in ways that best indicate—to you—the nature of their content.

7 THE ONE-SAMPLE *t* TEST

Earlier, the one-sample and two-sample *z* tests were described. However, these are rarely used, because the standard deviation in the population is rarely known, so the relevant standard error cannot be calculated.

Instead *t* tests are used. These follow exactly the same logic, but to ascertain the critical values of *t*, a family of sampling distributions known as *t* distributions is used. Although these all resemble the normal distribution, there is a different *t* distribution for different sample sizes or, more specifically, for each of the different **degrees of freedom** (*df*). The concept of degrees of freedom is not easy to grasp. It refers to the number of independent pieces of information that are needed to determine a statistic. For example, to work out a mean you need *N* pieces of information; that is, you need to know every score in the sample. If in a sample of three scores I tell you that two of them are 5 and 7, you are still unable to work out the mean until you know the third score. The mean has *N* degrees of freedom. However, if I told you the mean was 5 you would be able to work out the third score (i.e., 3). Any statistic that uses the sample mean can have no more than $N - 1$ degrees of freedom. Do not worry too much about this, although you may want to read up on it in textbooks.

When do you use the one-sample *t* test?

The one-sample *t* test is used when you have **scores from a single sample or group of participants** and you want to know if the mean of the population from which the sample is drawn is the same as some hypothesised mean. In order to use the one-sample *t* test, the following assumptions need to be satisfied, as the accuracy of the test depends on them.

Assumptions

1. The scores should be at the interval or ratio level of measurement.
2. The scores are randomly sampled from the population of interest
3. The scores are independent of each other.
4. The scores are normally distributed in the population.
   When *N* is large (30+) the *t* test is considered to be **robust**; that is, violation of the normality assumption has little effect on its accuracy.

Example of a one-sample *t* test

Suppose at the end of your three-year degree you get a job in the Human Resource Management department of the local branch of a large national company. The headquarters of your company is located in Victoria and they are so concerned about their poor safety record that they have just conducted a survey of the work safety knowledge of all their employees. The instrument measuring work safety has a possible range of 0-40; the higher the score, the greater the knowledge. In particular, they have been disturbed to find that the level of knowledge among supervisory staff, who have a key role to play in work safety, is only 15.9.

Suppose budgetary restraints preclude a similar full-scale survey among the local workforce. Instead, the branch manager asks you to select a random sample of supervisory staff, and to ascertain if the local mean is the same as or different from the headquarters mean.

Manual calculation of a one-sample *t* test

Step 1: Specify hypotheses and decision criteria

H_0: $\mu = 15.9$
H_1: $\mu \neq 15.9$
$\alpha = .05$, Two-Tailed
Retain H_0 if absolute value of observed $t <$ critical t
Reject H_0 if absolute value of observed $t \geq$ critical t

Step 2: Set down data table

[***Note:*** *X* represents the work safety knowledge score for each of 16 participants in a random sample of the local branch supervisory staff.]

ID	*X*	X^2
1	19	361
2	29	841
3	23	529
4	10	100
5	17	289
6	7	49
7	30	900
8	25	625
9	25	625
10	32	1024
11	8	64
12	22	484
13	15	225
14	26	676
15	12	144
16	22	484
$N = 16$	$\sum X = 322$	$\sum X^2 = 7420$

Step 3: Calculate the mean

$$\overline{X} = \frac{\Sigma X}{N}$$

$$= 322/16$$

$$= 20.125$$

$$= \mathbf{20.13}$$

Step 4: Calculate the standard deviation

$$s = \sqrt{\frac{\Sigma X^2 - \frac{(\Sigma X)^2}{N}}{N-1}}$$

$$= \sqrt{\frac{7420 - \frac{(322)^2}{16}}{16-1}}$$

$$= \sqrt{\frac{7420 - \frac{103684}{16}}{15}}$$

$$= \sqrt{\frac{7420 - 6480.25}{15}}$$

$$= \sqrt{\frac{939.75}{15}}$$

$$= \sqrt{62.65}$$

$$= 7.9152$$

$$= \mathbf{7.92}$$

Step 5: Calculate the estimate of the standard error of the mean

$$s_{\overline{X}} = \frac{s}{\sqrt{N}}$$

$$= \frac{7.9152}{\sqrt{16}}$$

$$= \frac{7.9152}{4}$$

$$= \mathbf{1.9788}$$

Step 6: Set down the appropriate formula, substitute the values, and calculate observed *t*

[***Note:*** The *t*-test formula is the same as the corresponding *z*-test formula, except that we do not know the standard error of the mean, and use the estimate instead.]

$$t = \frac{\overline{X} - \mu}{s_{\overline{X}}}$$

$$= \frac{20.125 - 15.9}{1.9788}$$

$$= \frac{4.225}{1.9788}$$

$$= 2.1351$$

$$= \mathbf{2.14}$$

Step 7: Calculate degrees of freedom

$$df = N - 1$$
$$= 16 - 1$$
$$= \mathbf{15}$$

Step 8: Look up a table of critical values of *t*

Internet site: http://www.danielsoper.com/statcalc/calc10.aspx
For a two-tailed test with $\alpha = .05$ and 15 *df* the critical value of *t* is 2.13145.

Step 9: Draw the statistical conclusion

The absolute value of the observed *t* is 2.1351 and this is > the critical *t* of 2.131; therefore **reject** H_0: $\mu = 15.9$, and **accept** H_1: $\mu \neq 15.9$.

Step 10: Specify the research conclusion

The mean work safety knowledge scores for the local branch supervisory staff is significantly higher, statistically, than the mean for supervisory staff in the Victorian headquarters.

Step 11: You may also want to calculate the confidence limits, as follows:

When $\alpha = .05$, we are interested in the corresponding 95% confidence interval.

$$\begin{array}{rcl} \overline{X} - (\text{critical } t)(s_{\overline{X}}) & \leq \mu \leq & \overline{X} + (\text{critical } t)(s_{\overline{X}}) \\ 20.125 - (2.131)(1.9788) & \leq \mu \leq & 20.125 + (2.131)(1.9788) \\ 20.125 - 4.2168 & \leq \mu \leq & 20.125 + 4.2168 \\ 15.9082 & \leq \mu \leq & 24.3418 \\ \mathbf{15.91} & \leq \mu \leq & \mathbf{24.34} \end{array}$$

Ninety-five per cent of the time a confidence interval calculated in this way from a randomly selected sample will include the population mean.

Because this confidence interval does not include the hypothesised population mean of 15.9, we can reject the null hypothesis (but **only just**), and can conclude that the mean work safety knowledge scores for local branch supervisory staff is higher than the mean for supervisory staff in the Victorian headquarters.

We can also be 95% confident that the local population mean is somewhere in the range 15.91 to 24.34 (or, by subtraction of 15.9, that the **difference** between the local and Victorian population means is somewhere in the range 0.01 to 8.44, in favour of the local branch). Note the extra information we obtain from confidence intervals. Here, they make us aware that the difference *could* be almost nonexistent.

Using SPSS to conduct the one-sample *t* test

Step 1: Enter data in SPSS and save data file

First, enter the data into SPSS and save it in a file called ***OneSampleTTest.sav***. There are two variables, with variable labels, and value labels as follows:

ID	Participant number	*Set **Measure** column to **Nominal***
Know	Safety knowledge NSW supervisors	*Set **Measure** column to **Scale***

Step 2: Screen data and check assumptions

Screen the data as described in the *Data Screening* chapter, checking the normality assumption carefully, given that the sample size is small. (You should find that the normality assumption is met.)

Step 3: Request the analysis

14 How to: Conduct a one-sample *t* test

Analyze
Compare Means ▸
One-Sample T Test...
[Select variables you want]
Safety knowledge local supervisors [know] ▸ **Test Variables(s):**
[In the box for **Test Value:** type] **15.9**
OK

Step 4: Examine and interpret the output

T-Test

One-Sample Statistics

	N	Mean	Std. Deviation	Std. Error Mean
Safety knowledge local supervisors	16	20.13	7.915	1.979

One-Sample Test

	Test Value = 15.9					
					95% Confidence Interval of the Difference	
	t	df	Sig. (2-tailed)	Mean Difference	Lower	Upper
Safety knowledge local supervisors	2.135	15	.050	4.225	.01	8.44

1

With SPSS you do not have to look up tables of critical values, because it gives you the **exact probability (*p*)** for your *t* test, but calls it the **Sig. (2-tailed)**.

You need only compare this *p* with your α:
If $p > \alpha$ then your *t* test is NOT significant
If $p \leq \alpha$ then your *t* test IS significant.

In this case p = .05 exactly; therefore the *t* test IS significant using the normal α of .05.

For a two-tailed test:
p = Sig. (2-tailed) = .05 in this case

For a one-tailed test:
p = Sig. (2-tailed) / 2 = .025 in this case
(i.e. where p = .05, any **one** tail has a p of .025)

2

This is the 95% confidence interval for the difference between the means.

It tells us that at the 95% confidence level the difference between the local and headquarters population means is somewhere between .01 and 8.44.

A note about E-notation:

Depending on how SPSS has been set up in the **Edit** then **Options...** menu you might encounter output that uses E-notation, or scientific notation, as a way of reporting small numbers. For example, the Lower limit of the 95% confidence interval in the **One-Sample Test** output above may be shown as **7.30E-03**, but what would this mean?

E-03 means **"move the decimal point three places to the left"**. Thus, 7.30E-03 becomes .0073 (which is .01 rounded to two decimal places). (E-04 would mean move four places to the left, so that a number of 4.2E-04 would become .00042, and so on).

Statistical significance versus practical importance

Statistical significance only means that there is very probably a **real** difference between the population means. However, it does not mean that this difference is "significant" in the sense of **important** from a practical point-of-view. With a large enough sample **any** difference, however trivial, will be statistically significant.

To assess the practical importance of a difference between means (or any other statistical result) you need to look at the magnitude of the difference—and the 95% confidence intervals—in the light of what is actually being measured by the scores. Then you have to ask yourself whether a difference of this magnitude is meaningful in a practical sense.

Effect size and Cohen's *d*

Increasingly, researchers are required to report another measure that speaks directly to the issue of importance, that is, some measure of **effect size**. There are a number of these, with **Cohen's *d*** (Cohen, 1988) being one of the most common. Cohen's d is simply a z score. For the one-sample t test it is the difference between means in units of standard deviation, and is calculated for this example as follows (see Green & Salkind, 2005, pp. 156-157):

$$d = \frac{Mean\ difference}{SD} = \frac{20.125 - 15.9}{7.9152} = \frac{4.225}{7.9152} = .5338 = \mathbf{.53}$$

OR
alternative formula

$$d = \frac{t}{\sqrt{N}} = \frac{2.135}{\sqrt{16}} = \frac{2.135}{4} = .5338 = \mathbf{.53}$$

Cohen (1988) proposed that d values of .2, .5, and .8 could be regarded as representing small, medium, and large effect sizes respectively. In this case the effect size is a medium one, with the local branch mean exceeding the headquarters mean by just over half a standard deviation. Effect size measures are useful as a standardised way of comparing results across different studies, and they can also be used in subsequent research to help determine the required sample size.

In addition to reporting in the results section, you might also comment on the **practical** importance of the findings in the Discussion section of a research report.

How to report statistical results in the Results section of a research report

Sample Results sections, written according to the style and content requirements of the *Publication manual of the American Psychological Association* (American Psychological Association, 2010), are provided throughout this text. They are not intended to be definitive examples, nor are they the only way to report the results. Rather, they provide an indication of appropriate contents for the Results section of a research report.

When reporting statistical results the exact *p* value is reported for the statistical test, for example, $p = .016$, or $p = .347$. Where the SPSS output gives a significance value of .000, report this as $p < .001$. (Note that an older convention was to simply report $p <$ or $> \alpha$; but this is no longer relevant now that statistical software gives exact values.) The previous edition of the *APA manual*[1] also required the alpha level to be specified, but that advice has been discontinued in favour of specifying 95% or 99% confidence intervals, which are more informative.

Statistical symbols (e.g., *t*) and the probability symbol (*p*) are **italicised** when reported in the Results section. In the past they were underlined, but now italics are **always** used instead of underlining.

After specifying the α level at the outset, a one-sample *t* test is reported as:

t(*df*) = observed *t* value to two decimal places, *p* = exact probability or significance value.

One-sample *t* test: Research report sample Results section

Results

To determine whether the mean work safety knowledge of supervisory staff in the local branch was the same as for the Victorian headquarters, a one-sample *t* test was conducted. The assumption of normality was met, and the result was statistically significant, $t(15) = 2.14$, $p = .05$. This indicated that the mean work safety knowledge score for local branch supervisory staff ($M = 20.13$, $SD = 7.92$) was significantly higher than the mean for supervisory staff in the Victorian headquarters ($M = 15.9$). A medium effect size (Cohen's $d = 0.53$) was evident in the mean difference of 4.23, 95% CI [0.01, 8.44].

[1] Although a common abbreviation for the *Publication manual of the American Psychological Association* this should **not** be used as a substitute for the full reference in manuscripts or student papers.

8 The Dependent-Samples *t* Test, and Wilcoxon Signed-Rank T Test Nonparametric Alternative

When do you use the dependent samples *t* test?

The dependent-samples *t* test (or paired- or correlated-samples *t* test) is used when there are data from **within-groups** or **repeated measures designs**, that is, when there is one group of participants, but there are two DV scores from those participants obtained under different levels of an IV.

Alternatively, there may be two groups of participants who are **related** or **matched** in some way (e.g., husbands and wives; fathers and sons; or matched-pairs, that is, pairs of participants who have been matched on some variable). The two groups receive or represent different levels of an IV, and the data are their scores on some DV.

The research question is whether the population means for the two sets of scores are the same or different; in other words, whether or not the **differences** between the scores are equal to zero.

The pretest-posttest **pre-experimental** design is a common use of the dependent *t* test. Here, scores are obtained on a DV before and after some intervention or manipulation, and the researcher wants to know if there is a difference between the means on the pretest and the posttest. (A **pre**-experimental design lacks a control group for comparison, so we are unable to rule out other causes for any observed difference).

With all within-subjects designs it is necessary to be alert to the potential problem of **order** effects. These occur when there is a carry-over effect from one score to the next that has nothing to do with the IV, for instance, when participants perform better at posttest simply because of the practise they received on the pretest.

In order to use the dependent-samples *t* test, the following assumptions need to be satisfied, as the accuracy of the test depends on them.

Assumptions

1. The data (scores) should be at the interval or ratio level of measurement.
2. The cases are randomly sampled from the population.
3. The dependent variable is normally distributed in the population at each level of the independent variable (and the **difference scores** are normally distributed). When n is large (30+) the dependent *t* test is considered to be **robust**; that is, violation of the normality assumption has little effect on its accuracy.
4. The difference scores are independent of one another.

Example of a dependent samples *t* test

Imagine that for a research project you want to investigate the hypothesis that waiting in a room with a fish tank will reduce systolic blood pressure. To conduct such a study you might randomly select a sample of first year psychology students to take part in "a study about exercise and stress". You might arrange for them to meet in a laboratory at a specified time, then when they are all present you might ask them to wait for 10 minutes (the control condition) while the study apparatus is set up. After 10 minutes you come in and explain you need to take two blood pressure readings 10 minutes apart in order to ascertain their resting stress levels. You take the first reading for each participant, then you might "collect a piece of apparatus" from the waiting room that has been concealing the fish tank. Participants wait a further 10 minutes in the presence of the fish tank before you take the second blood pressure reading. After this, you debrief them about the true purpose of the study.

As an aside, take a few moments to consider what design faults might be present in this study. Are there likely to be order effects? Also, do you envisage any ethical problems?

Manual calculation of a dependent samples *t* test

The dependent-samples *t* test is based on **difference scores**; and the formula is used to compare the **mean difference score** with the hypothesised mean difference of 0, which is the null hypothesis of no difference between the two sets of scores.

Step 1: Specify hypotheses and decision criteria

H_0: $\mu_1 = \mu_2$ or $\mu_1 - \mu_2 = 0$
H_1: $\mu_1 \neq \mu_2$ or $\mu_1 - \mu_2 \neq 0$

Suppose you want to be extra sure about the presence of any difference, so you decide to use:

α = **.01**, two-tailed (two-tailed, because you do not want to rule out the possibility of a difference in the other direction)

Retain H_0 if absolute value of observed $t <$ critical t
Reject H_0 if absolute value of observed $t \geq$ critical t

[Note that the dependent *t* test does **not** have to have $\alpha = .01$. Choice of $\alpha = .05$ or $\alpha = .01$ depends on your research requirements, and how certain of an effect you want to be. If it is only possible to obtain a very small sample, where you know power will be low, a case may even be made for using a more liberal alpha level of .10 or even .15, although there is still reluctance to do this, because of the .05 tradition.]

Step 2: Set down data table

[***Note***: X_1 represents the DV scores under the control condition; X_2 represents the DV scores under the fish tank condition. D is the **difference** between X_1 and X_2.]

ID	X_1	X_2	D	D^2
1	120	114	6	36
2	115	110	5	25
3	125	125	0	0
4	130	120	10	100
5	135	123	12	144
6	110	112	-2	4
7	120	112	8	64
8	135	130	5	25
9	140	133	7	49
10	125	120	5	25
11	138	130	8	64
12	112	110	2	4
13	110	115	-5	25
14	115	112	3	9
15	130	122	8	64
16	132	125	7	49
$N = 16$	$\sum X_1 = 1992$	$\sum X_2 = 1913$	$\sum D = 79$	$\sum D^2 = 687$

Step 3: Calculate the means

$$\bar{X}_1 = \frac{\Sigma X_1}{N}$$
$$= 1992/16$$
$$= 124.5$$
$$= \mathbf{124.5}$$

$$\bar{X}_2 = \frac{\Sigma X_2}{N}$$
$$= 1913/16$$
$$= 119.5625$$
$$= \mathbf{119.56}$$

Step 4: Calculate the standard deviations

$$s = \sqrt{\frac{\Sigma X^2 - \frac{(\Sigma X)^2}{N}}{N-1}}$$

In order to calculate these you need to extend the data table to have columns for X_1^2 and X_2^2, so that you can calculate $\sum X_1^2$ and $\sum X_2^2$ for the above formula. Rather than do this here, the standard deviations will be obtained from the SPSS output.

Step 5: Calculate the mean of the difference scores

$$\overline{D} = \frac{\Sigma D}{N}$$

$$= 79/16$$

$$= 4.9375$$

$$= \mathbf{4.94}$$

Step 6: Calculate the standard deviation of the difference scores

$$s_D = \sqrt{\frac{\Sigma D^2 - \frac{(\Sigma D)^2}{N}}{N-1}}$$

$$= \sqrt{\frac{687 - \frac{(79)^2}{16}}{16-1}}$$

$$= \sqrt{\frac{687 - \frac{6241}{16}}{15}}$$

$$= \sqrt{\frac{687 - 390.0625}{15}}$$

$$= \sqrt{\frac{296.9375}{15}}$$

$$= \sqrt{19.7958}$$

$$= \mathbf{4.4493}$$

Step 7: Calculate the estimate of the standard error of the difference mean

$$s_{\overline{D}} = \frac{s_D}{\sqrt{N}}$$

$$= \frac{4.4493}{\sqrt{16}}$$

$$= \frac{4.4493}{4}$$

$$= \mathbf{1.1123}$$

Step 8: Set down the appropriate formula, substitute the values, and calculate observed *t*

Note how this formula is just a variation on the formula for the one-sample *t* test.

$$t = \frac{\overline{D} - \mu}{s_{\overline{D}}}$$

$$= \frac{4.9375 - 0}{1.1123}$$

$$= \frac{4.9375}{1.1123}$$

$$= 4.439$$

$$= \mathbf{4.44}$$

Step 9: Calculate degrees of freedom

$$df = N - 1$$

$$= 16 - 1$$

$$= \mathbf{15}$$

Step 10: Look up a table of critical values of *t*

Internet site: http://www.danielsoper.com/statcalc/calc10.aspx
For a two-tailed test with $\alpha = .01$ and 15 *df* the critical value of *t* is 2.947.

Step 11: Draw statistical conclusion

The absolute value of the observed *t* is 4.44 and this is > the critical *t* of 2.947, therefore **reject** H_0: $\mu_1 = \mu_2$ or $\mu_1 - \mu_2 = 0$, and **accept** H_1: $\mu_1 \neq \mu_2$ or $\mu_1 - \mu_2 \neq 0$.

Step 12: Specify the research conclusion

The mean systolic blood pressure was significantly lower, statistically, after participants had waited for 10 minutes in the presence of a fish tank, then after they had waited 10 minutes in a normal resting state.

Using SPSS to conduct the dependent-samples *t* test

Step 1: Enter data in SPSS and save data file

First, enter the data into SPSS and save them in a file called ***DependentTtest.sav***.

ID	Participant number	*Set **Measure** column to **Nominal***
Control	Control blood pressure	*Set **Measure** column to **Scale***
Fish	Fishtank blood pressure	*Set **Measure** column to **Scale***

Step 2: Screen data and check assumptions

Screen the data for the two DVs, Control and Fish, as described in Chapter 6. Check the normality assumption for the **difference scores** carefully, given that the sample size is small. To do this use **Compute** to create a new variable that is the **difference** between the Control and Fish scores as shown below, then perform data screening for normality on this difference variable

15 **How to: Use Transform to compute new variables**

Transform
  Compute Variable...
  [In space for **Target Variable** type **Difference** or whatever the new variable is to be called]
  [In the space for **Numeric Expression** select variables as follows]
    Control blood pressure [control] ▸ Numeric Expression
    [Then select or type minus sign **-**]
    Fishtank blood pressure [fish] ▸ Numeric Expression
  OK

Step 3: Request the analysis

16 **How to: Conduct a dependent-samples *t* test**

Analyze
  Compare Means ▸
    Paired-Samples T Test...
    [Select **both** variables you want simultaneously]
    Control blood pressure [control]
    Fishtank blood pressure [fish] ▸ Paired Variables:
    [Click on] **Options...**
      [For **Confidence Interval:** type] **99** [The 99% CI is appropriate when $\alpha = .01$]
      Continue
  OK

Note that from Version 18 you can repeat this step if you have many variables and want to perform a number of different paired *t* tests.

Step 4: Examine and interpret the output

Output: **Dependent *t* test**

Paired Samples Statistics

		Mean	N	Std. Deviation	Std. Error Mean
Pair 1	Control blood pressure	124.50	16	10.191	2.548
	Fishtank blood pressure	119.56	16	7.668	1.917

Paired Samples Correlations

		N	Correlation	Sig.
Pair 1	Control blood pressure & Fishtank blood pressure	16	.914	.000

Paired Samples Test

		Paired Differences					t	df	Sig. (2-tailed)
					99% Confidence Interval of the Difference				
		Mean	Std. Deviation	Std. Error Mean	Lower	Upper			
Pair 1	Control blood pressure - Fishtank blood pressure	4.938	4.449	1.112	1.660	8.215	4.439	15	.000

1

The mean **difference** between the two group means.

2

The **99%** confidence interval.

99% of the time this interval (1.66 to 8.22) will contain the true difference between the population **means.**

3

t value; degrees of freedom; and the **Sig. (2-tailed)**, which is the *p* value.

The *p* value is < α, which we had set at .01 in this study.

Therefore, the observed *t* value **IS significant.**

Statistical significance versus practical importance

Remember: Statistical significance only means that there is very probably a **real** difference between the population means. However, it does not mean that this difference is "significant" or **important** in the practical sense. With a large enough sample any difference, however trivial, will be statistically significant.

To assess the practical importance of a difference between means (or any other statistical result) you need to look at the magnitude of the difference—and the 95% confidence intervals—in the light of what is actually being measured by the scores. Then you have to ask yourself whether a difference of this magnitude is meaningful in a practical sense.

Effect size: Cohen's *d*

You also need to consider the effect size, which is calculated for a dependent *t* test as follows (see Green and Salkind (2005, p.163):

$$d = \frac{\overline{D}}{S_D} = \frac{4.9375}{4.4493} = 1.1097 = \mathbf{1.11}$$

OR alternative formula

$$d = \frac{t}{\sqrt{N}} = \frac{4.439}{\sqrt{16}} = \frac{4.439}{4} = 1.1097 = \mathbf{1.11}$$

This tells us that the mean difference between the two sets of scores is 1.11 times the standard deviation of the difference scores (i.e., the "average" amount of difference across individuals), and this is a large effect using Cohen's (1988) guidelines of .2, .5, and .8 for small, medium, and large effects.

Effect size: The correlation coefficient, and proportion of variance

An alternative to using *z* scores (e.g., Cohen's *d*) as a measure of effect size, is to use correlation-based measures. The correlation coefficient (symbol *r*), as a statistic of relationship, is dealt with in a later chapter, but it also has a more general application as a measure of the strength of an effect (although, remember, we can only assume an effect is **causal** if we have an **experimental** research design).

Never lose sight of the fact that the whole reason for doing research is to try to identify and understand the factors that produce **variance**. Why, in this case, is there variation in blood pressure? Here, we are pursuing a theory that the extent to which one is relaxed will be associated with blood pressure, and on the assumption that the fish tank will induce greater relaxation we are expecting that blood pressure will be lower in the fish tank condition than in the control condition. We want to find out if relaxation associated with the fish tank is associated with a **difference** in blood pressure over-and-above what can be attributed to chance.

Correlation **squared** (r^2) tells us the **proportion of variance** in one variable that can be predicted by another. In **experimental** designs, such as the one we have here, r^2 is the proportion of variance in the DV, the **differences in blood pressure** (from control to fish tank conditions), that can be **explained** by the IV; the presence of the fish tank. Field (2005, p. 32) makes a case for preferring the correlation coefficient as a measure of effect size, because it has a fixed possible range from 0 (indicating no effect at all) to 1 (indicating a complete effect that explains 100% of the variance)

Using Cohen's guidelines, the following relationships exist between Cohen's *d*, and *r* (Cohen, 1988, pp. 82-83):

Effect size	Cohen's *d*	*r*	r^2
Small	.2	.10	.01 (1%)
Medium	.5	.30	.09 (9%)
Large	.8	.50	.25 (25%)

The r^2 effect size squared is calculated as follows for this example:

$$r^2 = \frac{t^2}{t^2 + df}$$

$$= \frac{(4.439)^2}{(4.439)^2 + 15}$$

$$= \frac{19.70402}{19.70402 + 15}$$

$$= \frac{19.70402}{34.70402}$$

$$= .567773$$

$$= \mathbf{.57}$$

There is clearly a large effect size in this example; the presence of the fish tank accounts for 57% of the difference in blood pressure across the two measures, after the variance due to individual differences has been removed.

You need to be aware that there is a good deal of confusion in the way correlation effect sizes are reported, for reasons beyond the scope of this guide. Cohen (1988) and Field (2005) have routinely used *r*, while Pallant (2007, p. 240) has used r^2, although by way of eta squared (to be discussed later). The distinction is not critical as one can easily be obtained from the other.

Once again, in addition to reporting effect size in the Results section, it generally warrants some attention in the Discussion section of a research report. In addition you might comment on the practical importance of the findings.

Dependent *t* test: Research report sample Results section

Results

A dependent *t* test was conducted on the mean systolic blood pressure readings under the two conditions (control, and fish tank present). Assumptions of normality were met and the result indicated a statistically significant difference between the control and fish tank conditions, $t(15) = 4.44, p < .001$. The mean systolic blood pressure of 119.56 ($SD = 7.67$) when the fish tank was present was lower than the mean of 124.50 ($SD = 10.19$) for the control condition when the fish tank was absent. A large effect size (r^2 = .57) was indicated by the mean difference of 4.94 between the two conditions, 99% CI [1.66, 8.22].

Wilcoxon Signed-Rank Test

When the assumptions of the dependent *t* test are not met, an alternative is to use the Wilcoxon signed-rank test, which is a nonparametric alternative.

Another nonparametric test, the Sign test is an even simpler alternative that can be used with related samples.

When do you use the Wilcoxon (Signed-Rank) T Test?

The Wilcoxon signed-rank or Wilcoxon T test is used in the same sort of circumstances as apply for the dependent samples *t* test, that is, when there are two DV scores under different levels of an IV from one group of participants, OR there are scores on a DV from **two related groups** that experience different levels of an IV. However, conditions appropriate to the use of nonparametric tests apply (see Chapter 6). The researcher wants to know if the population distributions on the DV are identical, or if they differ in central tendency.

Assumptions

The Wilcoxon T test assumes random sampling, and, when used to test differences in central tendency, it assumes that the two distributions have similar shape and variability.

Example of Wilcoxon (Signed-Rank) T test

An environmental psychologist is interested in how much importance people attach to environmental protection. She is aware that most people tend to say it is very important, but suspects that **relative** to concerns about economic prosperity it is not so important. As a preliminary test of this hypothesis she randomly selects a sample of 11 people from her suburb, and asks them to **rank** the following societal goals in order of importance (1 being the most important and 5 being the least important):

Economic prosperity
Environmental protection
Law and order
Quality education for all
Quality health care for all

The resulting data are at the **ordinal** level of measurement. The psychologist's research hypothesis is that the median rank for environmental protection will be significantly lower than that for economic prosperity, although she does not have strong enough grounds to specify a one-tailed test.

The two sets of ranks are on the next page (ECON is the variable name for economic prosperity and ENVIRON is the variable name for environmental protection).

ID	ECON Rank	ENVIRON Rank	Difference in Ranks	Signed Rank of Difference
1	2	4	**-2**	-5
2	3	5	**-2**	-5
3	1	3	**-2**	-5
4	2	4	**-2**	-5
5	4	3	**1**	1
6	3	5	**-2**	-5
7	3	5	**-2**	-5
8	2	4	**-2**	-5
9	1	5	**-4**	-10.5
10	4	1	**3**	9
11	1	5	**-4**	-10.5

Manual calculation of a Wilcoxon T Test

Step 1: Specify hypotheses and decision criteria

H_0: The two distributions are identical: $\mu_1 = \mu_2$
H_1: The two distributions differ in central tendency: $\mu_1 \neq \mu_2$

$\alpha = .05$, two-tailed

Retain H_0 if absolute value of observed T > critical T
Reject H_0 if absolute value of observed T ≤ critical T

Note that T tables work opposite to usual. H_0 is rejected when the observed T is **less than or equal to** the critical value of T.

Step 2: Set down data table (as above)

To calculate T, the two extra columns are needed. The first is simply the **Difference** in the economic prosperity and environmental protection rankings (ECON – ENVIRON).

The last column has the **Signed Rank of the Difference**. Difference scores are **ranked** and given the same **sign** as the difference score.

Note that tied scores are given an average rank. For example, there are 7 scores of -2 that need to share ranks 2, 3, 4, 5, 6, 7, and 8:

$2 + 3 + 4 + 5 + 6 + 7 + 8 = 35$
$35 / 7 = 5$
Signed rank = **-5.**

Similarly, there are 2 scores of –4 that need to share ranks 10 and 11:

$10 + 11 = 21$
$21 / 2 = 10.5$
Signed rank = **-10.5**

Step 3: Calculate the sum of the positive ranks (T+) and the sum of the negative ranks (T-)

T+ = Σ of positive signed ranks = **10**

T- = Σ of negative signed ranks = **-56**

Step 4: Look up a table of critical values of T

From Howell (2002, pp. 715-716), for a **two-tailed** test with α = .05 and N = 11, the critical value for the **smaller** of either T+ or T- is 10.

Step 5: Draw statistical conclusion

The smallest observed T is 10 and this is equal to the critical T of 10; therefore **reject** H_0: $\mu_1 = \mu_2$ and **accept** H_1: $\mu_1 \neq \mu_2$.

Step 6: Specify the research conclusion

There was a statistically significant difference in the rankings of economic prosperity and environmental protection, with economic prosperity being ranked higher than environmental protection.

Using SPSS to conduct the Wilcoxon T test

Step 1: Enter data in SPSS and save data file

First, enter the data into SPSS and save them in a file called ***Wilcoxon.sav***.

ID	Participant number	*Set **Measure** column to **Nominal***
Econ	Economic prosperity	*Set **Measure** column to **Ordinal***
Environ	Environmental protection	*Set **Measure** column to **Ordinal***

Step 2: Screen data and check assumptions

Although there are no assumptions to formally test you should always screen the data to check accuracy of entry. A visual inspection of the data is all that is required here. You should also run the **Explore** procedure to obtain full descriptive statistics, including the median which is the more appropriate measure of central tendency with ordinal data.

Step 3: Request the analysis

17 **How to: Conduct a Wilcoxon T test**

Analyze
Nonparametric Tests ▸
Legacy Dialogs ▸
2 Related Samples...

- Older versions of SPSS do not have the **Legacy Dialogs** option.
- See Appendix 1 for the new style of nonparametric tests in PASW 18.

[Select **both** variables you want simultaneously]
Economic prosperity [econ]
Environmental protection [environ] ▸ Test Pair(s) List:

[From] **Test Type** [Select] ☑ **Wilcoxon**

[Click on the **Options** button and select as follows]
Statistics
☑ **Descriptive**
Continue
OK

Step 4: Examine and interpret the output

Output: **Wilcoxon Signed-Rank test**

Descriptive Statistics

	N	Mean	Std. Deviation	Minimum	Maximum
Economic prosperity	11	2.36	1.120	1	4
Environmental protection	11	4.00	1.265	1	5

Ranks

		N	Mean Rank	Sum of Ranks
Environmental protection - Economic prosperity	Negative Ranks	2[a]	5.00	10.00
	Positive Ranks	9[b]	6.22	56.00
	Ties	0[c]		
	Total	11		

a. Environmental protection < Economic prosperity

b. Environmental protection > Economic prosperity

c. Environmental protection = Economic prosperity

1

CARE! These are not literally the Mean Ranks, but the Mean **Signed** Ranks. The mean ranks are in the first table headed **Descriptive Statistics**.

Test Statistics[b]

	Environmental protection - Economic prosperity
Z	-2.105[a]
Asymp. Sig. (2-tailed)	.035

a. Based on negative ranks.

b. Wilcoxon Signed Ranks Test

2

SPSS uses the *z* score normal approximation to the Wilcoxon signed-rank test.

The **Asym. Sig. (2-tailed)** is the *p* value. It is $< \alpha$ of .05; therefore, the test result **IS significant**.

Effect size for Wilcoxon (Signed-Rank) T test

An effect size in terms of r (then r^2) can be obtained for the Wilcoxon signed-rank test using the following formula:

$$r = \frac{z}{\sqrt{N}} \text{, where } N = \text{number of observations, not cases}$$

$$= \frac{-2.105}{\sqrt{22}}$$

$$= \frac{-2.105}{4.6904}$$

$$= -.4488$$

$$= -\mathbf{.45}$$

Wilcoxon (Signed-Rank) T test: Research report sample Results section

Results

Participant rankings of economic prosperity and environmental protection were compared using the Wilcoxon Signed-Rank test. As expected, the median rank for environmental protection (*Mdn* = 4.00, Range = 4.00) was significantly lower than the median rank for economic prosperity (*Mdn* = 2.00, Range = 3.00), with a medium to large effect size, $z\ (N = 11) = 2.11, p = .035, r^2 = .20$.

Notes:

Unless using a one-tailed test there is no need for a minus sign to be used when reporting *z* or *t* results.

Strictly speaking, with ordinal data, the median and range are more appropriate descriptive statistics than the mean and standard deviation. These can be obtained from the **Explore** procedure during data screening.

9 THE INDEPENDENT-SAMPLES *t* TEST, AND MANN-WHITNEY *U* TEST NONPARAMETRIC ALTERNATIVE

When do you use the independent samples *t* test?

The independent-samples *t* test is used in **between-subjects** research designs, or what are now more likely to be called, **between-groups** designs; where there are data from **two different samples or groups of participants** that have experienced different levels of an IV. The data are participants' scores on some DV and the researcher wants to know if the means of the populations from which the samples are drawn are the same or different on that DV.

In order to use the independent samples *t* test, the following assumptions need to be satisfied, as the accuracy of the test depends on them.

Assumptions

1. The data should be at the interval or ratio level of measurement.
2. The scores are randomly sampled from the populations of interest (or, in true experiments available participants have been randomly assigned to the different groups).
3. The groups are **independent**, that is, participants appear in one and only one group and the groups are not related to one another (an example of violation of the independence assumption would be if one group consisted of mothers, while the second group consisted of sons).
4. The scores on the DV are normally distributed in each population. When *n* is relatively large (30+ in each group) the *t* test is considered to be **robust**, that is, violation of the normality assumption has little effect on its accuracy. When both groups are reasonably large (20-30 participants) and $n_1 = n_2$ departure from normality can be tolerated (unless there is extreme skewness or extreme nonnormal kurtosis). With even larger samples the normality assumption can be ignored.
5. There is **homogeneity of variance**, that is, the variances in the populations of the two groups are equal (the sample variances need to be approximately equal). This assumption is really only of concern when group sizes are unequal. In any event, SPSS comes to the rescue. It uses the **Levene Test for Equality of Variances** to test the null hypothesis that the variances are equal. If the probability for this test is significant ($\leq .05$), then the null hypothesis of equal variances is rejected, and the alternative hypothesis of unequal variances is accepted. The *t* value appropriate for when groups have unequal variances is then used.

   The decision rule is:

   Use *t* value, *df*, and probability labelled **equal** if the Levene test is **not** significant.

   Use *t* value, *df*, and probability labelled **unequal** if the Levene test **is** significant.

When the assumptions underlying the *t* test cannot be met, or the DV is measured on an ordinal scale, yet one still has independent groups, the nonparametric alternative to the independent *t* test should be used. This is the Mann-Whitney *U* Test (the Mann-Whitney for short). It is also known as the Wilcoxon Rank Sum test.

Example of an independent samples *t* test

Here, we will use the example from the data screening chapter, which was as follows.

Imagine you a school psychologist studying the effect of mild exercise on mental alertness. You begin your research by conducting a **true experiment** in which you randomly assign 34 student volunteers to one of two groups: A Control group and an Exercise group. Exercise is the IV and the DV is the number of errors on a proofreading task. Students in the Control group (no exercise) spend the morning in sedentary activities, reading their course materials and taking notes. Between 11:00am and 11:30am they perform the proofreading task. Students in the Exercise condition follow exactly the same procedure, except that they engage in a light aerobics workout between 10:00am and 10:30am.

Remember that participant number 11 in the Control group with a raw score of 18 was identified as an outlier. The decision was taken to change this raw score, according to recommendations of Tabachnick and Fidell (2007), to 7 which was one unit larger than the next most extreme score of 6.

Manual calculation of an independent samples t test

Step 1: Specify hypotheses and decision criteria

Let Group 1 = Control group and Group 2 = Exercise group

H_0: $\mu_1 = \mu_2$

H_1: $\mu_1 \neq \mu_2$

$\alpha = .05$, two-tailed

Retain H_0 if absolute value of observed $t <$ critical t

Reject H_0 if absolute value of observed $t \geq$ critical t

Step 2: Set down data table

Note: X_1 represents the DV scores (number of errors) in the Control group, and X_2 represents the DV scores in the Exercise group.

ID	X_1	X_1^2	ID	X_2	X_2^2
1	4	16	18	5	25
2	3	9	19	4	16
3	5	25	20	4	16
4	2	4	21	4	16
5	5	25	22	2	4
6	4	16	23	1	1
7	6	36	24	3	9
8	4	16	25	2	4
9	3	9	26	3	9
10	6	36	27	2	4
11	7	49	28	3	9
12	3	9	29	5	25
13	2	4	30	2	4
14	4	16	31	3	9
15	5	25	32	2	4
16	6	36	33	3	9
17	5	25	34	5	25
$n_1 = 17$	$\sum X_1 = 74$	$\sum X_1^2 = 356$	$n_2 = 17$	$\sum X_2 = 53$	$\sum X_2^2 = 189$

Step 3: Calculate the means

$$\bar{X}_1 = \frac{\Sigma X_1}{n_1} \qquad \bar{X}_2 = \frac{\Sigma X_2}{n_2}$$

$$= 74/17 \qquad = 53/17$$

$$= 4.3529 \qquad = 3.1176$$

$$= \mathbf{4.35} \qquad = \mathbf{3.12}$$

Step 4: Calculate the standard deviations

$$s_1 = \sqrt{\frac{\Sigma X_1^2 - \frac{(\Sigma X_1)^2}{N_1}}{N_1 - 1}} \qquad s_2 = \sqrt{\frac{\Sigma X_2^2 - \frac{(\Sigma X_2)^2}{N_2}}{N_2 - 1}}$$

$$= \sqrt{\frac{356 - \frac{(74)^2}{17}}{17-1}} \qquad = \sqrt{\frac{189 - \frac{(53)^2}{17}}{17-1}}$$

I am sure I can leave you to complete these calculations, and when you do, you should have:

$$s_1 = 1.4552 \qquad s_2 = 1.2187$$

$$= \mathbf{1.46} \qquad = \mathbf{1.22}$$

Step 5: Set down the appropriate formula, substitute the values, and calculate observed *t*

Note: The *t*-test formula is the same as the corresponding *z*-test formula, except that we do not know the standard error of the difference between means in the denominator, and must estimate it from the sample standard deviations instead. The formula is relatively simple with equal *n*. A more complex formula that can deal with unequal *n* is given in some text books.

$$t_{\bar{X}_1 - \bar{X}_2} = \frac{\bar{X}_1 - \bar{X}_2}{\sqrt{\frac{s_1^2 + s_2^2}{n}}}$$

$$= \frac{4.3529 - 3.1176}{\sqrt{\frac{1.4552^2 + 1.2187^2}{17}}}$$

$$= \frac{1.2353}{\sqrt{\frac{2.1176 + 1.4852}{17}}}$$

$$= \frac{1.2353}{\sqrt{\frac{3.6028}{17}}}$$

$$= 1.2353/\sqrt{0.2119}$$

$$= 1.2353/0.4604$$

$$= 2.6831$$

$$= \mathbf{2.68}$$

Step 6: Calculate degrees of freedom

$$df = n_1 + n_2 - 2$$

$$= 17 + 17 - 2$$

$$= \mathbf{32}$$

Step 7: Look up a table of critical values of t

Internet site: http://www.danielsoper.com/statcalc/calc10.aspx

For a two-tailed test with α = .05 and 32 *df* the critical value of *t* is somewhere between 2.042 and 2.021 according to most tables that only give critical *t* values for 30 *df* and 40 *df*. However, from the internet site we obtain the exact value of 2.036933.

Step 8: Draw statistical conclusion

The absolute value of the observed *t* is 2.68 and this is > any of the *t* values at Step 7; therefore **reject** H_0: $\mu_1 = \mu_2$ and **accept** H_1: $\mu_1 \neq \mu_2$.

Step 9: Specify the research conclusion

The number of errors on the proofreading task was significantly less, statistically, for students who engaged in half an hour of light aerobics exercise than for students who did not.

Step 10: Calculate the effect size

Note that an effect size calculator for independent *t* tests is available at:
http://web.uccs.edu/lbecker/Psy590/escalc.htm

$$r^2 = \frac{t^2}{t^2 + df}$$

$$= \frac{(2.683)^2}{(2.683)^2 + 32}$$

$$= \frac{7.198489}{7.198489 + 32}$$

$$= \frac{7.198489}{39.19849}$$

$$= .183642$$

$$= \mathbf{.18}$$

Using SPSS to conduct the independent-samples *t* test

Steps 1 and 2 for data entry and data screening have already been conducted in Chapter 6.

Step 1: Enter data in SPSS and save data file

First, enter the data into SPSS and save them in a file called ***IndTtest.sav***.

ID	Participant number	*Set **Measure** column to **Nominal***
Group	Experimental group 0 = Control 1 = Exercise	*Set **Measure** column to **Nominal***
Errors	Number of proofreading errors	*Set **Measure** column to **Scale***

Make sure the outlier (participant number 11 in the Control group) has the revised score of 7.

Step 2: Screen data and check assumptions

Screen the data for the DV, separately for each group, as described in Chapter 6. Check the normality assumption carefully given that the sample size is small. Make sure you have screened the data **after** dealing with the outlier.

Step 3: Request the analysis

18 How to: Conduct an independent-samples *t* test

Analyze
Compare Means ▸
Independent-Samples T Test...
[Select variables you want]
Number of proofreading errors [errors] ▸ Test Variables(s):
Experimental group [group] ▸ Grouping Variable:

[Click on **Define Groups...** and enter the values for **Group]**
[Enter **Group 1]** **0** [Tab]
[Enter **Group 2]** **1**
Continue

OK

Note the **Options** button.This allows you to specify the Confidence Interval, and it has the 95% CI as the default. You can also change how any missing values are handled under **Options**.

Step 4: Examine and interpret the output

Output: **Independent *t* test**

T-Test

Group Statistics

	Experimental group	N	Mean	Std. Deviation	Std. Error Mean
Number of proofreading errors	Control	17	4.35	1.455	.353
	Exercise	17	3.12	1.219	.296

1

Examine the means and standard deviations. What do they suggest about group differences?

Independent Samples Test

		Levene's Test for Equality of Variances		t-test for Equality of Means						
									95% Confidence Interval of the Difference	
		F	Sig.	t	df	Sig. (2-tailed)	Mean Difference	Std. Error Difference	Lower	Upper
Number of proofreading errors	Equal variances assumed	.780	.384	2.683	32	.011	1.235	.460	.298	2.173
	Equal variances not assumed			2.683	31.044	.012	1.235	.460	.296	2.174

2

Remember the Levene's Test decision rule:

If Levene's is **NOT** significant (Sig. > .05), use the **top** row of figures—*t* value, *df*, and Sig. (2-tailed)—for **Equal variances assumed**.

If Levene's **IS** significant (Sig. ≤ .05),use the **bottom** row of figures—*t* value, *df*, and Sig. (2-tailed)—for **Equal variances not assumed**.

3

Look at appropriate *t* test result. **Sig. (2-tailed)** is the two-tailed probability.

The *t* test is **NOT** significant if this is > α (usually .05).

The *t* test **IS** significant if this is ≤ α (usually .05).

4

The 95% confidence interval.

Effectively, we can be 95% confident that the difference between the population means is somewhere within this range.

Here, at Sig. = .384 (> .05) it is not significant; the homogeneity of variance assumption is met, so use the *t* test for Equal variances assumed.

Here, it **is** significant at Sig. (2-tailed) = .011 (<.05).

Independent-samples *t* test: Research report sample Results section

Results

An independent *t* test was conducted on the number of errors. Following the recommendations of Tabachnick and Fidell (2007, p. 77), one outlier (with an unusually high score) in the control group was changed to have a score one unit larger than the next most extreme score in the group. Assumptions of normality and homogeneity of variance were met, and the *t* test indicated a statistically significant difference between the control and exercise groups, $t(32) = 2.68$, $p = .011$. The mean number of errors was lower in the exercise group ($M = 3.12$, $SD = 1.22$) than in the control group ($M = 4.35$, $SD = 1.46$); the difference in means of 1.23, 95% CI [0.30, 2.17] representing a medium to large effect ($r^2 = .18$).

Notes:

Were the assumption of homogeneity of variance to be violated one might report something along the following lines:

> Using the *t* test for unequal variances because of violation of the assumption of homogeneity of variance, a statistically significant difference was found between the control and exercise groups, $t(31.04) = 2.68$, $p = .012$.

The manually calculated effect size of .18 is between Cohen's (1988) guidelines for a medium effect (.09) and a large effect (.25).

Mann-Whitney *U* Test (Wilcoxon Rank-Sum)

(Nonparametric alternative to Independent *t* test

When do you use the Mann-Whitney *U* Test?

The Mann-Whitney *U* Test is used in the same sort of circumstances as apply for the independent samples *t* test, that is, there are scores on a DV from two different groups that experience two different levels of the IV. However, conditions appropriate to the use of nonparametric tests apply (see Chapter 6). The researcher wants to know if the population distributions for the two groups on the DV are identical, or if they differ in central tendency.

Assumptions

The assumption of independence is necessary for the Mann-Whitney *U* Test. That is, participants appear in one and only one group, and the groups are not related in any way.

It is also assumed that the distributions have similar shape and variability.

Example of Mann-Whitney *U* Test

Suppose you are running a program to enhance the coping skills of parents of severely intellectually and physically disabled children. Small sample sizes tend to be a problem in areas like this. Suppose 13 sets of parents volunteer for your program and you are able to randomly assign seven of them to an Experimental condition where they are exposed to a newly developed program. The other six are assigned to the Control group, which receives the standard program.

At the end of six weeks training both groups complete a coping competence inventory with a possible range of 0-40 (the higher the score, the greater the competence). The literature suggests scores on the inventory tend to be negatively skewed (i.e., the majority of people have scores that are relatively high, but there are always some participants who return relatively poor scores).

The scores for the two groups are as follows:

Control	Experimental
21	12
5	26
20	24
15	28
25	2
22	26
	10

You could analyse these results using an independent *t* test, but satisfaction of the assumptions is dubious to say the least. There are very small and uneven sample sizes where it is difficult to screen the data adequately. Just "eyeballing" the histograms indicates skewness and the presence of somewhat extreme scores in the Experimental group; however, none of the formal tests can confirm this as the sample sizes are just too small.

Manual calculation of a Mann-Whitney *U* test

A word of warning: Do not be surprised to find different formulae used in different text books. The formula that follows is taken from Field (2005, p. 526; see also pp. 522-525).

Step 1: Specify hypotheses and decision criteria

Let Group 1 = Control group and Group 2 = Experimental group

H_0: The two distributions are identical

H_1: The two distributions differ in central tendency: $\mu_1 \neq \mu_2$

$\alpha = .05$, two-tailed

Retain H_0 if two-tailed probability > .05

Reject H_0 if two-tailed probability $\leq .05$

Step 2: Set down data table

X_1 represents the DV scores in the Control group, X_2 represents the DV scores in the Experimental group.

List the scores **in order** for each group. (Note that is this case ID refers to the "**set** of parents", that is, mother and father.)

Rank **all** the scores in ascending order from 1 to 13. Tied scores are given tied ranks. For example, there are two scores of 26 that correspond to ranks 11 and 12. These ranks are averaged (11 + 12 / 2) and the scores of 26 share the same tied rank of 11.5.

ID	X_1	Rank	ID	X_2	Rank
1	5	2	7	2	1
2	15	5	8	10	3
3	20	6	9	12	4
4	21	7	10	24	9
5	22	8	11	26	11.5
6	25	10	12	26	11.5
			13	28	13
$n_1 = 6$	$\Sigma X_1 = 108$	$R_1(\Sigma\text{Ranks}) = 38$	$n_2 = 7$	$\Sigma X_2 = 128$	$R_2(\Sigma\text{Ranks}) = 53$

Step 3: You may wish to calculate the means if you have interval level data (i.e., to compare with medians).

$$\bar{X}_1 = \frac{\Sigma X_1}{n_1} = 108/6 = \mathbf{18.00}$$

$$\bar{X}_2 = \frac{\Sigma X_2}{n_2} = 128/7 = 18.2857 = \mathbf{18.29}$$

Step 4: Set down and calculate the appropriate formula

$$U_1 = n_1 n_2 + \frac{n_1(n_1+1)}{2} - R_1 \qquad U_2 = n_1 n_2 + \frac{n_2(n_2+1)}{2} - R_2$$

$$= (6x7) + \frac{6(6+1)}{2} - 38 \qquad = (6x7) + \frac{7(7+1)}{2} - 53$$

$$= 42 + \frac{6x7}{2} - 38 \qquad = 42 + \frac{7x8}{2} - 53$$

$$= 42 + \frac{42}{2} - 38 \qquad = 42 + \frac{56}{2} - 53$$

$$= 42 + 21 - 38 \qquad = 42 + 28 - 53$$

$$= \mathbf{25} \qquad = \mathbf{17}$$

Step 5: Look up a table of critical values of *U*

Not all statistics books contain *U* tables, and they can appear in different forms. It is all very confusing, but no longer critical, as we would normally rely on the SPSS.

A website that will calculate the probability value for you is:
http://elegans.swmed.edu/~leon/stats/utest.html

You need to specify the two sample sizes ($n_1 = 6$, $n_2 = 7$), and the **smaller** of the two *U* values ($U = 17$)
The *p* value returned for a two-tailed test is .628

Step 6: Draw statistical conclusion

As the observed two-tailed probability of .628 is greater than α of .05 retain H_0: The two distributions are identical; and **reject** H_1 that the two distributions differ in central tendency, $\mu_1 \neq \mu_2$.

Step 7: Specify the research conclusion

There was no statistically significant difference in the coping competence of the two groups (Control and Experimental) at the end of the six-week training program.

Step 10: Calculate the effect size

Using the *z*-score approximation (from the SPSS output to follow):

$$r = \frac{z}{\sqrt{N}} \text{ , where } N = \text{total number of cases} \qquad = \frac{-0.572}{3.60555}$$

$$= \frac{-0.572}{\sqrt{13}} \qquad = -.1586$$

$$= \mathbf{-.16}$$

Using SPSS to conduct the Mann-Whitney *U* test

Step 1: Enter data in SPSS and save data file

First, enter the data into SPSS and save them in a file called ***Mann.sav***.

ID	Participant number	*Set **Measure** column to **Nominal***
Group	Experimental group 1=Control 2=Experimental	*Set **Measure** column to **Nominal***
Coping	Coping skill score	*Set **Measure** column to **Scale***

Step 2: Screen data and check assumptions

Although there are no assumptions to formally test you should always screen the data to check accuracy of entry. A visual inspection of the data is all that is required here. You should also run the **Explore** procedure to obtain full descriptive statistics, including the median which is the more appropriate measure of central tendency with ordinal data.

Step 3: Request the analysis

19 How to: Conduct a Mann-Whitney *U* test

Analyze
 Nonparametric Tests ▸
  Legacy Dialogs ▸
   2 Independent Samples...

- Older versions of SPSS do not have the **Legacy Dialogs** option.
- See Appendix 1 for the new style of nonparametric tests in PASW 18.

[Select variables you want]
Coping skill score [coping] **▸ Test Variables List:**
Experimental group [group] **▸ Grouping Variable:**

[Click on **Define Groups...** and enter the values for **Group]**
 [Enter **Group 1] 1** [Tab]
 [Enter **Group 2] 2**
 Continue

[From] **Test Type** [Select] **☑ Mann-Whitney U**

OK

Step 4: Examine and interpret the output

Output: **Mann-Whitney *U* test**

NPar Tests

Mann-Whitney Test

Ranks

	Experimental group	N	Mean Rank	Sum of Ranks
Coping skill score	Control	6	6.33	38.00
	Experimental	7	7.57	53.00
	Total	13		

1

CARE!
These are the mean **ranks**, not mean scores.

Test Statistics[b]

	Coping skill score
Mann-Whitney U	17.000
Wilcoxon W	38.000
Z	-.572
Asymp. Sig. (2-tailed)	.567
Exact Sig. [2*(1-tailed Sig.)]	.628[a]

a. Not corrected for ties.

b. Grouping Variable: Experimental group

2

SPSS uses the *z* score normal approximation to the Mann-Whitney *U* test.

The **Exact. Sig. [(2*(1-tailed Sig.)]** is the *p* value.

It is > α of .05, therefore, the test result is **NOT significant.**

Mann-Whitney *U* test: Research report sample Results section

Results

In view of small sample sizes and indications that coping competence was negatively skewed, a Mann-Whitney nonparametric test was conducted on the coping competence scores for the control and experimental groups. It showed no statistically significant difference between the groups, z (N = 13) = 0.57, p = .63, r^2 = .03. Descriptive statistics are reported in Table 1.

Table 1

Descriptive Statistics for Control and Experimental Groups

Group	*n*	*M*	*Mdn*	*SD*	Range
Control	6	18.00	20.50	7.16	20
Experimental	7	18.29	24.00	10.16	26

Notes:

Unless using a one-tailed test there is no need for a minus sign to be used when reporting *z* or *t* results.

You could report either *U* or *z*.

The descriptive statistics can be obtained from the **Explore** procedure during data screening.

Strictly speaking only medians (and range) are appropriate for nonparametric statistics, however, if using interval level data you may wish to report means and standard deviations (recall with skewed data, the mean will be drawn away from the median towards the tail of the distribution). If only reporting the median and range there would be no need for a Table. Instead, the text might read something like:

> There was no statistically significant difference between the control group (*Mdn* = 20.50, Range = 20, *n* = 6) and the experimental group (*Mdn* = 24.00, Range = 26, *n*=7), *U* = 17.00, *p* = .63, r^2= .03.

10 One-Way Analysis of Variance (Anova), And The Kruskal-Wallis Nonparametric Alternative

When do you use a one-way ANOVA?

The one-way analysis of variance (ANOVA) can be thought of as an extension of the independent-samples *t* test, where instead of comparing two independent groups, three or more independent groups are compared.

A one-way ANOVA is used in **between-groups** research designs, when there are data from **three or more different samples** or groups of participants that have experienced different levels of one IV. The data are participants' scores on some DV and the researcher wants to know if the means of the populations from which the samples are drawn are the same or different on that DV.

The samples or groups may come from a true experimental design, in which case it can be inferred that the IV causes the DV (provided it is a well-designed experiment).

Groups may also come from a subject-variable design (also known as a **natural group design**). This type of design could arise with survey data. For example, a researcher might obtain a stress measure from a stress scale included in a survey questionnaire. This might constitute the "DV". Respondents might then be grouped according to whether they are married or in a partnered relationship (Group 1), separated or divorced (Group 2), widowed (Group 3), or never married (Group 4). Marital status would be the "IV", and the research question would be about whether the mean stress score is different in the different marital status populations. Of course, finding a significant difference would not allow causation to be inferred, as these types of design are really correlational in nature, with all the problems of causation that face correlational designs. Note also that the terms IV and DV are used for convenience in natural group designs, but technically this is not correct, because these terms are themselves causal in nature.

In order to use the one-way ANOVA, the following assumptions need to be satisfied, as the accuracy of the test depends on them. The assumptions are similar to those for the independent t test.

Assumptions

1. The DV should be at the interval or ratio level of measurement.
2. The scores are randomly sampled from the populations of interest (or, in true experiments available participants have been randomly assigned to the different groups).
3. The scores are **independent**, that is, participants appear in one and only one group; the groups are not related to one another; and the score for each participant is independent of the scores for all the other participants.
4. The scores on the DV are normally distributed in each population.
5. There is **homogeneity of variance**, that is, the variances in the group populations are equal (the sample variances need to be approximately equal to satisfy this assumption).

When n is large (20-30+ in each group) ANOVA is considered to be **robust**, that is, violation of the normality assumption has little effect on its accuracy. Some sources consider that with as little as 10-20 participants in each group moderately non-normal distributions pose little problem.

Now that computers can perform the calculations there is no necessity to have equal numbers of participants in each group, although it is desirable to achieve equal group sizes wherever possible.

With large and **equal** groups ANOVA is also robust against heterogenous variances. In fact, differences in group size not exceeding 1:1.5 (smallest to largest) can be tolerated.

The assumptions of normality and homogeneity of variance are only really of concern when group sizes are unequal, or worse still when there are **small** and **unequal** group sizes (e.g., with groups of 6, 12, and 70 participants the use of ANOVA would be questionable). SPSS can compute the Levene Test for Equality of Variances. If the probability for this test is **not** significant (>.05), then the null hypothesis of equal variances is retained.

When the assumptions underlying ANOVA are important (i.e., with small and unequal n) and cannot be met, or the DV is measured on an ordinal scale, the nonparametric alternative to the one-way ANOVA should be used. This is the Kruskal-Wallis test.

The *F* statistic

In ANOVA, the F statistic is calculated, as opposed to the t statistic in the t test. In fact, the t test is the simplest member of the ANOVA family ($t^2 = F$).

The null hypothesis in ANOVA is that the population means for the groups are equal; the alternative hypothesis is that they are unequal.

F is a simple **ratio** of:

$$\frac{\text{the variance } \textbf{between} \text{ groups}}{\text{the variance } \textbf{within} \text{ groups (or } \textbf{error} \text{ variance)}}$$

If the variation **between** groups is of the same order of magnitude as the naturally occurring level of variation among individuals **within** groups (i.e., individual differences or error variance), the F ratio will be close to 1, and the null hypothesis of no (real) difference between the groups will be retained.

On the other hand, if the variation **between** groups is much greater than the naturally occurring variation among individuals **within** groups, the F ratio will be much greater than 1, and the null hypothesis of no difference between the groups will be rejected.

Take note that the variance within groups is also known as the **error** variance.

Analytical comparisons

A significant overall or **omnibus** F simply indicates that there is a difference **somewhere** among the groups. In order to find out exactly which groups differ from which **post hoc analytical comparisons** must be conducted.

There are a number of **post hoc** comparisons (as well as **planned** comparisons that can be performed **instead** of overall F). These are discussed in most texts, and will be addressed in more detail in later chapters.

Why not just do a series of *t* tests among the groups? The problem of familywise error

A question that is often asked is "why do we need ANOVA, why not just do a series of *t* tests?" comparing every group with every other group. The answer has to do with the fact that a series of *t* tests increases the chance of a **Type I error** occurring **somewhere**, resulting in what is known as **familywise error** (or what used to be called experiment-wise error).

For example, with three groups (A, B, C) three *t* tests are possible (A-B, B-C, A-C). Were these *t* tests independent it would be possible to work out the Type I error rate across the *family* of three *t* tests and adjust the alpha level to control for it. With three *t* tests the **familywise error rate**, using an alpha of .05 for each test is approximately .05 x 3 = .15 (more precise formulae are given in many texts). This increase in the familywise error rate can be controlled by setting a more stringent alpha level for each test, using what is known as a **Bonferroni adjustment** (or correction) of .05 divided by the number of tests, that is, .05/3 = .017. Note that a slightly more powerful version of the Bonferroni test is the **Sidak test** (pronounced "Sh"idak).

A further complication is that these *t* tests are not independent, so the adjustment only allows approximate control of the familywise error rate. The *t* tests would not be independent, because if it was found that A > B, and B > C, then it must be the case that A > C. Performing an ANOVA first, and analytical comparisons only if the ANOVA is significant controls this problem, at least to some extent, although it is a controversial issue.

Example of a one-way ANOVA

Imagine you are studying the effectiveness of different kinds of "Help" systems for experienced computer users. A new word processing package is being developed. You randomly assign 40 experienced word processor operators to one of four groups:

> The **Nohelp** control group has no online help, but works from hardcopy user manuals.
>
> The **Menu** group uses menu driven Help screens.
>
> The **Options** group uses a Help system that advises the user of the options they have at a particular point.
>
> The **Index** group uses a Help Index system.

After two weeks working with the packages for equivalent periods of time in equivalent conditions, the operators are given a word processing test with a possible score range of 0 to 40; the higher the score the better the performance. This score constitutes the DV.

This study involves manipulation of an IV, random assignment, and controlled conditions; hence it is an experimental research design.

If we find a significant *F*, we can use the **Tukey's HSD** (Honestly Significant Difference) **Test**, which is the most appropriate post hoc test for performing pairwise comparisons among all possible pairs of means. This will tell us which groups are significantly different from which.

Note: In many research methods courses you are required to perform a manual calculation of a one-way ANOVA; but in this age of computers there is little to be gained by such an exercise. However, later in the chapter a **conceptual** calculation will be performed, as it **is important that you understand, logically, what ANOVA is doing**.

Using SPSS to conduct one-way ANOVA

Step 1: Enter data in SPSS and save data file

Suppose there are 10 participants in each group, and the test scores for each group are as follows:

Nohelp	Options	Menu	Index
21	23	36	33
20	34	28	24
27	32	29	32
21	26	28	21
24	24	28	23
26	30	25	19
30	32	33	32
19	33	35	28
28	24	37	26
29	26	35	30

Of course you would **not** enter the data into SPSS in these four columns. Rather, you would need to enter the data into **three columns** for the following **three variables**, with variable labels and value labels appropriately defined:

ID — Participant number — *Set **Measure** column to **Nominal***

Group — Experimental group — *Set **Measure** column to **Nominal***
1=Nohelp
2=Options
3=Menu
4=Index

Score — Word processing test score — *Set **Measure** column to **Scale***

Remember to save the data in a file that you might call ***Oneway.sav***.

Step 2: Screen data and check assumptions

Screen the data for the DV, separately for each group, as described in Chapter 6. Check the normality assumption carefully given that the sample sizes are small.

Step 3: Request the analysis

20 How to: Conducting a one-way ANOVA

Analyze
    Compare Means ▸
        One-Way ANOVA...
            [Select variables you want]
            Word processing test score [score] ▸ **Dependent List**
            Experimental group [group] ▸ **Factor:**

            Options... →
                [From **Statistics** select]
                ☑ **Descriptive**
                ☑ **Homogeneity-of-variance**

                [Select] ☑ **Means Plot**
                Continue

            Post Hoc...
                [From **Equal Variances Assumed** select]
                ☑ **Tukey**
                Continue
            OK

There are options here for tests that are more appropriate than *F* if the assumption of homogeneity-of-variance is violated. You might select them if this were to be the case:
☑ **Brown-Forsythe**
☑ **Welch**

Note:
If you find the variances are **not** equal, select a **Post Hoc** test from **Equal Variances Not Assumed** (e.g., select ☑ **Games-Howell**).

Copying SPSS Output

To copy output from SPSS to a Word document, click on each section of output you want, then:

Select **Copy Objects** from the **Edit** menu in SPSS to transfer it to a Word document, where you choose **Paste Special** from the **Word Edit menu**.

Or, select **Copy** in SPSS; then **Paste Special**, then **Picture** in Word (or simply select **paste** if you want to be able to edit the output in Word).

Step 4: Examine and interpret the output

Oneway

Descriptives

Word processing test score

	N	Mean	Std. Deviation	Std. Error	95% Confidence Interval for Mean		Minimum	Maximum
					Lower Bound	Upper Bound		
Nohelp	10	24.50	4.035	1.276	21.61	27.39	19	30
Options	10	28.40	4.222	1.335	25.38	31.42	23	34
Menu	10	31.40	4.248	1.343	28.36	34.44	25	37
Index	10	26.80	4.962	1.569	23.25	30.35	19	33
Total	40	27.78	4.917	.777	26.20	29.35	19	37

1

Examine the means and standard deviations. What do they suggest about group differences?

A plot of the means is also produced at the end of the output, but will not be included here.

Test of Homogeneity of Variances

Word processing test score

Levene Statistic	df1	df2	Sig.
.288	3	36	.834

2

The Levene's test is **NOT** significant (Sig. > .05); therefore, we retain the null hypothesis that the variances are equal. Thus, the homogeneity-of-variance assumption is met.

ANOVA

Word processing test score

	Sum of Squares	df	Mean Square	F	Sig.
Between Groups	252.075	3	84.025	4.378	.010
Within Groups	690.900	36	19.192		
Total	942.975	39			

3

The *F* value **IS** significant (Sig. < α, which is .05 unless otherwise specified). There IS a difference somewhere among the groups.

Post Hoc Tests

Multiple Comparisons

Dependent Variable: Word processing test score
Tukey HSD

(I) Experimental group	(J) Experimental group	Mean Difference (I-J)	Std. Error	Sig.	95% Confidence Interval Lower Bound	95% Confidence Interval Upper Bound
Nohelp	Options	-3.900	1.959	.210	-9.18	1.38
	Menu	-6.900*	1.959	.006	-12.18	-1.62
	Index	-2.300	1.959	.647	-7.58	2.98
Options	Nohelp	3.900	1.959	.210	-1.38	9.18
	Menu	-3.000	1.959	.430	-8.28	2.28
	Index	1.600	1.959	.846	-3.68	6.88
Menu	Nohelp	6.900*	1.959	.006	1.62	12.18
	Options	3.000	1.959	.430	-2.28	8.28
	Index	4.600	1.959	.106	-.68	9.88
Index	Nohelp	2.300	1.959	.647	-2.98	7.58
	Options	-1.600	1.959	.846	-6.88	3.68
	Menu	-4.600	1.959	.106	-9.88	.68

*. The mean difference is significant at the .05 level.

4

Tukey HSD post hoc test results show which groups differ from which.
The **Mean Difference** column shows the difference between the means in the columns to the left; the (I) column minus the (J) column. Where there is an **asterisk** beside the mean difference, the two means are **significantly different**. This will be supported by the **Sig.** value and the **95% confidence interval**.

These tables are a little confusing because they double-up. Nohelp – Menu gives the same result as Menu – Nohelp, with just a change in sign.

Note that an equal variance post hoc test **is** appropriate here because the Levene's test was **not** significant, indicating that the assumption of homogeneity of variance (equal variances) was met.

The only significant difference is between the Nohelp and Menu groups. This is also shown by the fact that they are not in overlapping subgroups below.

Homogeneous Subsets

Word processing test score

Tukey HSD[a]

Experimental group	N	Subset for alpha = .05 1	Subset for alpha = .05 2
Nohelp	10	24.50	
Index	10	26.80	26.80
Options	10	28.40	28.40
Menu	10		31.40
Sig.		.210	.106

Means for groups in homogeneous subsets are displayed.

a. Uses Harmonic Mean Sample Size = 10.000.

Understanding conceptually the ANOVA output, and what ANOVA does

ANOVA is really very simple, and it does exactly what its name suggests: it analyses the variance in the scores to determine the source of that variance. In this example there are 40 scores that vary from the grand mean of the entire sample, which is 27.775. Why do they vary? Is the variation just due to chance individual differences (error), or is it due to the effect of the independent variable manipulation (the different Help systems in this example)?

The main **ANOVA** summary table in the output has the following information.

Sum of Squares (*SS*)

To answer the question of "why the variation?" ANOVA calculates the sum of the squared deviations from the mean (**sum of squares**), which were introduced in Chapter 3 in the calculation of variance and standard deviation.

The ANOVA Summary Table in the SPSS output lists the **Sum of Squares** on which ANOVA is based.

The **Total sum of squares** is a measure of the variability in all 40 scores. It is the sum of the squared deviations of each score from the Grand Mean of all 40 scores.

The **Total** sum of squares is made up of two parts:

> **The Within-Groups or Error sum of squares.**
> This is a measure of the chance (or error) variability among individuals within each group. It is the sum of the squared deviations of each score from its Group Mean.
>
> **The Between-Groups sum of squares.**
> This is a measure of the variability between groups. It is the sum of the squared deviations of each score's Group Mean from the Grand Mean.

To illustrate, take the score for the first participant in the Nohelp Group.

- We can see from the Descriptives that the Total mean or **Grand Mean** is 27.775 (27.78).
- The **group mean** for Nohelp is 24.5.
- The **score** for the first participant in the Nohelp Group is 21.

This participant's total variability is:	21 - 27.775	= -6.775
It is made up of two parts:		
The Within-Groups part	21 - 24.5	= -3.5
The Between-Groups part	24.5 - 27.775	= -3.275

If these values are squared and summed for ALL the participants (as shown on the next page), the Sum of Squares figures in the SPSS output ANOVA Summary Table are obtained. You never need to do these calculations, but you should study the table to make sure you understand what is happening—and how *commonsensical* it really is.

This is the **deviation score** approach to calculating ANOVA that provides conceptual understanding of what ANOVA is about. However, **computational formulae** based on raw scores are used in the manual calculation of ANOVA that is shown in most text books.

		Grand Mean		**27.775**		
		Nohelp Group Mean		**24.5**		
		Options Group Mean		**28.4**		
		Menu Group Mean		**31.4**		
		Index Group Mean		**26.8**		

Score	Score - Grand Mean	Score - Group Mean	Group Mean - Grand Mean	Score - Grand Mean Squared	Score - Group Mean Squared	Group Mean - Grand Mean Squared
No-Help Group:						
21	-6.775	-3.5	-3.275	45.900625	12.25	10.725625
20	-7.775	-4.5	-3.275	60.450625	20.25	10.725625
27	-0.775	2.5	-3.275	0.600625	6.25	10.725625
21	-6.775	-3.5	-3.275	45.900625	12.25	10.725625
24	-3.775	-0.5	-3.275	14.250625	0.25	10.725625
26	-1.775	1.5	-3.275	3.150625	2.25	10.725625
30	2.225	5.5	-3.275	4.950625	30.25	10.725625
19	-8.775	-5.5	-3.275	77.000625	30.25	10.725625
28	0.225	3.5	-3.275	0.050625	12.25	10.725625
29	1.225	4.5	-3.275	1.500625	20.25	10.725625
Options Group:						
23	-4.775	-5.4	0.625	22.800625	29.16	0.390625
34	6.225	5.6	0.625	38.750625	31.36	0.390625
32	4.225	3.6	0.625	17.850625	12.96	0.390625
26	-1.775	-2.4	0.625	3.150625	5.76	0.390625
24	-3.775	-4.4	0.625	14.250625	19.36	0.390625
30	2.225	1.6	0.625	4.950625	2.56	0.390625
32	4.225	3.6	0.625	17.850625	12.96	0.390625
33	5.225	4.6	0.625	27.300625	21.16	0.390625
24	-3.775	-4.4	0.625	14.250625	19.36	0.390625
26	-1.775	-2.4	0.625	3.150625	5.76	0.390625
Menu Group:						
36	8.225	4.6	3.625	67.650625	21.16	13.140625
28	0.225	-3.4	3.625	0.050625	11.56	13.140625
29	1.225	-2.4	3.625	1.500625	5.76	13.140625
28	0.225	-3.4	3.625	0.050625	11.56	13.140625
28	0.225	-3.4	3.625	0.050625	11.56	13.140625
25	-2.775	-6.4	3.625	7.700625	40.96	13.140625
33	5.225	1.6	3.625	27.300625	2.56	13.140625
35	7.225	3.6	3.625	52.200625	12.96	13.140625
37	9.225	5.6	3.625	85.100625	31.36	13.140625
35	7.225	3.6	3.625	52.200625	12.96	13.140625
Index Group:						
33	5.225	6.2	-0.975	27.300625	38.44	0.950625
24	-3.775	-2.8	-0.975	14.250625	7.84	0.950625
32	4.225	5.2	-0.975	17.850625	27.04	0.950625
21	-6.775	-5.8	-0.975	45.900625	33.64	0.950625
23	-4.775	-3.8	-0.975	22.800625	14.44	0.950625
19	-8.775	-7.8	-0.975	77.000625	60.84	0.950625
32	4.225	5.2	-0.975	17.850625	27.04	0.950625
28	0.225	1.2	-0.975	0.050625	1.44	0.950625
26	-1.775	-0.8	-0.975	3.150625	0.64	0.950625
30	2.225	3.2	-0.975	4.950625	10.24	0.950625
		Sum of Squares		**942.9750**	**690.9000**	**252.0750**
				SS Total	*SS* within groups	*SS* between groups

Mean square (variance)

The **Mean Square** is actually the **variance** and is obtained by dividing the Sum of Squares by the Degrees of Freedom (*df*).

The Between Groups *df* $= k - 1$ where k = the number of groups
The Within Groups *df* $= N - k$ where N = the *total* number of participants
The Total *df* $= N - 1$

The *F* value (*F* ratio)

As mentioned earlier, *F* is the ratio of:

$$\frac{\text{between-groups variance (Mean square)}}{\text{within-groups (error) variance (Mean square)}} = \frac{84.025}{19.192} = 4.378$$

The Sig. value

This is the probability (p) of obtaining the observed F -ratio if the null hypothesis is true, and it is always a two-tailed probability. There cannot be a one-tailed test when there are more than two groups.

If $p > \alpha$, the test is **not** significant, and the null hypothesis of no difference between the population means is retained.

If $p \leq \alpha$, the test **is** significant. The null hypothesis is rejected, and the alternative hypothesis that there is a difference **somewhere** among the population means is accepted.

Alpha (α) is usually set at .05 unless there is a particular reason for doing otherwise.

Homogeneity of variance assumption

The Levene Test is a test of the null hypothesis that the population variances are equal.

If it is **not** significant, the null hypothesis is retained, and the homogeneity of variance assumption is met. Post hoc tests for **Equal Variances Assumed** can then be used.

If it **is** significant, the null hypothesis is rejected, and the assumption is violated. It is then necessary to choose post hoc tests for **Equal Variances Not Assumed**.

Analytical (post hoc) comparisons

The significant *F* indicates that there is a difference **somewhere** among the groups, but it does not specify which groups differ from which.

The Tukey HSD test is a post hoc analytical test of all pairwise comparisons to ascertain where the differences lie. It is appropriate when the homogeneity of variance assumption is met.

If the homogeneity of variance assumption is not met, one of the SPSS post hoc comparisons for unequal variances should be used instead (e.g., Games-Howell).

Post hoc tests are **only** conducted if the omnibus *F* is significant.

Using the General Linear Model (GLM) procedure in SPSS: A better way to conduct a one-way ANOVA

Steps 1 and 2 are as before

Step 3: Request the analysis using GLM

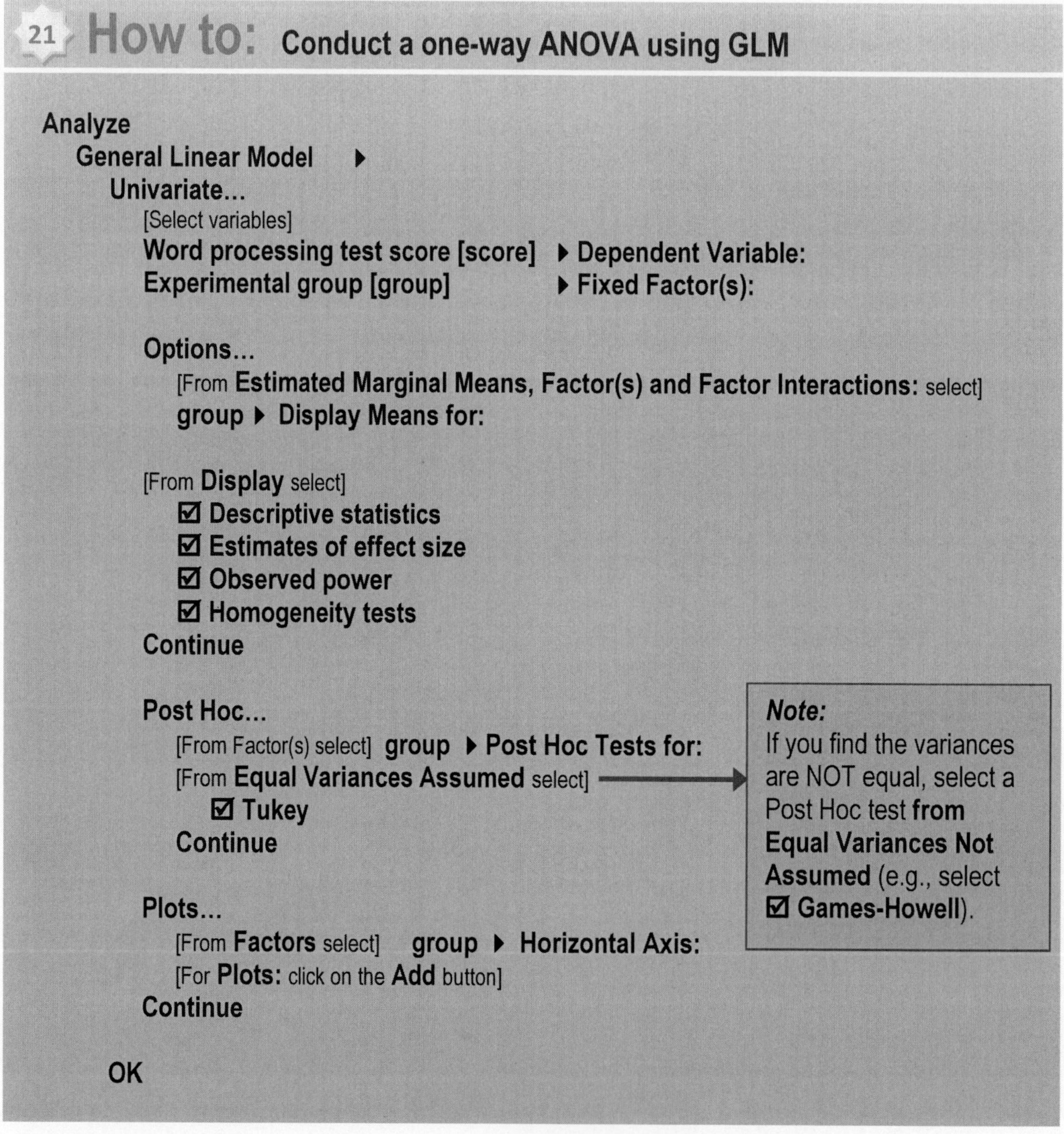

21 How to: Conduct a one-way ANOVA using GLM

Analyze
General Linear Model ▸
Univariate...
[Select variables]
Word processing test score [score] ▸ Dependent Variable:
Experimental group [group] ▸ Fixed Factor(s):

Options...
[From **Estimated Marginal Means, Factor(s) and Factor Interactions:** select]
group ▸ Display Means for:

[From **Display** select]
☑ **Descriptive statistics**
☑ **Estimates of effect size**
☑ **Observed power**
☑ **Homogeneity tests**
Continue

Post Hoc...
[From Factor(s) select] **group ▸ Post Hoc Tests for:**
[From **Equal Variances Assumed** select]
☑ **Tukey**
Continue

Note: If you find the variances are NOT equal, select a Post Hoc test **from Equal Variances Not Assumed** (e.g., select ☑ **Games-Howell**).

Plots...
[From **Factors** select] **group ▸ Horizontal Axis:**
[For **Plots:** click on the **Add** button]
Continue

OK

As well as all the output provided by the one-way procedure, GLM, gives additional information, the most important of which is as follows.

Step 4: Examine and interpret the additional output provided by GLM

Output: **One-way between-groups ANOVA using GLM**

Univariate Analysis of Variance

Group is the same as Between Groups, and Error is the same as Within Groups.

(You would normally ignore the **Intercept** which in ANOVA tests the null hypothesis that the grand mean is zero.)

These are estimates of the power of the test, which should ideally be .80 or above.

In this case, power is satisfactory at .833.

1

3

Tests of Between-Subjects Effects

Dependent Variable: Word processing test score

Source	Type III Sum of Squares	df	Mean Square	F	Sig.	Partial Eta Squared	Noncent. Parameter	Observed Power[a]
Corrected Model	252.075[b]	3	84.025	4.378	.010	.267	13.135	.833
Intercept	30858.025	1	30858.025	1607.887	.000	.978	1607.887	1.000
group	252.075	3	84.025	4.378	.010	.267	13.135	.833
Error	690.900	36	19.192					
Total	31801.000	40						
Corrected Total	942.975	39						

a. Computed using alpha = .05

b. R Squared = .267 (Adjusted R Squared = .206)

2

Partial Eta Squared, which has the symbol η^2 in research reports, is another measure of **effect size**. Here, it indicates that 26.7% (.267) of the total variance is explained by the independent variable. This is quite a large effect.

Effect size: Eta squared, and proportion of variance

Partial eta squared (η^2) which is the same as eta squared in **one-way** ANOVA designs is the effect size measure used by SPSS, although not everyone agrees it is the best measure (see Field, 2005, pp. 357-359). It is a generalisation of r^2, and is the same as r^2 when there are only two groups involved.

Eta squared is calculated as the sum of squares between groups, divided by the sum of squares total.

In this example that works out as follows:

$$\frac{\text{Sum of Squares}_{\text{Between groups}}}{\text{Sum of Squares}_{\text{Corrected Total}}} = \frac{252.075}{942.975} = .267$$

Cohen (1988, pp. 284-288) provided the following guidelines for eta squared:

Effect size	η^2
Small	.01 (1%)
Medium	.06 (6%)
Large	.14 (14%)

One-way ANOVA: Research report sample Results section

Results

The word processing test scores for the four types of help group were analysed with a one-way analysis of variance (ANOVA). Test assumptions of normality and homogeneity of variance were satisfactory, and the result was statistically significant, $F(3, 36) = 4.38$, $p = .01$, $\eta_p^2 = .27$. Tukey HSD post hoc comparisons revealed that the only significant difference was between the no help and menu groups; a higher score being achieved in the menu group than in the no help group ($M_{Diff} = 6.90$, $p = .006$, 95% CI [1.62,12.18]). Descriptive statistics are shown in Table 1.

Table 1

Mean Word Processing Test Scores

Group	*M*	*SD*
Nohelp	24.50	4.04
Options	28.40	4.22
Menu	31.40	4.25
Index	26.80	4.96

Note: $n = 10$ for each group.

Notes:

The convention for reporting a one-way ANOVA result is as follows:
F(*df* between groups, *df* error or within groups) = F ratio, p = Sig. value, η_p^2 = Partial Eta Squared value (for one-way designs).

Where cell sizes are equal, a note underneath the table is sufficient. With unequal cells include an n column.

Kruskal -Wallis test

(Nonparametric alternative to one-way ANOVA)

The Kruskal-Wallis is similar to the Mann-Whitney test described in the previous chapter, except that it is not limited to two groups, but can accommodate three or more groups. Like the Mann-Whitney it uses **ranks**, not raw scores, in its calculation.

Should you need to perform the Kruskal-Wallis test you can do so in SPSS as follows.

Using SPSS to conduct the Kruskal-Wallis test

Step 1: Enter data in SPSS and save data file

For convenience, the same data will be used as for the one-way ANOVA.

Step 2: Screen data and check assumptions

Although there are no assumptions to formally test you should always screen the data to check accuracy of entry. A visual inspection of the data is all that is required here. You should also run the **Explore** procedure to obtain full descriptive statistics, including the median which is the more appropriate measure of central tendency with ordinal data.

Step 3: Request the analysis

22 How to: Conduct a Kruskal-Wallis test

Analyze
Nonparametric Tests ▶
Legacy Dialogs ▶
K Independent Samples...
[Select variables you want]
Word processing test score [score] ▶ **Test Variable List:**
Experimental group [group] ▶ **Grouping Variable:**

[Click on **Define Groups...** and enter the values for **Group]**
[Enter **Minimum]** **1** [Tab]
[Enter **Maximum]** **4**
Continue

[From] **Test Type** [Select] ☑ **Kruskal-Wallis H**
OK

- Older versions of SPSS do not have the **Legacy Dialogs** option.
- See Appendix 1 for the new style of nonparametric tests in PASW 18.

Step 4: Examine and interpret the output

NPar Tests

Kruskal-Wallis Test

Ranks

	Experimental group	N	Mean Rank
Word processing test score	Nohelp	10	13.10
	Options	10	21.75
	Menu	10	28.85
	Index	10	18.30
	Total	40	

1

CARE!
These are the mean **ranks**, not mean scores.

Test Statistics[a,b]

	Word processing test score
Chi-Square	9.637
df	3
Asymp. Sig.	.022

a. Kruskal Wallis Test

b. Grouping Variable: Experimental group

2

SPSS uses a chi-square approximation to the Kruskal-Wallis test.

The **Asymp. Sig**. is the *p* value.

Here, it is < α of .05; therefore, the test result **IS significant**.

Effect size for the Kruskal-Wallis test

An eta squared effect size is computed for the Kruskal-Wallis test from the chi-square value, using the following formula (Green & Salkind, 2005, p. 386):

$$\eta^2 = \frac{\chi^2}{N-1}$$

$$= \frac{9.637}{40-1}$$

$$= \frac{9.637}{39}$$

$$= 0.2471$$

$$= \mathbf{.25}$$

Post hoc comparisons for the Kruskal-Wallis test

SPSS does not perform post hoc comparisons for the Kruskal-Wallis test. One option is to perform a series of Mann-Whitney tests on each of the possible pairs of groups, or ideally, on a subset of possible pairs that represent the **minimum number that are theoretically meaningful** (see Field, 2005, pp. 550-553). In this example, all possible pairs of comparisons, as we saw from the Tukey test earlier, involves six tests. You can always work out the number of possible pairs of tests, according to the following formula:

$k(k-1)/2$, where k = the number of groups.

Here, that is: (4 x (4-1))/2 = 4x3/2 = 12/2 = 6

But remember . . . if we want to keep **familywise error** to .05 across these six tests, we must use a **Bonferroni adjusted alpha level** for each test of .05 divided by the number of tests (i.e., .05/6 = .008)

If you perform these six Mann-Whitney tests, you will again find that the only significant difference (at adjusted α = .008) is between the Nohelp and Menu groups.

Kruskal-Wallis: Research report sample Results section

Results

A Kruskal-Wallis nonparametric test showed a significant difference in word processing test scores among the four types of help group, χ^2 (3, N = 40) = 9.64, p = .02, η^2 = .25. Six post hoc comparisons between pairwise means were conducted using the Mann-Whitney test, and a Bonferroni adjusted alpha of .008. The only significant difference was between the no help and menu groups, where a higher median score was achieved in the menu group, z (N = 20) = 2.73, p = .005, r^2 = .37.

Descriptive statistics are shown in Table 1.

Table 1

Descriptive Statistics for Word Processing Test Scores

Group	*M*	*Mdn*	*SD*	Range
Nohelp	24.50	25.00	4.04	11
Options	28.40	28.00	4.22	11
Menu	31.40	31.00	4.25	12
Index	26.80	27.00	4.96	14

Note: n = 10 for each group.

Notes:

Where cell sizes are equal, a note underneath the table is sufficient. With unequal cells include an n column. You would only include means and standard deviations when scores are at interval or ratio levels of measurement.

The general format for reporting chi-square (symbol χ^2) is:

χ^2 (*df*, *N* = number of cases) = chi-square value, *p* = probability value.

11 One-Way Repeated Measures, Anova And Friedman Nonparametric Alternative

When do we use repeated measures ANOVA?

Just as one-way between-groups ANOVA can be thought of as an extension of the independent *t* test where three or more independent groups are involved, so **repeated measures ANOVA** can be thought of as an extension of the **dependent *t* test** where three or more **related** groups of scores are involved. It usually involves the **same group of participants** being measured repeatedly under different levels of an independent variable.

Repeated measures ANOVA is also known as within-groups or **within-subjects ANOVA**, to distinguish it from **between-groups ANOVA** that involves different and independent groups of participants.

Advantages and disadvantages

The advantages of repeated measures over between-groups designs are that:

1. They require fewer participants.
2. They are more **powerful**, because error due to individual differences is removed from the error term. The smaller the error term (the denominator of the *F* ratio) the easier it is to detect a treatment effect.

The main disadvantage is that **carry-over** or **order** effects can occur, whereby the performance of participants is affected by their exposure to prior levels of the independent variable (e.g., there may be practise or learning effects, or participants may suffer from fatigue). Carry-over effects can often be addressed by some form of **counterbalancing**, whereby the order of conditions is randomised or balanced. There are a number of problems that can arise, however, and before using counterbalancing it is necessary to read up on these design issues in an appropriate research methods text book (e.g., Martin, 2008).

Assumptions

The usual ANOVA assumptions that were listed for the dependent *t* test apply; namely interval or ratio level data, random selection of participants, normality of scores and difference scores, and independence of observations across individuals. In addition, repeated measures ANOVA has the important assumption of **sphericity** (refer Field, 2005, p. 428). This, in essence, refers to homogeneity of variance in the **difference** scores between pairs of treatment levels.

It is tested by a statistic known as **Mauchly's test of sphericity**, which tests the null hypothesis that there **is** sphericity. If the Mauchly sphericity test is **not** significant (i.e., > .05) the null hypothesis is retained, and we conclude that the assumption has been satisfactorily met.

If the assumption is violated a correction is applied to the degrees of freedom, and the significance of the *F* ratio is ascertained using the new degrees of freedom. The correction is obtained by multiplying the original degrees of freedom by a value (provided by SPSS) known as **epsilon**. Two sets of adjusted figures are then provided by SPSS to be used when sphericity is violated. The first is the **Greenhouse-Geisser correction**, which tends to be conservative; and the second is the **Huynh-Feldt correction**, which tends to be more liberal. The two often lead to the same conclusion (significant or not), in which case the Greenhouse-Geisser result is normally reported. When they do **not** agree, there are various, rather confusing, recommendations about what to do (see Field, 2005, pp. 430-431, and 444-445).

One simple recommendation is to use the Huynh-Feldt correction when the epsilon values are around .75 or above, and to use the Greenhouse-Geisser correction when they are definitely below .75. However, an increasingly popular recommendation is to abandon univariate ANOVA when sphericity is violated, and to use the **multivariate** option instead. Multivariate ANOVA (MANOVA) results are always provided in the repeated-measure output. MANOVA itself is discussed in a later chapter.

For now, we will not pay too much attention to the multivariate option, but if you ever find yourself writing a journal article when you have a sphericity problem, you need to thoroughly read the relevant references on the issue. You should also look at other articles that have been published in your intended journal to ascertain whether the journal favours the univariate or multivariate approach.

Nonparametric alternative: Friedman test

The Friedman nonparametric test for one-way repeated measures designs is available in SPSS, under **Nonparametric Tests ▸, Legacy Dialogs ▸, K Related Samples...** . It is evaluated as a chi-square, and so is interpreted and written up in the same way as the Kruskal-Wallis test. For this reason, it will not be demonstrated in this chapter. Post hoc comparisons can be conducted if necessary using the Wilcoxon signed-rank test, in the same way as Mann-Whitney can be used for post hoc comparisons following a significant Kruskal-Wallis.

(See Appendix 1 for the new style of nonparametric tests in SPSS. Post hoc tests are performed automatically, as shown in the Kruskal-Wallis example in Appendix 1.)

Example of a repeated measures ANOVA

A new recreational drug is sweeping the world. It is even being touted as a possible solution to the illegal drug problem. Easy to make, and completely nonaddictive, it goes under the popular name of *Bliss*, because of the feeling of blissful wellbeing it produces. Most impressive of all, it has no detrimental effects on cognitive functioning, and, indeed, no negative side effects at all . . . until now! A bright young psychology PhD student, with a double-degree in medicine, has detected an unusual side effect. *Bliss* seems to have a very detrimental effect on mathematical ability that appears some twelve hours after consumption and is present for around six hours, by which time the drug has been cleared from the body.

In the first of a series of experiments, the student wants to test the effect of different quantities of *Bliss* on mathematical ability, and to do so on people who have not yet been exposed to it. They prove difficult to find, but eventually she tracks down eleven of them. Over a series of four days the participants take different quantities of the drug at 10pm each night, then at 11am the next day they are given a speed test on mathematics. Four equivalent versions of the test are used, and the score is the number of errors (including failures to complete) in the fixed 10-minute period allowed for the test.

To control for order effects (i.e., practise effects), partial counterbalancing is used, and the four conditions are: A control condition where the drug is not taken, 10mg of *Bliss*, 20 mg of *Bliss*, and 30 mg of *Bliss*. [Please note that this is an entirely fictitious example.]

Using SPSS to conduct one-way repeated measures ANOVA

Step 1: Enter data in SPSS and save data file

To demonstrate partial counterbalancing, the order in which the different quantities of the drug were taken by each participant is shown in the columns to the left of the ID column. For example, the participant with ID 4 took 10 mg on the first night, 20 mg on the second night, 30 mg on the third night, and did not take they drug on the fourth night.

The resulting data (number of errors) for the different quantities of the drug (regardless of the order in which it was taken) are in the columns to the right:

Night1 mg	*Night2 mg*	*Night3 mg*	*Night4 mg*	**ID**	**Control (No drug)**	**10 mg**	**20 mg**	**30 mg**
0	*10*	*20*	*30*	1	37	31	37	43
30	*0*	*10*	*20*	2	34	25	38	45
20	*30*	*0*	*10*	3	36	30	35	40
10	*20*	*30*	*0*	4	25	26	40	41
0	*10*	*20*	*30*	5	26	28	36	37
30	*0*	*10*	*20*	6	27	29	39	33
20	*30*	*0*	*10*	7	23	31	38	41
10	*20*	*30*	*0*	8	26	32	33	36
0	*10*	*20*	*30*	9	39	25	32	34
30	*0*	*10*	*20*	10	28	27	33	41
20	*30*	*0*	*10*	11	30	30	30	37

The first thing to note about repeated measures ANOVA, is that you **do** enter the data in four columns. The data comprise **four** scores (variables) from each of 11 participants (cases), unlike the situation in a between-groups design where there would be one score (variable) from 44 participants (cases).

The variable and value labels used in this guide are:

ID	Participant number	*Set **Measure** column to **Nominal***
Control	No drug control	*Set **Measure** column to **Scale***
Mg_10	10 mg	*Set **Measure** column to **Scale***
Mg_20	20 mg	*Set **Measure** column to **Scale***
Mg_30	30 mg	*Set **Measure** column to **Scale***

Save the data in a file that you might call ***Repeated.sav***.

Step 2: Screen data and check assumptions

Screen the data, check for outliers, and perform normality assumption checks (given the small sample size) for each of the four variables.

Step 3: Request the analysis

23 How to: Conduct a one-way repeated measures ANOVA

Analyze
  General Linear Model
    Repeated Measures...

[In the box for **Within-Subject Factor Name** type a name for the repeated measures factor:] **Drug**

[In the box for **Number of Levels** type:] **4**
[Click on the **Add** button]

[Click on **Define** button]

[Select the **Within-Subjects Variables** in the **CORRECT ORDER**, as follows:]
No drug control [Control] ▶ _?_ [1]
10 mg [Mg_10] ▶ _?_ [2]
20 mg [Mg_20] ▶ _?_ [3]
30 mg [Mg_30] ▶ _?_ [4]

Options...
[From **Estimated Marginal Means, Factor(s) and Factor Interactions** select:]
Drug ▶ **Display Means for:**

☑ **Compare main effects**
[From Confidence interval adjustment: select] **Sidak**[1] ▼

[From **Display** select:]
☑ **Descriptive Statistics**
☑ **Estimates of Effect Size**
☑ **Observed power**
Continue

OK

> ***Note:***
> The Sidak post hoc test is preferred, because it is very slightly more powerful than Bonferroni.

---

[1] Recall that the Sidak correction is a very slightly more powerful version of the Bonferroni adjustment, and is recommended for that reason.

Step 4: Examine and interpret the output

The essential output is shown below.

Descriptive Statistics

	Mean	Std. Deviation	N
No drug control	30.09	5.486	11
10 mg	28.55	2.505	11
20 mg	35.55	3.205	11
30 mg	38.91	3.780	11

1

Examine the means and standard deviations. What do they suggest might be happening?

Multivariate Tests[c]

Effect		Value	F	Hypothesis df	Error df	Sig.	Partial Eta Squared	Noncent. Parameter	Observed Power[a]
Drug	Pillai's Trace	.855	15.770[b]	3.000	8.000	.001	.855	47.309	.997
	Wilks' Lambda	.145	15.770[b]	3.000	8.000	.001	.855	47.309	.997
	Hotelling's Trace	5.914	15.770[b]	3.000	8.000	.001	.855	47.309	.997
	Roy's Largest Root	5.914	15.770[b]	3.000	8.000	.001	.855	47.309	.997

a. Computed using alpha = .05

b. Exact statistic

c.

Design: Intercept
Within Subjects Design: Drug

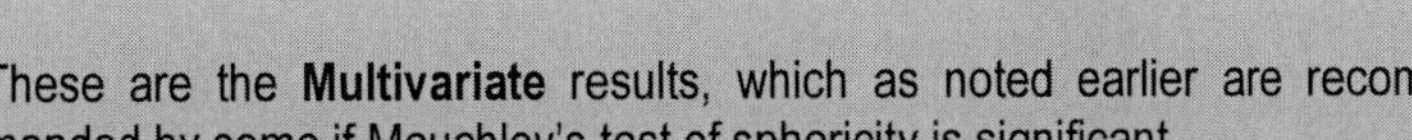

These are the **Multivariate** results, which as noted earlier are recommended by some if Mauchley's test of sphericity is significant.

If you were using them, they would indicate there is a significant difference among the scores (Pillai's Trace Sig. = .001, and **multivariate** η^2 = .855).

Note, there are four variants of the multivariate test, which are often the same or similar (as here). Pillai's is generally regarded as the more robust, although you often find that Wilkes Lambda is selected. The choice between them is not critical, but should they ever yield a different conclusion Wilkes Lambda would only be preferred if all assumptions had been clearly met.

Mauchly's Test of Sphericity[b]

Measure: MEASURE_1

Within Subjects Effect	Mauchly's W	Approx. Chi-Square	df	Sig.	Epsilon[a]		
					Greenhouse-Geisser	Huynh-Feldt	Lower-bound
Drug	.573	4.851	5	.437	.728	.939	.333

Tests the null hypothesis that the error covariance matrix of the orthonormalized transformed dependent variables is proportional to an identity matrix.

a. May be used to adjust the degrees of freedom for the averaged tests of significance. Corrected tests are displayed in the Tests of Within-Subjects Effects table.

b.
Design: Intercept
Within Subjects Design: Drug

2

Mauchly's tests of sphericity is **not** significant (Sig. = .437, > .05), therefore, we retain the null hypothesis, and conclude that the assumption of sphericity has been met.

Tests of Within-Subjects Effects

Measure: MEASURE_1

Source		Type III Sum of Squares	df	Mean Square	F	Sig.	Partial Eta Squared	Noncent. Parameter	Observed Power[a]
Drug	Sphericity Assumed	763.455	3	254.485	16.004	.000	.615	48.011	1.000
	Greenhouse-Geisser	763.455	2.184	349.625	16.004	.000	.615	34.947	.999
	Huynh-Feldt	763.455	2.817	271.000	16.004	.000	.615	45.085	1.000
	Lower-bound	763.455	1.000	763.455	16.004	.003	.615	16.004	.949
Error(Drug)	Sphericity Assumed	477.045	30	15.902					
	Greenhouse-Geisser	477.045	21.836	21.846					
	Huynh-Feldt	477.045	28.172	16.933					
	Lower-bound	477.045	10.000	47.705					

a. Computed using alpha = .05

3 4

Because Mauchley's test was **not** significant, we can use the results for **Sphericity Assumed**. The *F* value **is** significant (Sig. = .000, which means < .001, and that is certainly less than our normal α of .05).

This tells us that there is a difference **somewhere** among the Drug conditions.

Note also the results for **Partial Eta Squared** and Observed Power (which is perfect in this example).

Note: Early versions of SPSS did not automatically show the Greenhouse-Geisser etc. corrections that you only need to use when the sphericity assumption is not met.

Pairwise Comparisons

Measure: MEASURE_1

(I) Drug	(J) Drug	Mean Difference (I-J)	Std. Error	Sig.[a]	95% Confidence Interval for Difference[a]	
					Lower Bound	Upper Bound
1	10mg	1.545	1.983	.973	-4.929	8.020
	20mg	-5.455	2.146	.163	-12.461	1.552
	30mg	-8.818*	1.901	.006	-15.024	-2.612
2	Control	-1.545	1.983	.973	-8.020	4.929
	20mg	-7.000*	1.300	.002	-11.245	-2.755
	30mg	-10.364*	1.435	.000	-15.048	-5.679
20mg	Control	5.455	2.146	.163	-1.552	12.461
	10mg	7.000*	1.300	.002	2.755	11.245
	30mg	-3.364	1.201	.107	-7.283	.555
30mg	Control	8.818*	1.901	.006	2.612	15.024
	10mg	10.364*	1.435	.000	5.679	15.048
	20mg	3.364	1.201	.107	-.555	7.283

Based on estimated marginal means

*. The mean difference is significant at the .05 level.

a. Adjustment for multiple comparisons: Sidak.

Note:
SPSS does not automatically label the various repeated conditions, but just **numbers** them (**in the order you specified**) from 1 to 4. You have to manually enter the labels if you want them, as I have done here for some. To do this, double-click on the table in SPSS, then click on the number and type in the appropriate label. Be sure not to get confused about which number belongs to which drug condition: 1=control, 2=10mg, 3=20mg, and 4=30mg.

The **Sidak** post hoc results show which drug conditions differ from which.

The Mean Difference column shows the value of the (I) column mean to the left **minus** the (J) column mean. Where there is an asterisk beside the mean difference, the two means are significantly different. This will be supported by the Sig. value and the 95% confidence interval. Remember, too, that these tables are a little confusing because they double-up.

The asterisks tell us that there are significant differences, whereby scores are higher (more errors) with 30mg of the drug compared to either no drug or 10 mg, and with 20mg compared to 10mg.

There are no significant differences between no drug and 10mg, no drug and 20mg; and 20mg and 30mg.

You might wonder how it is possible to conclude there is **no** significant difference between no drug and 10 mg, and no drug and 20 mg, while at the same time concluding there **is** a significant difference between 10 mg and 20 mg.

There is a very important lesson here. If something is **statistically significant**, we conclude there **is** a real difference, but if something is **not** significant we do not really conclude there is no difference, but rather that ***we cannot rule out the possibility of no difference***, in other words that there ***might be*** no difference, as opposed to there ***is*** no difference. Results like this can, nonetheless, make interpretation a little difficult.

One-way repeated measures ANOVA: Research report sample Results section

Results

The error scores on the mathematics speed test for the four drug conditions were analysed with a one-way repeated measures analysis of variance (ANOVA. Assumptions of normality were satisfactory, as was Mauchley's test of sphericity. The result, $F(3, 30) = 16.00$, $p < .001$, $\eta_p^2 = .62$, indicated a statistically significant difference among the drug condition means, which are shown in Table 1. Sidak post hoc comparisons revealed that 30 mg of *Bliss* resulted in more errors than the no drug condition ($M_{\text{Diff}} = 8.82$, Sidak 95% CI [2.61,15.02]), and also the 10 mg condition ($M_{\text{Diff}} = 10.36$, Sidak 95% CI [5.68,15.05]). With 20 mg of *Bliss* there were more errors than with 10 mg ($M_{\text{Diff}} = 7.00$, Sidak 95% CI [2.76,11.25]), but 20 mg did not differ significantly from the control condition ($M_{\text{Diff}} = 5.46$, Sidak 95% CI [-1.52,12.46]). No other pairwise differences achieved statistical significance.

Table 1

Mean Number of Errors on the Mathematics Speed Test

Drug Condition	*M*	*SD*
No drug control	30.09	5.49
10mg	28.55	2.51
20mg	35.55	3.21
30mg	38.91	3.78

Note: $N = 11$.

Note: The convention for reporting a one-way repeated measures ANOVA is:
$F(df\text{ IV}, df\text{ error}) = F$ ratio, p = Sig. value, η_p^2 = Partial Eta Squared value (for one-way designs).

12 Factorial Between-Groups ANOVA

This chapter gets into quite complex territory that many students (and researchers) find daunting. It is not really that difficult, but can be overwhelming, especially when first encountered. Because the concepts and procedures of factorial ANOVA are of fundamental importance, it is essential to invest the time and effort needed to understand them, and then to retain that understanding **forever** in your **permanent** memory store. You must grasp this chapter in order to ever really be able to cope with analysis of variance.

To assist in this process, a chapter table of contents is provided, not only as a "find-quick", but also as a review facility, whereby you can tick off each section once you are satisfied you understand it. You need to work through the concepts, then follow the analysis systematically to see them illustrated. You should then review the concepts in the light of what you learn from the analysis.

Chapter contents

When do we use factorial ANOVA?

Between-groups factorial ANOVA is the next level of complexity in the ANOVA family of parametric statistics. It is used when there are **two or more IVs**; also known as **factors**. The groups or **cells** correspond to the various combinations of those IVs, and all the cells are independent (i.e., they are made up of different groups of unrelated participants).

The simplest **two-way** factorial ANOVA with two IVs is referred to as an A x B design (factor A with factor B); a **three-way** with three IVs is referred to as an A x B x C design, and so on.

In addition, the IVs can have various levels, so that factorial designs can also be written as 3 x 4 (a two-way design with three levels of factor A and four levels of factor B), or 2 x 5, or 3 x 2 x 3, and so on. The very simplest factorial design is the 2 x 2 design with only 4 cells (i.e., a two-way design with two levels of factor A and two levels of factor B). A 3 x 4 design would have 12 cells; a 2 x 5 design 10 cells; and a 3 x 2 x 3 design 18 cells. Throughout this chapter, factorial ANOVA will be illustrated using the hypothetical example that follows shortly. It is 3 x 2 design with 6 cells.

Assumptions

Assumptions for factorial between-groups ANOVA are the same as for one-way between-groups ANOVA:

1. The DV should be at the interval or ratio level of measurement.
2. The scores are randomly sampled from the populations of interest (or, in true experiments available participants have been randomly assigned to the different groups or cells).
3. The scores are **independent**. That is, participants appear in one and only one cell; the cells are not related to one another, and the score for each participant is independent of the scores for all the other participants.
4. The scores on the DV within each cell are **normally distributed** in the respective populations.
5. There is **homogeneity of variance**, that is, the variances in the cell populations are equal (the sample variances need to be approximately equal to satisfy this assumption).

As with the one-way ANOVA, factorial ANOVA is robust against violation of assumptions of normality and homogeneity of variance when cell sizes are large and equal. Assumptions are of concern mainly when cell sizes are small (i.e., <20-30) and unequal. However, heterogeneity of variance is a problem for analytical comparisons, although there are alternatives available to deal with it.

All ANOVA designs are less complicated if they are **balanced**, that is, there are equal numbers of participants in each cell. Aim for a balanced design wherever possible. **Unbalanced** designs, with unequal numbers in each cell, are possible, but more problematic, and homogeneity of variance becomes even more critical.

There are no nonparametric equivalents for the more complex ANOVAs.

Example of a two-way (3 x 2) factorial between-groups ANOVA

The psychology department of a large university runs a day care centre for children. You are a psychologist interested in motivation, and you identify a group of children who enjoy working with the computers provided at the centre.

You introduce a new computer game to the children and identify a group of 30, 6- to 7-year-olds, who enjoy playing the game. You obtain the parents' consent for these children to take part in an experiment to examine the effect of extrinsic rewards on intrinsic motivation. Intrinsic motivation is where satisfaction comes from performing an activity for its own sake; extrinsic motivation is where an activity is performed to gain extrinsic rewards such as praise, grades, or money.

You randomly allocate the children to one of three groups. The children in one group serve as the no-reward control group and are allowed to play the game spontaneously, while those in the second group are told that from now on they will gain or lose points as they play to tell them how well they are performing (an informational extrinsic reward). The children in the third group are told that from now on they will gain or lose points as they play, and that they must achieve a certain number of points in order to continue to be allowed to use the computers (a controlling extrinsic reward, or to avoid confusion with control group, you decide to call it a managing extrinsic reward, that is, one intended to control or manage behaviour).

After the first week you arrange for it so that each child gets to be left alone in a room with the computer game and other toys for a period of 15 minutes. You observe how many minutes they spend playing the computer game during that period.

To complete the experiment you manipulate one other variable, task difficulty, because there is literature to suggest that the effects of extrinsic rewards may differ for different levels of task difficulty. The computer game is set up so children play either an easy version or a hard version. Half the children in each group are randomly allocated to the easy version, while the other half are allocated to the hard version.

Schematic representation

A very useful first step in planning any factorial ANOVA is to draw a schematic representation of the design. In this example, it is a two-way (A x B), or more specifically, a 3 x 2 design, where three levels of factor A are **fully crossed** with two levels of factor B, producing a between-groups ANOVA with 6 cells. The dependent variable is the number of minutes of (spontaneous) playing time.

In addition to ensuring your understanding of the design, a schematic representation immediately draws attention to the question of sample size. Ideally, a minimum of 20 participants is needed in each cell, so 6 x 20 means that this design requires 120 participants if it is to be robust with respect to assumptions.

With only 30 children (5 assigned to each) cell, we know we are facing a less than ideal situation, where assumptions need to be taken seriously, and where there is a likelihood of low power.

	Factor (IV) B: Task Difficulty	
Factor (IV) A: Reward Type	Easy task	Hard task
No reward (control)		
Informational reward		
Managing reward		

The six **cells** of the design.

Imagine that the playing time in minutes (the DV) for each participant is as follows:

Factor (IV) A: Reward type	Factor (IV) B: Task difficulty Easy	 Hard	Marginal (Row) Means
No reward (Control)	7 10 8 8 9 Cell Mean: 8.40	9 8 9 7 8 Cell Mean: 8.20	8.30
Informational	8 8 9 10 11 Cell Mean: 9.20	12 10 13 12 10 Cell Mean: 11.40	10.30
Managing	10.5 9 11.5 9 10 Cell Mean: 10.00	5 6.5 7 6 6 Cell Mean: 6.10	8.05
Marginal (Column) Means	9.20	8.57	Grand Mean: 8.88

Main effects and interaction

In a **two-way** design such as this there is not just one effect and one *F* statistic, but three, one each for:

the **main effect** of Factor A;
the **main effect** of Factor B; and
the **interaction** of Factors A and B.

These are critically important concepts that you must understand and retain in your **permanent** knowledge base.

The Factor A main effect is the difference in the overall (row) means for the levels of Factor A (the three levels of reward), combined or collapsed across the levels of Factor B (the two levels of task difficulty). The **null hypothesis** is that there is no difference in these population means, that is, the mean playing time is the same when there is no reward, as it is for an informational reward, as it is for a managing reward[1].

The Factor B main effect is the difference in the overall (column) means for the levels of Factor B (the two levels of task difficulty), combined or collapsed across the levels of Factor A (the three levels of

---

[1] Significant main effects involving more than two levels are followed by analytical comparisons to determine exactly which levels differ from which.

reward). The **null hypothesis** is that there is no difference in these population means, that is, the mean playing time is the same for the easy task as for the hard task.

The A x B interaction is the extent to which differences in the levels of Factor A are **the same** at each level of Factor B, and vice versa. It concerns differences among the **cell means** (the six cell means in this case). A significant interaction takes priority over main effects, because it qualifies their interpretation, as the effects of factors are **conditional** on each other (i.e., the effect of Factor A **depends** on what level of Factor B is involved, and vice versa). If the interaction is significant in this example, it will mean that the effect of type of reward will **depend** on task difficulty (and vice versa). The **null hypothesis** is that the effect for each factor is the same at the levels of the other factor.

Types of interaction, and their implications for main effects

Whenever a significant interaction occurs it is necessary to plot the cell means as a first step toward its interpretation, and in particular, to establish whether it is an **ordinal** or **disordinal** interaction, as these have different implications for main effects.

In a two-way ANOVA, it is possible to produce two interaction plots, one with factor A on the horizontal axis (*x*-axis), and another with factor B on the *x*-axis.

No interaction

To appreciate the appearance of significant interactions, it is useful to look first at an example of **no interaction**. Here is a plot of one, as produced by SPSS.

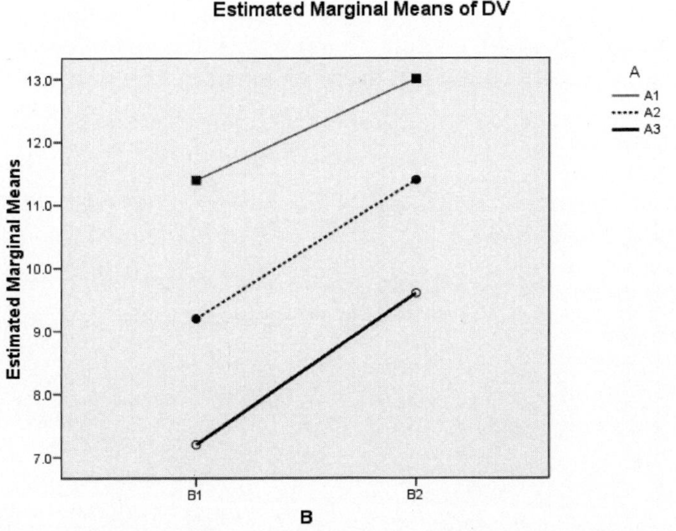

Factor B is plotted on the *x*-axis, and the lines plot the means for the three levels of factor A. The bold A1 line at the bottom, for example, plots a mean of just over 7 at B1, and around 9.5 at B2.

The more **parallel** the lines the more indicative they are of no interaction. There is clearly no interaction here, as the effect of A does not depend on B. The means for A1 are higher than for A2, and those for A2 are higher than those for A3, and the differences are maintained across both levels of B.

When there is **no interaction**, the main effect or main effects (if more than one is significant) **tell the whole story**, and are the focus of attention.

Ordinal interactions

Now, suppose the interaction plot had appeared as follows.

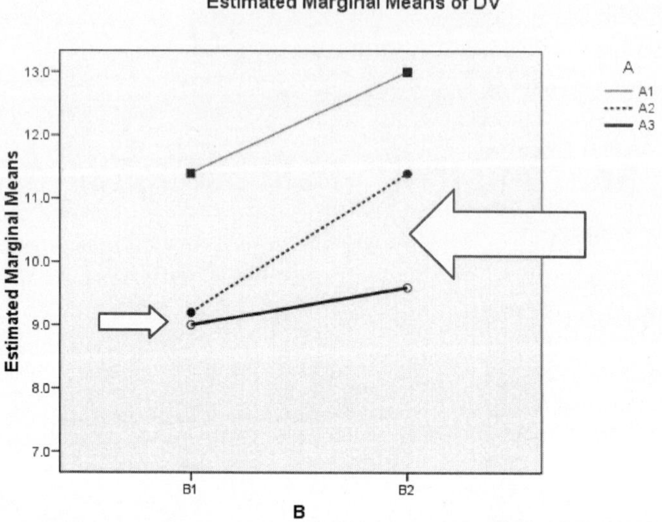

This shows an **ordinal** interaction, apparent when the lines are not parallel, but also do not cross over. The means for A1 are still higher than for A2, and those for A2 are higher than those for A3, but the latter differences are not maintained across both levels of B. The difference between A2 and A3 is very small at B1, but considerable at B2. (Note there is still no interaction involving A1 and A2.)

With an ordinal interaction the main effects remain important; they just do not tell the whole story, because the exact nature of an effect **depends** on the other factor involved.

Disordinal interactions

The plot to illustrate a disordinal interaction is one of the ones from the example we are using: the one with factor B (task difficulty) on the *x*-axis, and separate lines for the three levels of reward.

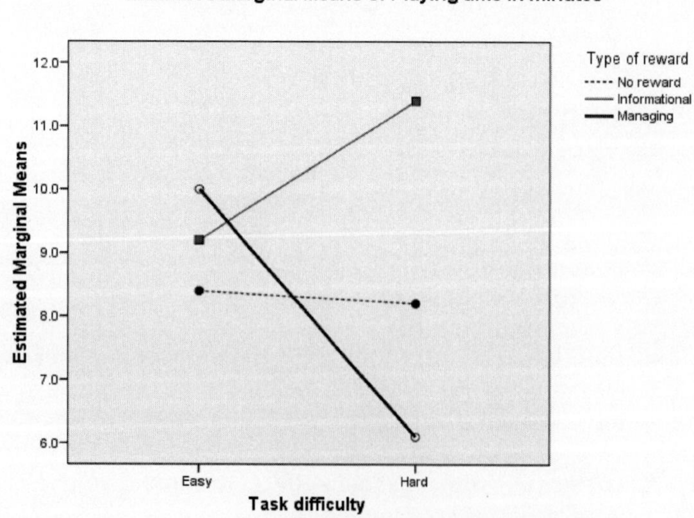

This shows a **disordinal** interaction, which is apparent when lines cross over one another.

With a disordinal interaction, the effect of one IV **depends** on the other IV, because things are working in opposite directions at levels of the other IV. This is apparent in the plot, so there can be no general statement about the effect of reward, because it utterly depends on what task difficulty is involved.

Disordinal interactions render main effects misleading, so that the main effects cease to be of any real interest. We will have more to say about this particular disordinal interaction a little later on.

Analysing a significant interaction

Before proceeding with a factorial ANOVA analysis, the researcher should decide on what to do if a significant interaction is found. There are a number of ways to break down an interaction to determine exactly what is going on; how **exactly** does one IV depend on the other IV? Unfortunately, this is a very controversial area, where there are different opinions on what should be done. At the heart of the disagreement is the issue of **familywise error**, which is **the most** controversial issue in the whole of statistics. Each researcher needs to make their own decision on how to proceed, based on the particular circumstances of the research being conducted, and also on any approach preferred by the journal to which they intend to submit the research report. Keppel and Wickens (2004) is a good source of advanced information and a variety of alternative approaches to this issue.

The approach that is most often recommended will be demonstrated in this chapter.

A simple effects approach to analysing interactions

A **simple effect** of a factor is its effect at **only one level** of the other factor in a factorial design. In our example, there are two simple effects of reward: the simple effect of reward for the easy task only; and the simple effect of reward for the hard task only.

There are three simple effects of task difficulty: The simple effect of task difficulty with no reward; an informational reward; and a managing reward, respectively.

You should be able to see that a simple effect analysis is just a one-way ANOVA, the only difference being that in the denominator of the *F* ratio it uses the **error mean square** from the full factorial ANOVA[1].

As with one-way ANOVA, if a simple effect is significant when there are more than two cells involved, it needs to be followed by analytical comparisons to pinpoint exactly which cells differ from which.

Do we need to perform all five of the simple effects in this example?

Sometimes we might want to, when both IV's are of equal theoretical importance, but the more tests performed, the more familywise error becomes an issue, and the more severe a Bonferroni (or Sidak) correction needs to be made, with a resulting loss of power. The goal, therefore, is to find an optimal compromise between controlling familywise error, and maintaining sufficient power, and this is where the disagreement comes in.

---

[1] If the assumption of **homogeneity of variance is violated**, a simple effect analysis should **not** use the error term from the full analysis. In this event, normal one-way ANOVAs are conducted, separately for each level of the second IV (using **Split File** in SPSS). Those that are significant are followed by analytical comparisons if they have more than two levels (using options for unequal variances if the homogeneity of variance assumption is violated for the one-way ANOVA).

In this example, the focus is on the effect of reward; task difficulty per se is not really of interest. Therefore, the first decision that could be taken is to only consider the simple effects of reward, which would mean just two simple effect analyses for each level of task difficulty.

Controlling familywise error for simple effects analyses

Given a decision to perform two simple effects analyses, the next question is what adjustment should be made for familywise error? Some sources advocate testing both at the conventional .05 alpha level, because it is already known from the factorial ANOVA that an interaction is present, and it is just a matter of clarifying it. Others, however, recommend strictly maintaining familywise error at .05, using a Bonferroni adjusted alpha of .05/2 = .025.

A third option, is to set a more liberal familywise error rate of .10 or even .15 to maintain reasonable power across the number of tests that are likely to be performed once analytical comparisons are added. This leads here to the same conclusion as the first option (using conventional .05). With only two simple effect analyses, a familywise error rate of .10, followed by a Bonferroni adjusted alpha for each one results in .10/2 = .05.

To make a decision, the researcher needs to consider what is important for the particular research project. Is it more important to avoid making a Type I error (e.g., deciding a reward type has an effect when it really doesn't)? If so, familywise error should be maintained at .05. On the other hand, if the research is more exploratory, seeking to maximise power, a familywise error of .10 or even .15 could be justified. The decision can also depend on many other things such as the overall power of the study given the same size, the complexity of the design, practical considerations, preferences of the journal to which the research is to be submitted, the cost/benefit of making either a Type I or Type II error, as well as the extent to which the research is likely to be replicated in an ongoing research program. In the latter event, maximising power might be more important. A more liberal familywise error might also be used if sample sizes are small, or if all the simple effects are to be examined.

Another decision, to which the same arguments apply, is what to do with a significant simple effect involving more than three levels, where further analytical or **simple comparisons** are needed? The most conservative approach is to test each comparison using the simple effect alpha level divided by the number of comparisons. For example, if the simple effect used an adjusted alpha of .025, and three analytical comparisons were needed, each would use an adjusted alpha of .025/3 = .008.

In our example, even though sample size is small, it might be the case that one does not want to promote the use of rewards for children unless very sure of their effects. With this priority the emphasis would be on avoiding Type I errors, and a familywise error rate of .05 would be used. Each simple effect would use a Bonferroni adjusted alpha of .05/2 = .025, with follow-up analytical comparisons using .025/3 = .008.

Using SPSS to conduct factorial ANOVA

Step 1: Enter data in SPSS and save data file

First, enter the data into SPSS and save it in a file called ***ANOVA.sav***.
In this example, there are three variables of interest (the IVs Reward and Task difficulty, and the DV Playing Time).

It is also a good idea to include two other variables, one for participant ID number, and another that specifies the "cells" of the design. This can either be typed in at the beginning or you can use the **Transform** menu in SPSS to create it later.

ID	Participant number	*Set **Measure** column to **Nominal***
Reward	Type of reward 0=No reward 1=Informational 2=Managing	*Set **Measure** column to **Nominal***
Task	Task difficulty 1=Easy 2=Hard	*Set **Measure** column to **Nominal***
Cell	1=None/Easy 2=None/Hard 3=Info/Easy 4=Info/Hard 5=Manage/Easy 6=Manage/Hard	*Set **Measure** column to **Nominal***
Time	Playing time in minutes	*Set **Measure** column to **Scale***

Step 2: Screen data and check assumptions

Screen the data and check for outliers as described in Chapter 6. Data screening needs to be conducted **separately for each cell** of the design, using the SPSS **Split File** option. Check the normality assumption carefully, given that the sample size is small.

Step 3: Request the analysis

24 How to: Conduct a factorial between-groups ANOVA

Analyze
General Linear Model
Univariate...
[Select variables, as follows:]
Playing time in minutes [time] ▸ Dependent Variable:

Type of reward [reward] }
Task difficulty [task] } ▸ Fixed Factor(s):

Options...
[From **Estimated Marginal Means, Factor(s) and Factor Interactions** select:]
reward }
task }
reward*task } ▸ Display Means for:
[Select] ☑ **Compare main effects**
[For **Confidence interval adjustment:** select]
Sidak[1]
[From **Display** select:]
☑ **Descriptive Statistics**
☑ **Estimates of Effect Size**
☑ **Observed power**
☑ **Homogeneity tests**
Continue

Plots...
[From **Factors:** select]
reward ▸ Horizontal Axis
task ▸ Separate Lines
[For **Plots:** Click on the **Add** button]
Continue
OK

> ***Note:***
> As an alternative to asking for Sidak main effect comparisons, it might be more powerful to choose from **Post Hoc . . .** tests, as demonstrated previously for One-way between-groups ANOVA using GLM.

[1] Recall, the Sidak test is a slightly more powerful version of the Bonferroni test.

Step 4: Examine and interpret the output

Univariate Analysis of Variance

Between-Subjects Factors

		Value Label	N
Type of reward	0	No reward	10
	1	Informational	10
	2	Managing	10
Task difficulty	1	Easy	15
	2	Hard	15

Descriptive Statistics

Dependent Variable: Playing time in minutes

Type of reward	Task difficulty	Mean	Std. Deviation	N
No reward	Easy	8.400	1.1402	5
	Hard	8.200	.8367	5
	Total	8.300	.9487	10
Informational	Easy	9.200	1.3038	5
	Hard	11.400	1.3416	5
	Total	10.300	1.7029	10
Managing	Easy	10.000	1.0607	5
	Hard	6.100	.7416	5
	Total	8.050	2.2292	10
Total	Easy	9.200	1.2790	15
	Hard	8.567	2.4412	15
	Total	8.883	1.9418	30

1

Examine the means and standard deviations, and consider what they suggest might be happening.

Levene's Test of Equality of Error Variances[a]

Dependent Variable: Playing time in minutes

F	df1	df2	Sig.
.967	5	24	.458

Tests the null hypothesis that the error variance of the dependent variable is equal across groups.

a. Design: Intercept+reward+task+reward * task

2

The Levene's test is **not** significant **(Sig. > .05)**; therefore, we retain the null hypothesis that the variances are equal. Thus, the homogeneity-of-variance assumption is met.

Tests of Between-Subjects Effects

Dependent Variable: Playing time in minutes

Source	Type III Sum of Squares	df	Mean Square	F	Sig.	Partial Eta Squared	Noncent. Parameter	Observed Power[a]
Corrected Model	80.642[b]	5	16.128	13.487	.000	.738	67.436	1.000
Intercept	2367.408	1	2367.408	1979.714	.000	.988	1979.714	1.000
reward	30.417	2	15.208	12.718	.000	.515	25.436	.992
task	3.008	1	3.008	2.516	.126	.095	2.516	.331
reward * task	47.217	2	23.608	19.742	.000	.622	39.484	1.000
Error	28.700	24	1.196					
Total	2476.750	30						
Corrected Total	109.342	29						

a. Computed using alpha = .05

b. R Squared = .738 (Adjusted R Squared = .683)

3

Examine the ANOVA summary table to establish which effects are significant, ignoring the Corrected Model and Intercept.

Here, the main effect for reward is significant (Sig. or $p < .001$), as is the reward*task interaction (Sig. or $p < .001$), but the main effect for task is not (Sig. or $p = .126$).

4

Be sure to stipulate **partial** eta squared for the effect size in **factorial** ANOVA.

Also note the **power** of each test. It is low in this example for the main effect of task.

Sum of Squares, Mean Square, and *F* Ratio

As with one-way ANOVA the Total sum of squares (the **Corrected Total** in this Table) is a measure of the variability in all 30 scores. It is the sum of the squared deviations of each score from the Grand Mean of all 30 scores. However, in two-way factorial ANOVA, the Total sum of squares is made up of four parts:

Within-Groups or **Error** sum of squares.
This is a measure of the variability among individuals within each group. It is the sum of the squared deviations of each score from its Cell Mean.

Factor A Between-Groups sum of squares (**reward** in this example).
This is a measure of the variability among Factor A levels. It is the sum of the squared deviations of each score's Factor A Row Mean from the Grand Mean.

Factor B Between-Groups sum of squares (**task** in this example).
This is a measure of the variability between Factor B levels. It is the sum of the squared deviations of each score's Factor B Column Mean from the Grand Mean.

Interaction sum of squares.
This is conceptually more difficult to understand, and is what is left in order to make up the corrected Total sum of squares. The more there is left, the more likely it is that an interaction is present.

Mean Squares and ***F* Ratio** are calculated as for the one-way ANOVA.

Degrees of freedom Total = N -1; for Factor A = a -1; for Factor B = b -1; and for the interaction = $(a -1)(b -1)$; where a and b are the number of levels in Factors A and B respectively.

Effect size: Partial Eta Squared

Partial eta squared does not equal eta squared when we move from one-way to factorial ANOVA.

Recall that eta squared is:

$$\frac{SS_{\text{Effect (Between-groups)}}}{SS_{\text{Corrected total}}}$$

Partial eta squared, however, is:

$$\frac{SS_{\text{Effect}}}{SS_{\text{Effect}} + SS_{\text{Error}}}$$

For example, here for reward it is :
30.417 / (30.417 + 28.700)
= 30.417 / 59.117
= .514522
= **.515**

See Tabachnick and Fidell (2007, pp. 54- 55) for a discussion of this issue, and for a generally preferred measure of effect size, although one that has to be calculated manually, namely **omega squared** (ω^2).

Estimated Marginal Means

1. Type of reward

Estimates

Dependent Variable: Playing time in minutes

Type of reward	Mean	Std. Error	95% Confidence Interval	
			Lower Bound	Upper Bound
No reward	8.300	.346	7.586	9.014
Informational	10.300	.346	9.586	11.014
Managing	8.050	.346	7.336	8.764

5

These are the **marginal means**. When cell sizes are **equal** (as they are here), marginal means are the same as the observed means reported in the **Descriptive Statistics** table at the beginning.

The Estimated Marginal Means tables also show the standard errors of the means, and their 95% confidence intervals.

What happens when cell sizes are Unequal?

Weighted means

When cell sizes are unequal the means in the **Descriptive Statistics** table are **weighted** means (i.e., they are biased toward larger cells). To illustrate, suppose we had two cells making up a main effect group, with details as follows:

	Cell 1	Cell 2	*(Weighted mean)* Group	*(Unweighted mean)* Group
M	4	24	8	14
n	20	5	25	25
ΣX	80	120		

The mean scores for the two cells combined would be $\Sigma X_1 + \Sigma X_2 / n_1 + n_2$ = (80+120) / (20+5) = 200/25 = 8.
You can see this group mean of 8 is biased in favour of the larger Cell 1; it is a **weighted mean**.

Unweighted means

When cells are unequal, the **marginal means** are **unweighted** means, and they differ from the Descriptive Statistics table. Unweighted means are simply the **average** of the contributing cell means, so each cell contributes equally, regardless of its size. The unweighted mean in our example is simply (4 + 24)/ 2 = 28/2 = 14.
This sits in the middle of the two cell means without being biased toward the larger cell; it is the **unweighted mean**.

Pairwise Comparisons

Dependent Variable: Playing time in minutes

(I) Type of reward	(J) Type of reward	Mean Difference (I-J)	Std. Error	Sig.[a]	95% Confidence Interval for Difference[a]	
					Lower Bound	Upper Bound
No reward	Informational	-2.000*	.489	.001	-3.255	-.745
	Managing	.250	.489	.942	-1.005	1.505
Informational	No reward	2.000*	.489	.001	.745	3.255
	Managing	2.250*	.489	.000	.995	3.505
Managing	No reward	-.250	.489	.942	-1.505	1.005
	Informational	-2.250*	.489	.000	-3.505	-.995

Based on estimated marginal means

*. The mean difference is significant at the .05 level.

a. Adjustment for multiple comparisons: Sidak.

6

Recall with one-way ANOVA, that because there are more than two groups we need to do analytical comparisons when there is a significant effect, in order to establish exactly which groups differ from which.

The same applies in factorial ANOVA when then there are **significant main effects that involve more than two levels**, as is the case here for type of reward.

The comparisons here (using Sidak adjustment to control familywise error across three tests) tell us that:

There is a significant difference between the no reward and informational conditions (informational reward has a higher mean than no reward).

There is a significant different between the informational and managing conditions (informational reward has a higher mean that the managing reward).

There is no significant difference between the no reward and managing reward cells.

Main effects are compromised by an interaction

We only focus on the main effects and their comparisons when there is **no** significant interaction.

In this case, because there **is** a significant interaction, these main effect comparisons, as we shall see, are actually misleading, and therefore they are of little interest.

2. Task difficulty

Estimates

Dependent Variable: Playing time in minutes

			95% Confidence Interval	
Task difficulty	Mean	Std. Error	Lower Bound	Upper Bound
Easy	9.200	.282	8.617	9.783
Hard	8.567	.282	7.984	9.149

Pairwise Comparisons

Dependent Variable: Playing time in minutes

					95% Confidence Interval for Difference[a]	
(I) Task difficulty	(J) Task difficulty	Mean Difference (I-J)	Std. Error	Sig.[a]	Lower Bound	Upper Bound
Easy	Hard	.633	.399	.126	-.191	1.457
Hard	Easy	-.633	.399	.126	-1.457	.191

Based on estimated marginal means

a. Adjustment for multiple comparisons: Sidak.

The top table shows the marginal means, standard errors, and 95% confidence intervals for the levels of the second main effect.

When there are only **two** levels involved in a main effect, as here for task difficulty, there is no real need to consider the table of **Pairwise Comparisons**. It will be give the same result as in the main ANOVA summary table (Tests of Between-Subjects Effects); namely, no significant difference (over the three reward levels combined) between the two levels of task difficulty

3. Type of reward * Task difficulty

Dependent Variable: Playing time in minutes

Type of reward	Task difficulty	Mean	Std. Error	95% Confidence Interval	
				Lower Bound	Upper Bound
No reward	Easy	8.400	.489	7.391	9.409
	Hard	8.200	.489	7.191	9.209
Informational	Easy	9.200	.489	8.191	10.209
	Hard	11.400	.489	10.391	12.409
Managing	Easy	10.000	.489	8.991	11.009
	Hard	6.100	.489	5.091	7.109

Profile Plots

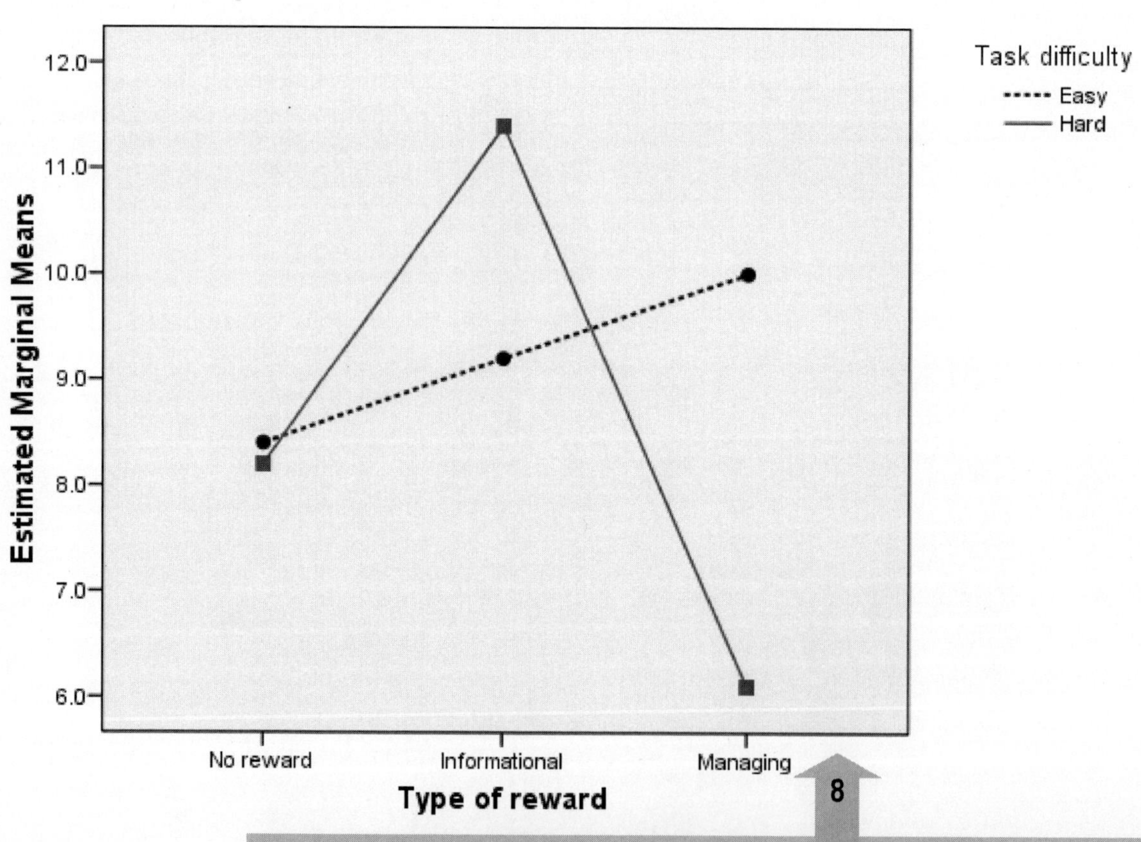

If there is a significant interaction, as there is here, it becomes very important to examine the plot of the means to help understand the nature of the interaction.

The interaction means in the table above also become important.

Step 5: Performing a simple effects analysis in SPSS following a significant interaction

Step 4 would have been the end of the analysis had the interaction **not** been significant.

With a significant interaction it is necessary to proceed to step 5: performing the two simple effect analyses previously decided upon in the event of an interaction (reward within the two levels of task).

Simple effects analysis in SPSS requires the use of command syntax. You might have noticed in your own work that the latest versions of SPSS include the command syntax for each analysis at the beginning of the output.

The easiest way to input the necessary command syntax is by using an older SPSS procedure, MANOVA (multivariate analysis of variance), which is extremely versatile and can do many things besides MANOVA. Here we will use it to perform simple effects analysis.

25 How to: Use SPSS command syntax to conduct an ANOVA simple effects analysis

[In the SPSS **Data View window** click on the following menu options]

File
  New
    Syntax

This opens a syntax window.
Type the following commands into the syntax window, exactly as they are written, then save the file (it will be given the extension .sps, and you might want to call it ***ANOVA.sps***)

```
MANOVA time BY reward(0,2) task(1,2)
   /ERROR = WITHIN
   /DESIGN = REWARD WITHIN task(1)
   /DESIGN = REWARD WITHIN task(2).
```

Do **not** forget this full stop.

To run this analysis, position the cursor at the beginning, or somewhere within the commands:

Then click on the run button ▶ located in the icon bar.
Alternatively, select from the **Run** option in the menu bar at the top.

Note:
The first line of the MANOVA command syntax specifies the name of the dependent variable, which is **time**; and the names and minimum and maximum levels of the independent variables, **reward(0,2)**, and **task(1,2)**.

The second line specifies that the **error term** to be used is the one from the main analysis.

Lines 3 and 4 ask for each of the simple effect analyses across the easy **task(1)**, then hard **task(2)**.

Step 6: Examine and interpret the simple effects analysis output

Output: **ANOVA simple effects analysis**

Manova

```
* * * * * * A n a l y s i s   o f   V a r i a n c e * * * * * *

        30 cases accepted.
         0 cases rejected because of out-of-range factor values.
         0 cases rejected because of missing data.
         6 non-empty cells.

         2 designs will be processed.

 - - - - - - - - - - - - - - - - - - - - - - - - - - - - - - - - - - - -

* * * * * A n a l y s i s   o f   V a r i a n c e -- design   1 * * * * *

 Tests of Significance for time using UNIQUE sums of squares
 Source of Variation          SS      DF        MS         F  Sig of F

 WITHIN CELLS              28.70      24      1.20
 REWARD WITHIN TASK(1)      6.40       2      3.20      2.68      .089

 - - - - - - - - - - - - - - - - - - - - - - - - - - - - - - - - - - - -
```

1

As of SPSS (version 15) MANOVA syntax still results in the old style of output. To be able to scroll through **all** the output you may need to double-click on it so it opens in its own window.

These are the results for the simple effect of reward within the **first** level of task difficulty. Refer back to where the data were entered and coded to confirm this was the **easy task**. The **Sig. of F** is **greater than** our Bonferroni adjusted alpha of .025, therefore, this simple effect is **not** significant.

This tells us there were no significant effects of reward on playing time when the children were playing the easy version of the computer game.

```
* * * * * A n a l y s i s   o f   V a r i a n c e -- design   2 * * * * *

 Tests of Significance for time using UNIQUE sums of squares
 Source of Variation          SS      DF        MS         F  Sig of F

 WITHIN CELLS              28.70      24      1.20
 REWARD WITHIN TASK(2)     71.23       2     35.62     29.78      .000
 - - - - - - - - - - - - - - - - - - - - - - - - - - - - - - - - - - - -
```

2

These are the results for the simple effect of reward within the **second** level of task difficulty; that for the **hard task**. The **Sig. of F** is **less than** our Bonferroni adjusted alpha of .025, therefore, this simple effect **is** significant.

This tells us that there were significant differences **somewhere** across the different levels of reward when the children were playing the hard version of the computer game.

Step 7: Performing simple comparisons in SPSS following a significant simple effect analysis

The simple effect analysis for the hard task was significant, but with three levels of reward we now need to perform simple effect comparisons to pinpoint exactly which reward conditions are different from which. These comparisons can also be conducted using MANOVA syntax, but the output is confusing. The easiest way to perform them is to specify the exact comparisons we want using **comparison coefficients** in the one-way ANOVA procedure.

When we have homogeneity of variance, as in this example, and all analyses require the mean square error from the main factorial analysis, we use all **six cells of the design as IV levels** in the one-way ANOVA, as shown below.

When homogeneity of variance is violated the simple effect analyses are just separate one-way ANOVAs anyway, where comparisons only need be made among the cells involved in each analysis.

Because we are performing yet more tests we need to make further adjustment for familywise error, by using the Bonferroni adjusted alpha used for the simple effect analysis (.025 in this case) and dividing it by the number of comparisons (three in this case), so our per-comparison adjusted alpha for the simple comparisons becomes .025/3 = .008.

Comparison coefficients

To perform comparisons (or **contrasts** as they are also called), weights or coefficients are used to indicate which cells are being compared. To avoid confusion in planning comparisons, it is always useful to draw up a table like the one below, which lists the six cells of the present example, **in the order in which they were coded.**

Here, we are only interested in comparing reward conditions for the hard task, but we need all six cells to retain the mean square error for the full design. To specify that none of the easy task cells are involved they are assigned coefficients of 0. **Note that coefficients must always sum to zero.**

Cell Code and Description	1 None/ Easy	**2 None/ Hard**	3 Info/ Easy	**4 Info/ Hard**	5 Manage/ Easy	**6 Manage/ Hard**	**SUM**
Mean	8.40	**8.20**	9.20	**11.40**	10.00	**6.10**	
Comparison 1	0	**-1**	0	**1**	0	0	**0**
Comparison 2	0	**1**	0	0	0	**-1**	**0**
Comparison 3	0	0	0	**1**	0	**-1**	**0**

To illustrate what the comparisons are "saying", consider Comparison 3 as an example. It says:

> Ignore cells 1, 2, 3, and 5, but take the mean of cell 4 and multiple it by 1 (11.40 x 1 = 11.40), and the mean of cell 6 and multiply it by -1 (6.10 x -1 = -6.10), then determine if the difference between them, 11.40 – 6.10 = 5.30, is significant or not.

It does not matter which of the two cells is assigned the negative coefficient, although to keep output consistent it is useful to always assign plus 1 to the cell with the larger mean.

Comparisons 1 and 2 are interpreted in the same way, such that they compare cells 2 and 4, and 2 and 6 respectively. Together, these three comparisons test all possible differences among reward conditions for the hard task.

The coefficients then need to be entered **in the correct order** into the SPSS one-way ANOVA procedure under **Contrasts...**, as shown on the next page.

26 How to: Conduct simple comparisons using the one-way ANOVA procedure

Analyze
Compare Means
One-Way ANOVA...
[Select variables you want]
Playing time in minutes [time] ▸ Dependent List:
Cell [cell] ▸ Factor:

Contrasts...
[Type in the coefficients for each Comparison, as follows]
[In the box for **Coefficients** type the Comparison 1 coefficients as follows:]
0 [Click on **Add** button]
-1 [Add]
0 [Add]
1 [Add]
0 [Add]
0 [Add]
[Click on the **Next button**, and type Comparison 2 Coefficients]
0 [Add]
1 [Add]
0 [Add]
0 [Add]
0 [Add]
-1 [Add]
[Click on the **Next button**, and type Comparison 3 Coefficients]
0 [Add]
0 [Add]
0 [Add]
1 [Add]
0 [Add]
-1 [Add]
Continue
OK

Step 8: Examine and interpret the simple comparisons output

Contrast Coefficients

Contrast	Cell					
	None/Easy	None/Hard	Info/Easy	Info/Hard	Manage/Easy	Manage/Hard
1	0	-1	0	1	0	0
2	0	1	0	0	0	-1
3	0	0	0	1	0	-1

1

Always double-check the table of **Contrast Coefficients** provided, to make sure the coefficients have been entered correctly, because if they haven't, the output will still look OK, but it will be completely wrong!

Contrast Tests

		Contrast	Value of Contrast	Std. Error	t	df	Sig. (2-tailed)
Playing time in minutes	Assume equal variances	1	3.200	.6916	4.627	24	.000
		2	2.100	.6916	3.036	24	.006
		3	5.300	.6916	7.663	24	.000
	Does not assume equal variances	1	3.200	.7071	4.525	6.702	.003
		2	2.100	.5000	4.200	7.886	[illegible]03
		3	5.300	.6856	7.731	6.236	.000

2

We know from the beginning, that we can assume equal variances, so we use the results from the top table. All three Sig. (2-tailed) results are less than our Bonferroni adjusted alpha level of .008.

From Contrast I we conclude that the mean playing time in the informational reward/hard task cell was significantly **higher** than in the no reward/hard task cell.
*The informational reward resulted in a **higher** mean than no reward.*

From Contrast 2 we conclude that the mean playing time in the no reward/hard task cell was significantly **higher** than the managing reward/hard task cell.
*In other words, the managing reward resulted in a **lower** mean than no reward.*

From Contrast 3 we conclude that the mean playing time in the informational/hard task cell was significantly **higher** than the managing reward/hard task cell.
*The informational reward resulted in a **higher** mean than the managing reward.*

The versatility of coefficients

Now is a good time to let you in on another way of doing things, one I personally find makes more sense. It is also an opportunity to introduce you to the control you have once you are able to use command syntax and coefficients effectively.

In the later versions of SPSS, you may already have noticed that output opens with the command syntax that SPSS created to run the analysis[1]. I have ignored it so far, but here is the command syntax SPSS created for the simple comparisons. I have just copied from SPSS and pasted into Word.

```
ONEWAY
  time BY cell
  /CONTRAST= 0 -1 0 1 0 0 /CONTRAST= 0 1 0 0 0 -1 /CONTRAST= 0 0 0 1 0 -1
  /MISSING ANALYSIS .
```

You can then rearrange it in Word to make it easier to follow, as in the left column at the bottom of the page.

Now, imagine in this example that a different decision had been taken, a decision to look at all five of the possible simple effects. This, could, in effect, be done by using nine simple comparisons instead, with a Bonferroni adjusted alpha for each one of .05/9 = .006.

All we need to do is copy and paste the three lines of syntax for the contrasts into a new version of the command syntax, then change the coefficients to specify the nine comparisons (but do this **very carefully**, as it is easy to make mistakes). The syntax would then look like the right hand column (do not forget the stop at the end):

- The first three contrasts remain the same; they are the three possible comparisons among the hard task cells.
- The next three contrasts have been changed to specify the same three comparisons, but this time among the easy task cells. Together, these six contrasts cover the two simple effects of reward.
- The final three contrasts are those for the three simple effects of task difficulty (easy versus hard with no reward; easy versus hard with informational reward; and easy versus hard with managing reward.)

```
ONEWAY
 time BY cell
 /CONTRAST= 0 -1 0 1 0 0
 /CONTRAST= 0 1 0 0 0 -1
 /CONTRAST= 0 0 0 1 0 -1
 /MISSING ANALYSIS .
```

```
ONEWAY
 time BY cell
 /CONTRAST= 0 -1 0 1 0 0
 /CONTRAST= 0 1 0 0 0 -1
 /CONTRAST= 0 0 0 1 0 -1
 /CONTRAST= -1 0 0 0 1 0
 /CONTRAST= 0 0 -1 0 1 0
 /CONTRAST= -1 0 1 0 0 0
 /CONTRAST= 1 -1 0 0 0 0
 /CONTRAST= 0 0 -1 1 0 0
 /CONTRAST= 0 0 0 0 1 -1
 /MISSING ANALYSIS .
```

[1] If not in your output, go to the **Data View** window, click on **Edit** menu, select **Options. . .** , then **Viewer** tab, then ☑ **Display commands in the log** (at bottom left of dialog box), then click **OK**.

Here is the output and it tells the whole story. I would argue this is the easiest way to do it, although you can find all manner of alternative and complicated syntax in various textbooks.

Contrast Coefficients

	Cell					
Contrast	None/Easy	None/Hard	Info/Easy	Info/Hard	Manage/Easy	Manage/Hard
1	0	-1	0	1	0	0
2	0	1	0	0	0	-1
3	0	0	0	1	0	-1
4	-1	0	0	0	1	0
5	0	0	-1	0	1	0
6	-1	0	1	0	0	0
7	1	-1	0	0	0	0
8	0	0	-1	1	0	0
9	0	0	0	0	1	-1

1

Be VERY SURE to check the coefficients because it is so easy to make a mistake. The potential for mistakes is the main disadvantage of this approach, but it is easily overcome with a little careful attention.

Contrast Tests

		Contrast	Value of Contrast	Std. Error	t	df	Sig. (2-tailed)
Playing time in minutes	Assume equal variances	1	3.200	.6916	4.627	24	.000
		2	2.100	.6916	3.036	24	.006
		3	5.300	.6916	7.663	24	.000
		4	1.600	.6916	2.313	24	.030
		5	.800	.6916	1.157	24	.259
		6	.800	.6916	1.157	24	.259
		7	.200	.6916	.289	24	.775
		8	2.200	.6916	3.181	24	.004
		9	3.900	.6916	5.639	24	.000

2

Simple effects of reward:

The first three results for **Assume equal variances** are the same as before for the hard task.

The next three, for the easy task, confirm no significant differences (at adjusted alpha = .006), consistent with the earlier simple effect analysis.

Simple effects of task difficulty:

The final three are the new information, the simple effects for task difficulty.
They show no significant difference between easy and hard tasks with no reward; a significantly longer playing time on the hard task with an informational reward; and a significantly longer playing time on the easy task with a managing reward.

Violation of the assumption of homogeneity of variance:

The other advantage here is that if homogeneity of variance has not been met, the results for **Does not assume equal variances** can be used instead.

Obtaining a bar graph with error bars from SPSS

Although interaction diagrams used to be shown as line graphs, and still are by SPSS, the previous edition of the *APA Manual* recommended they be shown as bar graphs when the independent variable is nominal or categorical. It is also increasingly recommended that graphs show error bars, which commonly represent either standard errors, or 95% confidence intervals.

27 How to: Obtain a bar graph with error bars using Legacy Dialogs

Graphs
  Legacy Dialogs ▸
    Bar...
      [Click mouse on box for **Clustered**]
      [From **Data in Chart Are** select]
        ⊙ **Summaries for groups of cases**
      Define
        [From **Bars Represent** select]
          ⊙ **Other statistic (e.g., mean)**
        [Select variable(s) you want]
          Playing time in minutes [time] ▸ **Bars Represent:**
          Type of reward [reward] ▸ **Category Axis**
          Task difficulty [task] ▸ **Define Clusters by:**

        Options...
          ☑ **Display error bars**
          [From Error **Bars Represent** select]
            ⊙ **Confidence intervals**
              [For **Level (%):** type] **95.0** [assuming it is not already entered]
          Continue
      OK

Notes:

1. The **Legacy Dialogs** option in **Graphs** is relatively easy to use, but it is likely to be discontinued in future versions of SPSS in favour of **Chart Builder**. See Appendix 2 for how to use **Chart Builder** to obtain this graph.
2. In earlier versions of SPSS it might just be a matter of selecting Graphs then Scatter... without needing to go into **Legacy Dialogs.**

The chart that is produced is still not in the style required by the *APA Manual*, but by double-clicking on it to open **Chart Editor** just about anything is possible. However, it would take pages to explain it all. Instead, practise clicking or double-clicking on everything in the chart or menus to see what appears, then explore the ways in which it can all be changed. (*Note:* If you have trouble positioning the legend for *Game Difficulty* using **Chart Editor** in the version of SPSS you are using, you might need to select **Hide Legend** and create it manually.)

Keep exploring until you have managed to produce a graph similar to the one in the Results section to follow (this graph was entirely produced in SPSS, albeit Version 15).

Note:
The way graphs are created and edited has changed considerably in different versions of SPSS. If using older versions, you might find things very different, but possibly a whole lot simpler.

Confidence intervals, error bars and significance: A caution

It is beyond the scope of this book, but be careful with confidence intervals and error bars as we move into complex designs. There is a common misconception that if confidence intervals do not overlap, the difference between means is significant, and if they do overlap it is *not* significant, but this is not necessarily the case. The two approaches are related but not identical. This blog site might be helpful: http://scienceblogs.com/cognitivedaily/2007/03/most_researchers_dont_understa.php.

Payton, Greenstone, and Schenker (2003), concluded that "caution should be exercised when results of an experiment are displayed with confidence or standard error intervals. Whether or not these intervals overlap does not imply the statistical significance of the parameters of interest." (p. 5.) In particular, "the method of examining overlap . . . mistakenly fails to reject the null hypothesis more frequently than does [significance testing] when the null hypothesis is false" (Schenker & Gentleman, 2001, p. 182). Advanced researchers should consult these references; and also Loftus and Masson (1994) for additional issues in the use of confidence intervals in repeated measures designs.

Go back and read p. 38 about confidence intervals. All decisions in inferential statistics are *probability judgements* not certainties, and confidence intervals are useful as another source of information for understanding the probabilistic nature of estimates of population means.

Factorial ANOVA: Research report sample Results section

Results

A 3 x 2 (type of extrinsic reward x game difficulty) between-groups ANOVA was performed on the time spent playing with the computer game. Assumptions of normality and homogeneity were satisfactory. The reward x difficulty interaction was significant, $F(2,24) = 19.74$, $p < .001$, partial $\eta_p^2 = .62$; as was the main effect of reward, $F(2, 24) = 12.72$, $p < .001$, partial $\eta_p^2 = .52$; but not the main effect of game difficulty, $F(1,24) = 2.52$, $p = .13$, partial $\eta_p^2 = .10$. Descriptive statistics are given in Table 1, and the interaction is shown in Figure 1.

Two simple effect analyses for reward across easy and hard games were conducted to further analyse the interaction, using a Bonferroni adjusted alpha level of .025 to maintain the familywise error rate at .05. The simple effect of reward was not significant for the easy version of the computer game, $F(2,24) = 2.68$, $p = .09$, but it was for the hard version, $F(2,24) = 29.78$, $p < .001$. Three pairwise comparisons among the means, using an adjusted alpha of .008, showed that when playing the hard version of the game children played for longer with an informational reward compared to no reward, $t(24) = 4.63$, $p < .001$, partial $r^2 = .47$[1]; and also compared with a managing reward, $t(24) = 7.66$, $p < .001$, partial $r^2 = .71$. They also played longer with no reward than they did with a managing reward, $t(24) = 3.04$, $p = .006$, partial $r^2 = .28$.

---

[1] The effect size for each of the comparisons is calculated manually using the effect size formula for r^2 of $t^2_{contrast}/(t^2_{contrast}+df)$.

Table 1

Mean Playing Time for Type of Extrinsic Reward by Difficulty of the Computer Game

	Easy Game		Hard Game		Total	
Type of reward	*M*	*SD*	*M*	*SD*	*M*	*SD*
No reward	8.40	1.14	8.20	0.84	8.30	0.95
Informational	9.20	1.30	11.40	1.34	10.30	1.70
Managing	10.00	1.06	6.10	0.74	8.05	2.23
Total	9.20	1.28	8.57	2.44	8.88	1.94

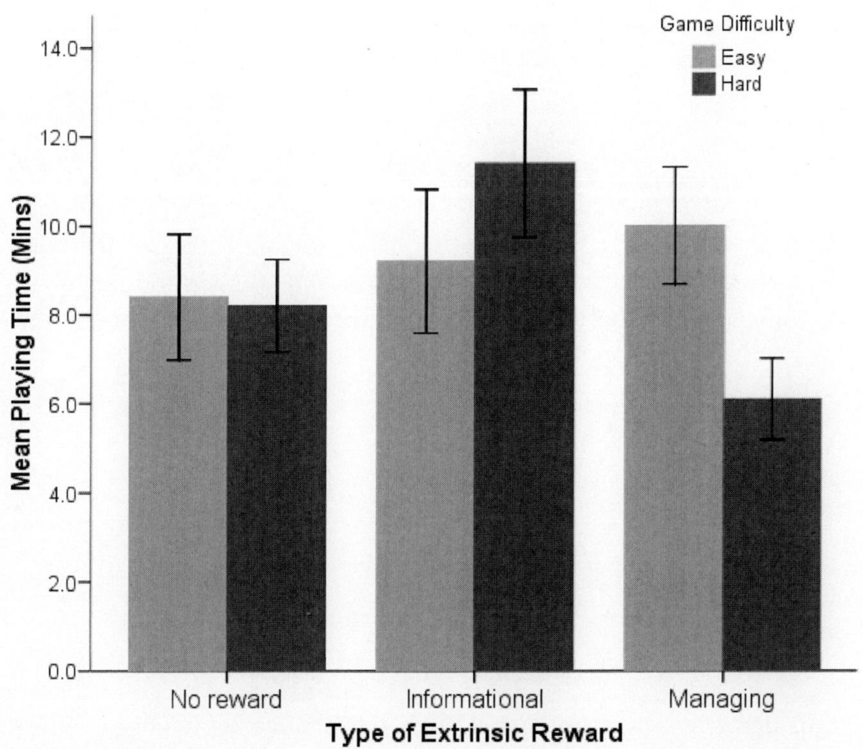

Figure 1. Mean playing time for type of extrinsic reward by difficulty of the computer game (error bars are 95% CI).

Notes:

1. The convention for reporting factorial ANOVA results is as follows:
   F(*df* effect, *df* error) = *F* ratio, *p* = Sig. value, effect size measure = value.

2. See page 130 of the *APA Manual* for how to show confidence intervals in a Table. This would certainly be required if error bars are not shown in a graph.

3. With multiway factorial ANOVA with a large number of effects, it may be preferable to report them in a table rather than in the text, as follows:

Table 2

Analysis of Variance Results

Source	*SS*	*df*	*MS*	*F*	*p*	η_p^2
Type of reward	30.42	2	15.21	12.72	<.001	.515
Game difficulty	3.01	1	3.01	2.52	.126	.095
Reward x Difficulty	47.22	2	23.61	19.74	<.001	.622
Error	28.70	24	1.20			
Total	109.34	29				

Multiway factorial between-groups ANOVA designs

Theoretically, there is no limit to the complexity of ANOVA designs. They can be three-way (three IVs), four-way (four IVs), five-way (five IVs), and so on. However, with more complex ANOVA designs, there are more effects to consider. For example, a three-way design with three independent variables (A, B, and C) has seven effects:

- Three main effects
  - The A main effect
  - The B main effect
  - The C main effect
- Three two-way interactions
  - The A x B two-way interaction
  - The A x C two-way interaction
  - The B x C two-way interaction
- One three-way interaction
  - The A x B x C three-way interaction.

Four-way and five-way designs become even more complicated. This complexity imposes a practical limit on the number of factors that can be considered. One avoids, if possible, going beyond a three-way design, and designs beyond four-way or five-way are not often seen in the literature. The more complex designs are only really feasible when there is a very strong theory to guide the interpretation of the results.

By way of illustration, suppose in the current example that all the children had been girls, and that we now obtained another sample of boys. This could become a three-way design with gender as a third IV. A significant three-way interaction would mean that the A x B two-way interaction would be **different** for the different levels of C. Here, it would mean that the two-way interaction between **reward** and **task** would be **different** for the different levels of **gender**.

The interaction line graphs might look like those on the next page. There is the significant two-way interaction for girls, but there does not appear to be a two-way interaction for boys. (The two-way interactions are **different** for boys and girls, because there is a three-way interaction.)

Girls

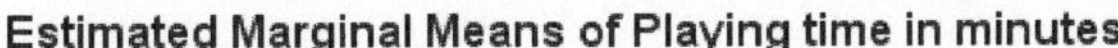

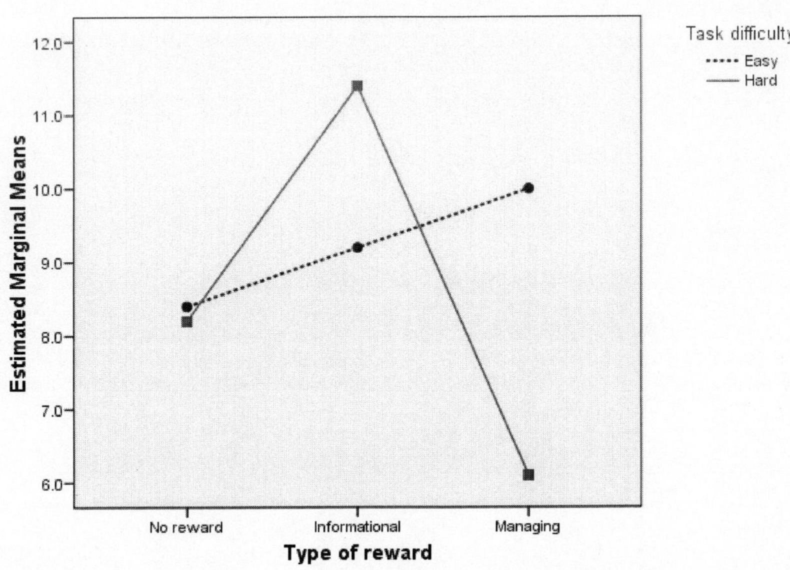

Boys

Estimated Marginal Means of Playing time in minutes

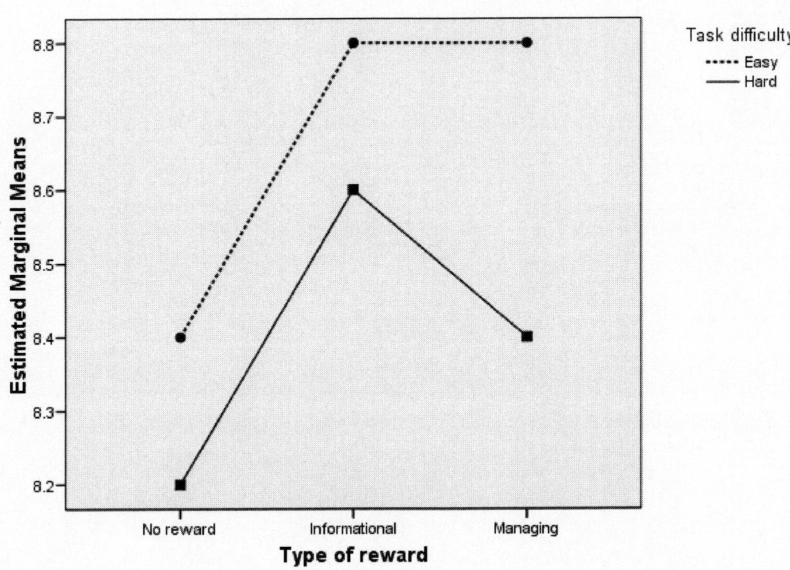

Missing cell designs

It is possible to perform ANOVA analyses when there are missing cells in a factorial design. However, in this case the **Type IV Sum of Squares** needs to be selected from the **Models...** option in the GLM procedure.

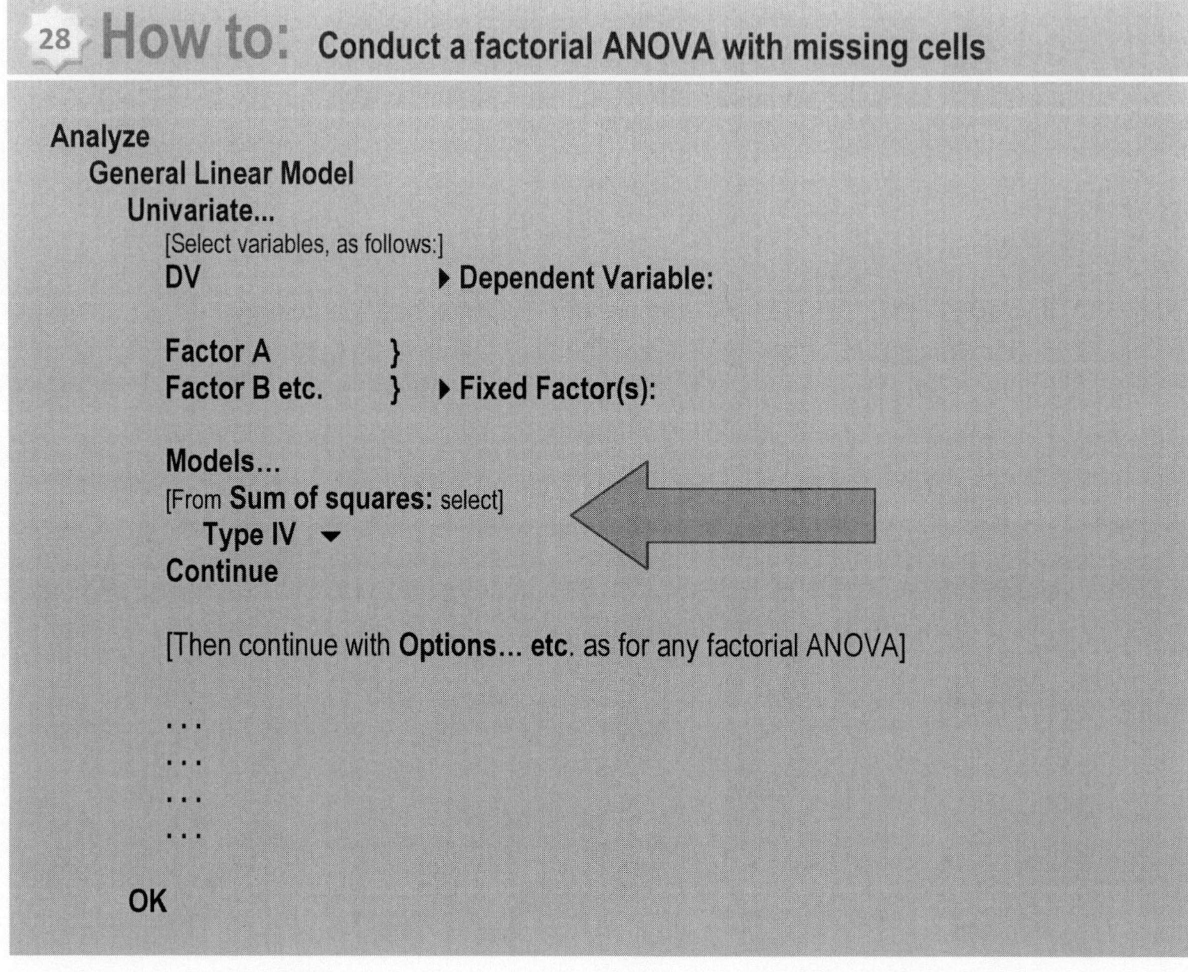

13 ANOVA ALTERNATIVES: PLANNED COMPARISONS, AND TREND ANALYSIS

The moment we moved away from comparing two groups, we encountered the problem of needing to do multiple tests to pinpoint the exact nature of differences, either for one-way analyses or main effects with more than two levels, or to analyse interactions; and with this came the central controversy about how to deal with **familywise error**. In the one-way ANOVA chapters, post hoc tests that compare everything with everything were used (e.g., Tukey's HSD), and they can also be used for main effects in factorial designs, because we frequently want to know how all the groups differ from one another.

However, even if we convert a factorial design to a one-way design, as in the previous chapter, post hoc tests that compare everything with everything are rarely appropriate to analyse interactions, because some of the comparisons are uninterpretable, as for example from the previous chapter, a comparison between an informational reward used with a hard game and a managing reward used with an easy game. If this comparison were significant there would be no way to tell if was due to the type of reward, or to the task difficulty.

A quick recap on familywise error, the Bonferroni test, and post hoc comparisons

The problem with doing multiple tests (of anything) is that of **familywise error**. If we set alpha (α) at .05, it holds the Type I error rate to .05 for **only one test**. When we conduct multiple tests, the Type I error rate increases across that "family" of tests. So, if four tests are conducted, the probability of **at least one** being significant by chance approaches .05 x 4, or .20. Familywise error, then, is the increase in the Type I error rate that occurs across multiple tests (i.e., the probability of at least one test being significant at the .05 level by chance alone).

Most authorities consider it necessary to make some adjustment to the α level to control familywise error across any large set of tests or comparisons, where "large" is often defined as more than the *df* in the design (although even the *df* may permit too many possibilities in complex designs with many effects).

The Bonferroni test

The most straightforward way to hold familywise error at a chosen α (e.g., .05, or perhaps a more liberal .10, or .15) for **any type of analysis or statistical test** is to use an **adjusted** α for each individual test or comparison that is the chosen α divided by the number of tests. For example, using the conventional α of .05 with five tests, the adjusted alpha becomes .05/5 or .01. This is the Bonferroni test[1]. Its main drawback is that it becomes very conservative (less powerful), and increasingly inappropriate, with a large number of tests. For this reason it is not generally recommended for post hoc comparisons that compare "everything with everything", and so generate many tests in even moderately complex designs.

---

[1] The Sidak test is a slightly more powerful version of the Bonferroni test that might be used in preference to Bonferroni.

Post hoc tests

There are a great variety of post hoc tests available in the one-way ANOVA procedure, or for main effects in the GLM procedure, and they are generally used only after a significant omnibus or overall *F*. Some are suited to specific situations. The Tukey HSD test, for instance, holds familywise error to the chosen alpha for all comparisons between pairs of means; the Dunnett test does so for comparisons between each of several groups and a control group; while the most conservative post hoc test of all, the Scheffé test, controls familywise error for all possible comparisons, including complex ones. **Complex comparisons** are those where several groups are combined and compared to another group or to another combination of groups.

Right clicking on the various post hoc options in the SPSS **Post Hoc** dialog window displays a short description of each test, but for more detailed information it is necessary to consult relevant text books, such as Howell (2002).

A priori or planned comparisons

Any ANOVA design can be dismantled into a set of pairwise or single *df* comparisons. These can be simple pairwise comparisons (comparing individual group means) or complex comparisons (of averaged means). They are all pairwise comparisons, because they only ever compare two means (simple or complex), and in effect, they are just *t* tests, except they use the combined error term from the full ANOVA (provided there is homogeneity of variance).

It is often the case, especially in factorial designs, that only a subset of the possible comparisons in the full design is of theoretical interest, in which case an argument can be made for performing only those comparisons **instead** of the omnibus ANOVA. This is the **planned comparisons** (or contrasts) approach, where the overall ANOVA is only used to supply the combined error term, which is a more accurate estimate of the population error variance (assuming homogeneity of variance).

Provided planned comparisons are limited in number (certainly no more than the *df*), most researchers accept that they can be conducted without any adjustment to the alpha level, because the choice of which ones to conduct is made **prior to collecting the data**, **not afterwards** when one knows what the data are like. Some adjustment to alpha should be made if there are a large number of comparisons, although it may be preferable in such cases to use a more liberal familywise alpha such as .10 or .15.

Planned comparisons that do not use an adjusted α are more powerful than the omnibus ANOVA (i.e., more likely to detect a real effect), because the ANOVA "wastes" power when only part of it is of theoretical interest. Because of this they are an acceptable alternative, **provided researchers are honest and make the decision beforehand of which ones to conduct.**

Example of planned comparisons

Imagine you are employed by a large national aged care provider to promote the physical and psychological health of residents in their aged care facilities. Based on emerging literature you are conducting research to test the health impact of two new programs: one in which residents undertake a gentle 20-minute weight training exercise session twice a week; and a second in which they have 10-minutes exposure to sunshine early in the morning for four days a week, in addition to the same weight-training program.

Forty-eight volunteers from low care facilities who meet minimum inclusion criteria for physical and mental health and mobility take part in the research over a 12-week period, and are randomly assigned to one of four groups as follows:

- **Control group 1**, the control group proper, that receives no exercise or sun exposure;

- **Control group 2** controls for the effect of general exercise, in that participants undertake a standard program of gentle movement and stretching exercises for 20-minutes twice a week, but have no sun exposure;
- **Experimental group 1**, where participants undertake the weights exercise program, but receive no sun exposure; and
- **Experimental group 2**, where participants receive both the weights program and sun exposure.

Each group consists of 6 women and 6 men, and fortunately all of them are able to complete the 12 weeks program, after which they are assessed on a 20-item index of physical and psychological health. The mean of the 20 items yields a general wellbeing score in the range 0 (poor) to 8 (excellent).

Assume there is a very strong theoretical basis to the study that supports the following hypotheses:

1. The combined experimental groups will have higher wellbeing than the combined control groups.
2. Experimental group 2 (weights and sunshine) will have higher wellbeing than experimental group 1 (weights only).

You decide to test these hypotheses by means of four planned comparisons (separately for women and men), conducted instead of the overall ANOVA, and because of the small number of tests, to do so using a conventional alpha level of .05 without any adjustment.

Step 1: Enter data in SPSS and save data file

The data resulting from the study are listed in the table below. You will recall, however, that data for a between-groups design are not entered in columns. Instead, you need to create variables as follows, and enter the data for the **48 participants** in **rows** accordingly (saving it in a file that you might call ***Planned.sav***):

ID	Participant number	***Nominal*** *variable*
Program	Health program 1= Control 1 2=Control 2 3=Experimental 1 4=Experimental 2	***Nominal*** *variable*
Gender	0=Women 1=Men	***Nominal*** *variable*
Cell	Code the 8 cells of the design according to the order in the coefficients table on the next page.	***Nominal*** *variable*
Wellness	Wellbeing score	***Scale*** *variable*

Men				Women			
Control 1	**Control 2**	**Exper 1**	**Exper 2**	**Control 1**	**Control 2**	**Exper 1**	**Exper 2**
5.67	5.83	6.00	7.00	6.17	5.67	6.55	6.30
5.00	5.50	5.67	6.67	5.00	5.67	7.20	6.80
4.83	5.17	5.83	6.83	5.67	6.17	6.53	7.00
5.50	5.00	6.67	6.33	6.00	5.17	6.37	7.13
6.00	5.50	6.00	6.17	6.33	4.67	7.03	7.10
4.67	5.17	7.33	6.00	5.00	6.00	6.20	7.70

Step 2: Construct the required coefficients

The first step in this analysis is to construct the necessary table of coefficients, according to the hypotheses (remember that coefficients must always sum to zero):

1. The combined experimental groups will have higher wellbeing than the combined control groups; tested for women and men separately.
2. Experimental group 2 (weights and sunshine) will have higher wellbeing than experimental group 1 (weights only); tested for women and men separately.

Cell Code Gender / Group	1 Women Con 1	**2** Men Con 1	3 Women Con 2	**4** Men Con 2	5 Women Exper 1	**6** Men Exper 1	**7** Women Exper 2	**8** Men Exper 2	SUM
Mean	5.70	5.28	5.56	5.36	6.65	6.25	7.01	6.50	
Comp. 1a	**-.5**	0	**-.5**	0	**.5**	0	**.5**	0	**0**
Comp. 1b	0	**-.5**	0	**-.5**	0	**.5**	0	**.5**	**0**
Comp. 2a	0	0	0	0	**-1**	0	**1**	0	**0**
Comp. 2b	0	0	0	0	0	**-1**	0	**1**	**0**

You should be able to see how **Comparisons 2a and 2b** compare experimental group 2 with experimental group 1, for women and men respectively, and so address hypothesis 2.

Comparisons 1a and 1b address hypothesis 1, but they are in the form of **complex comparisons,** where cells are combined. **Comparison 1a** only involves the women. It "averages" the two experimental groups by combining their means each multiplied by .5, then compares this new combined mean to a similarly combined mean for the two control groups. **Comparison 1b** does the same thing, but only for the men.

Whenever coefficients involve fractions it is a good idea to convert them to whole numbers by multiplying by their common denominator (.5 = ½; common denominator = 2; .05 x 2 = 1), so the coefficients become:

Cell Code Gender / Group	1 Women Con 1	**2** Men Con 1	3 Women Con 2	**4** Men Con 2	5 Women Exper 1	**6** Men Exper 1	**7** Women Exper 2	**8** Men Exper 2	SUM
Mean	5.70	5.28	5.56	5.36	6.65	6.25	7.01	6.50	
Comp. 1a	**-1**	0	**-1**	0	**1**	0	**1**	0	**0**
Comp. 1b	0	**-1**	0	**-1**	0	**1**	0	**1**	**0**
Comp. 2a	0	0	0	0	**-1**	0	**1**	0	**0**
Comp. 2b	0	0	0	0	0	**-1**	0	**1**	**0**

Step 3: Request the analysis using SPSS syntax

29 How to: Conduct planned comparisons using ANOVA syntax

[In the SPSS **Data View window** click on the following menu options]

File
New
Syntax

Type the following commands into the syntax window, exactly as they are written, then save the file (it will be given the extension .sps, and you might want to call it ***Planned.sps***)

```
ONEWAY wellness BY cell
  /STATISTICS DESCRIPTIVES HOMOGENEITY
  /CONTRAST= -1 0 -1 0 1 0 1 0
  /CONTRAST= 0 -1 0 -1 0 1 0 1
  /CONTRAST= 0 0 0 0 -1 0 1 0
  /CONTRAST= 0 0 0 0 0 -1 0 1
  /MISSING ANALYSIS .
```

Do **not** forget the full stop.

To run this analysis, position the cursor at the beginning, or somewhere within the commands:

Then click on the run button located in the icon bar.

Alternatively, select from the **Run** option in the menu bar at the top.

For advanced readers:

The GLM procedure can also handle comparisons, and can do so without having to convert to a one-way design using the cells of the analysis. However, the syntax is more complicated. It is demonstrated here only to allow the two types of output to be shown.

```
GLM wellness BY gender program
  /METHOD = SSTYPE(3)
  /LMATRIX "experimental vs control for women"
    program*gender -1 -1 1 1 0 0 0 0 program -1 -1 1 1
  /LMATRIX "experimental vs control for men"
    program*gender 0 0 0 0 -1 -1 1 1 program -1 -1 1 1
  /LMATRIX "exper 2 vs exper 1 for women"
    program*gender 0 0 -1 1 0 0 0 0  program 0 0 -1 1
  /LMATRIX "exper 2 vs exper 1 for men"
    program*gender 0 0 0 0 0 0 -1 1 program 0 0 -1 1.
```

In GLM the order of the cells is as specified by **Line 1.** Here, the cells are listed in order of **gender**, with the four cells of the different **programs** listed **within** each of the two genders; so the coefficients have a different pattern to the one-way procedure. There is an additional set of coefficients on each line that specifies the comparisons only within **program**. Anything enclosed in quotation marks is a comment, not a command. More explanation of this method of performing comparisons is provided in the MANOVA chapter

Step 4: Examine and interpret the output

Output: **ANOVA Planned comparisons from One-Way**

Descriptives

Wellbeing score

	N	Mean	Std. Deviation	Std. Error	95% Confidence Interval for Mean Lower Bound	95% Confidence Interval for Mean Upper Bound	Minimum	Maximum
Con1-Women	6	5.6950	.58113	.23724	5.0851	6.3049	5.00	6.33
Con1-Men	6	5.2783	.52381	.21384	4.7286	5.8280	4.67	6.00
Con2-Women	6	5.5583	.55391	.22613	4.9770	6.1396	4.67	6.17
Con2-Men	6	5.3617	.30394	.12408	5.0427	5.6806	5.00	5.83
Exper1-Women	6	6.6467	.38785	.15834	6.2396	7.0537	6.20	7.20
Exper1-Men	6	6.2500	.62938	.25694	5.5895	6.9105	5.67	7.33
Exper2-Women	6	7.0050	.45755	.18679	6.5248	7.4852	6.30	7.70
Exper2-Men	6	6.5000	.39385	.16079	6.0867	6.9133	6.00	7.00
Total	48	6.0369	.76198	.10998	5.8156	6.2581	4.67	7.70

1

The cell means are always of fundamental importance to the interpretation of any ANOVA analysis.

Test of Homogeneity of Variances

Wellbeing score

Levene Statistic	df1	df2	Sig.
.869	7	40	.539

2

The Levene test is not significant (Sig. > .05), meaning the assumption of homogeneity of variance is met. This assumption is always important when comparisons are conducted.

Contrast Coefficients

Contrast	Cell Con1-Women	Con1-Men	Con2-Women	Con2-Men	Exper1-Women	Exper1-Men	Exper2-Women	Exper2-Men
1	-1	0	-1	0	1	0	1	0
2	0	-1	0	-1	0	1	0	1
3	0	0	0	0	-1	0	1	0
4	0	0	0	0	0	-1	0	1

3

Check the contrast coefficients to make very sure they are correct.

Contrast Tests

		Contrast	Value of Contrast	Std. Error	t	df	Sig. (2-tailed)
Wellbeing score	Assume equal variances	1	2.3983	.40024	5.992	40	.000
		2	2.1100	.40024	5.272	40	.000
		3	.3583	.28301	1.266	40	.213
		4	.2500	.28301	.883	40	.382
	Does not assume equal variances	1	2.3983	.40913	5.862	18.363	.000
		2	2.1100	.39115	5.394	15.913	.000
		3	.3583	.24487	1.463	9.739	.175
		4	.2500	.30311	.825	8.395	.432

4

Using the results for **Assume equal variances** (because homogeneity of variance was met), the first hypothesis is supported. The experimental groups in combination have a significantly higher mean than the control groups in combination for both women (Contrast 1, Sig. < .001) and men (Contrast 2, Sig. < .001).

However, the second hypothesis is not supported. The mean for experimental group 2 is not significantly higher than the mean for experimental group 1, either for women (Contrast 3, Sig. = .213), or for men (Contrast 4, Sig. = .382). There is no significant difference between the weights only program and the weights with sunshine program.

Custom Hypothesis Tests #1

Contrast Results (K Matrix)[a]

Contrast			Dependent Variable: Wellbeing score
L1	Contrast Estimate		2.398
	Hypothesized Value		0
	Difference (Estimate - Hypothesized)		2.398
	Std. Error		.400
	Sig.		.000
	95% Confidence Interval for Difference	Lower Bound	1.589
		Upper Bound	3.207

a. Based on the user-specified contrast coefficients (L') matrix: experimental vs control for women

GLM 1

Comparison label

Test Results

Dependent Variable: Wellbeing score

Source	Sum of Squares	df	Mean Square	F	Sig.
Contrast	8.628	1	8.628	35.907	.000
Error	9.611	40	.240		

GLM 2

The **Contrast Results** above are the GLM version of the first contrast, to illustrate how they provide the added information of confidence intervals. The **Test Results** give the *F* value rather than *t*, but remember $t^2 = F$ (i.e., $5.992^2 = 35.91$). Only the first of the four contrasts is shown here for illustration purposes. Note also where the comparison label (based on comments in the syntax) appears in GLM.

Output: **ANOVA planned comparisons from GLM**

Tests of Between-Subjects Effects

Dependent Variable: Wellbeing score

Source	Type III Sum of Squares	df	Mean Square	F	Sig.
Corrected Model	17.677[a]	7	2.525	10.510	.000
Intercept	1749.305	1	1749.305	7280.064	.000
Gender	1.721	1	1.721	7.164	.011
Program	15.803	3	5.268	21.923	.000
Gender * Program	.153	3	.051	.212	.888
Error	9.611	40	.240		
Total	1776.594	48			
Corrected Total	27.289	47			

a. R Squared = .648 (Adjusted R Squared = .586)

GLM 2

The GLM procedure also produces the full ANOVA summary table. It appears at the beginning of the GLM output, but is not relevant to the planned comparisons, other than to demonstrate that the **Error Sum of Squares** is the same as that used for the contrast **Test Results**.

It will be used here to also demonstrate one other point about the versatility of contrast coefficients.

More on the versatility of coefficients

A coefficients approach can replace ANOVA altogether, as both main effects and interactions can be specified by coefficients. In the example we are using, the main effect of gender is easy to work out because it has only two levels, so the coefficients are those that compare women with men (each combined across program conditions), as follows:

Cell Code Gender / Group	1 Women Con 1	**2** Men Con 1	3 Women Con 2	**4** Men Con 2	5 Women Exper 1	**6** Men Exper 1	7 Women Exper 2	**8** Men Exper 2	SUM
Mean	5.70	5.28	5.56	5.36	6.65	6.25	7.01	6.50	
Gender main	**.25**	-.25	**.25**	-.25	**.25**	-.25	**.25**	-.25	**0**
Or, when multiplied by 4 to convert to whole numbers:									
Gender main	**1**	-1	**1**	-1	**1**	-1	**1**	-1	**0**

Output: **ANOVA Main effect contrasts**

Contrast Coefficients

Contrast	Cell							
	Con1-Women	Con1-Men	Con2-Women	Con2-Men	Exper1-Women	Exper1-Men	Exper2-Women	Exper2-Men
1	1	-1	1	-1	1	-1	1	-1

Contrast Tests

		Contrast	Value of Contrast	Std. Error	t	df	Sig. (2-tailed)
Wellbeing score	Assume equal variances	1	1.5150	.56602	2.677	40	.011
	Does not assume equal	1	1.5150	.56602	2.677	34.251	.011

The resulting main effect contrast for **gender** is in the form of a *t* test in procedure **One-way**, but it gives the same result as the ANOVA summary table above, with $t^2 = F$ (i.e., $2.677^2 = 7.16$), Sig. = .011.

Although only really of interest to advanced readers, the main effect for program can also be determined using coefficients, but it is more complicated because there are more than two levels. This main effect requires a **set** of three orthogonal comparisons (i.e., *k*-1, where *k* is the number of levels, or in other words, the *df*). The mean *F* value of the three comparisons equals the *F* value for the program main effect. Known as a **Helmert contrast** (which is available automatically under **Contrasts...** in the **GLM** procedure), the set compares each level of the main effect with the remainder. In our example, this is control group 1 with the other three groups combined; then control group 2 with the two experimental groups combined, then experimental group 1 versus experimental group 2.

Planned comparisons and orthogonality

There are two restrictions on the use of planned comparisons: the first we have already mentioned, namely, that the comparisons **must be determined before the study is conducted**, and must be strongly grounded in theory or prior research. Above all, you **must never** change to planned procedures after you know what the data are like, in order to obtain a significant difference.

The second restriction is that planned comparisons should ideally be **orthogonal**, meaning independent of one another, with no overlapping information.

Only *k* - 1 orthogonal comparisons are possible, where *k* = the number of means being compared. Hence, with four means, there are only three possible orthogonal contrasts (the same as mentioned above for the set of comparisons needed to test the main effect of program).

The test for orthogonality among pairs of comparisons is made by **multiplying their coefficients**. If the sum of these multiplied coefficients is **zero** the comparisons are orthogonal; if it is not zero, they are not orthogonal.

The planned comparisons used in this example were all orthogonal as demonstrated below:

Cell Code Gender / Group	1 Women Con 1	**2** Men Con 1	3 Women Con 2	**4** Men Con 2	5 Women Exper 1	**6** Men Exper 1	7 Women Exper 2	**8** Men Exper 2	SUM
Mean	5.69	5.28	5.56	5.36	6.65	6.25	7.01	6.50	
Comp. 1a	**-1**	0	**-1**	0	**1**	0	**1**	0	**0**
Comp. 1b	0	**-1**	0	**-1**	0	**1**	0	**1**	**0**
Comp. 2a	0	0	0	0	**-1**	0	**1**	0	**0**
Comp. 2b	0	0	0	0	0	**-1**	0	**1**	**0**
1a x 1b	0	0	0	0	0	0	0	0	**0**
1a x 2a	0	0	0	0	**-1**	0	**1**	0	**0**
1a x 2b	0	0	0	0	0	0	0	0	**0**
1b x 2a	0	0	0	0	0	0	0	0	**0**
1b x 2b	0	0	0	0	0	**-1**	0	**1**	**0**
2a x 2b	0	0	0	0	0	0	0	0	**0**

Sometimes, meaningful comparisons that a researcher wants to perform are not orthogonal with one another. The general advice is that they can still be conducted, as meaningfulness is ultimately of most importance, but because they are not orthogonal some will partly reflect others and so will not be fully interpretable in their own right (see Keppel, 2004). The recommendation is to **try** to use orthogonal comparisons wherever possible.

Trend analysis

Where there is a **quantitative** independent variable, that is, it has levels that differ in a quantitative fashion (e.g., drug quantity: 25 ml, 50 ml, 75 ml, 100 ml; time allowance: 10 min, 15 min, 20 min, 25 min, 30 min; etc.), **trend analysis** might be performed in preference to comparisons among individual means. That is, one looks to see if there are significant linear, quadratic, cubic, and/or higher order trends in the relationship between the IV and the DV.

The **SPSS one-way procedure** can automatically perform trend analysis, using the **Contrasts option**, as follows:

30 How to: Conduct a trend analysis using One-Way ANOVA

Contrasts...
[Select] ☑ Polynomial
[From **Degree:** select the type(s) of trend analysis required]

Alternatively, one can use coefficients to specify different types of trends. For example, an IV with four levels, consisting of the age groups of children, as shown in the graph below, has linear coefficients of -3 -1 1 3; and quadratic coefficients of 1 -1 -1 1.

The bold line plots the actual means; the grey line represents the extent to which there is a linear trend in the means; and the dotted line represents the extent to which there is a quadratic trend. Just as interactions take priority over main effects, so higher order trends take priority over lower order ones. Here, there is a significant linear trend, but it does not convey the whole story. There is also a significant and overriding quadratic trend, whereby the means increase across the younger age groups, but reverse direction after age 10 years. It is necessary to refer to relevant statistics texts for a detailed explanation of trend analysis, and tables of trend coefficients (e.g., Howell, 2002, pp. 408-415).

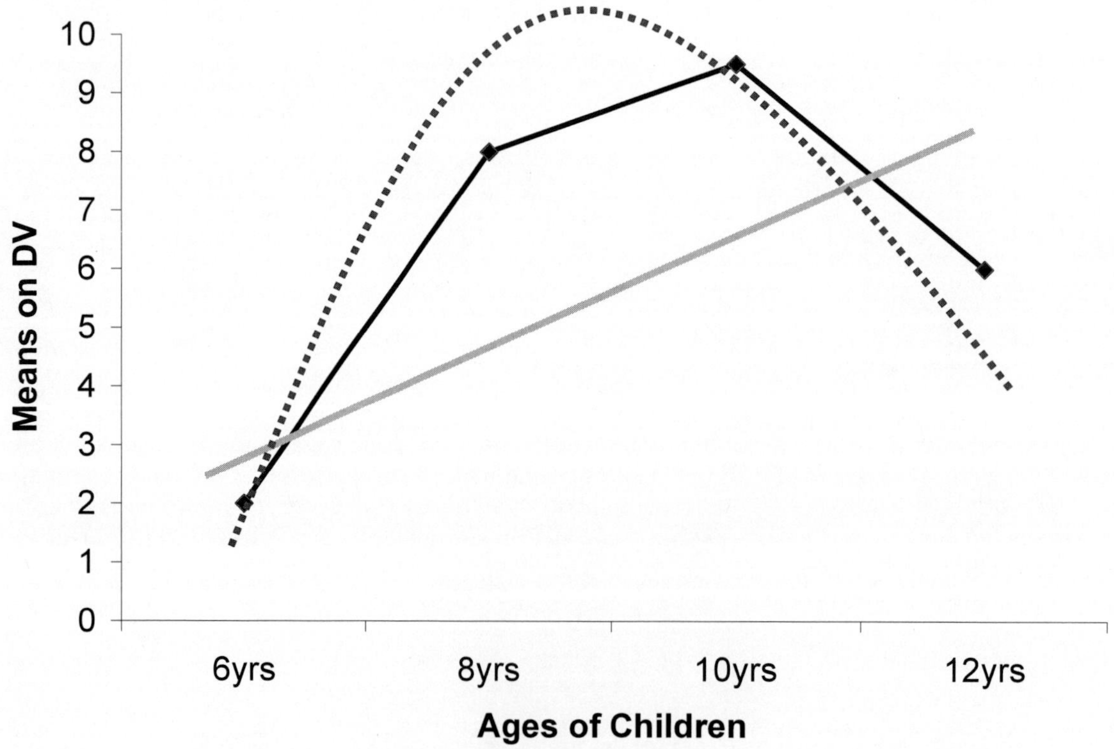

14 Factorial And Mixed Repeated Measures ANOVA

Just as between-groups ANOVA can have multiple IVs in factorial designs, so repeated measures ANOVA can have factorial designs involving more than one repeated measures factor.

The notation often used for repeated measures is a little different to between-groups because it includes "S" within parenthesis to indicate that the same "subjects" (participants) serve in all conditions. So (A x S) would indicate a one-way design with factor A repeated across subjects (S), while (A x B x S) would indicate a two-way design with factors A and B both repeated across subjects.

There is also an additional possibility, and that is a **mixed**, or what is sometimes called **split-plot** (or SPANOVA) design, where there are mixed within- and between-groups factors. The simplest mixed design is an A x (B x S) design. The notation indicates one between-groups factor (A) and one repeated measures factor (B).

All sorts of possibilities exist, for example:

A x B x (C x S)	Three factors, two between-groups (A, B); one repeated (C)
A x (B x C x S)	Three factors, one between-groups (A); two repeated (B and C)
A x B x (C x D x S)	Four factors, two between-groups (A and B); two repeated (C and D).

The various factors can also have different numbers of levels, but the same caution applies as with factorial ANOVA, in that the more complex the design the more difficult and problematic can be the interpretation.

Assumptions

1. The data (scores) should be at the interval or ratio level of measurement.
2. The cases are randomly sampled from the population.
3. The dependent variable is normally distributed in the population at each level of the independent variable (and the **difference scores** are normally distributed). When *n* is large (30+) the dependent *t* test is considered to be **robust**; that is, violation of the normality assumption has little effect on its accuracy. With extreme non-normality even larger samples may be required to provide robustness.
4. The difference scores are independent of one another.
5. There is **sphericity**, that is, homogeneity of variance in the **difference** scores between pairs of repeated levels. This is assessed by Mauchly's test of sphericity, as discussed in the one-way repeated measures chapter, and only applies when there are more than two levels of the repeated measures factors.

When there is also a between-groups factor, the following additional assumptions apply, although they are not critical when sample sizes are equal or close to.

6. There is homogeneity of variance (i.e., Levene's test).
7. There is homogeneity of covariance, that is, the pattern of covariances (or correlations) between pairs of repeated measures scores is the same across the groups. This is assessed by a statistic known as Box's *M*, using an alpha level of .001 because it is notoriously sensitive and also affected by non-normality (see Tabachnick & Fidell, 2007; or Stevens, 2002). The assumption is met if Box's *M* is not significant (> .001), or on some occasions if Box's *M* is not computed because there are fewer than two nonsingular cell covariance matrices.

Example of factorial, mixed, repeated measures ANOVA

Imagine you are making some attempt to evaluate the effectiveness of a rehabilitation program at a juvenile detention centre for young male offenders. Shortly after their entry to the centre you obtain a pretest score for each youth on some measure of socially desirable behaviour. Imagine this measure is on an interval scale with a possible range of 0 to 20, with low scores indicating antisocial behaviour and high scores indicating socially desirable behaviour. Rather than relying purely on the juvenile's self-report, you also have a staff member (e.g., their case worker) complete an assessment of the juvenile's behaviour on the same instrument.

Imagine post-test scores are then obtained for each juvenile just before they leave the detention centre after a six-month rehabilitation program. Finally, imagine you are able to obtain follow-up scores for the juveniles on the same measure some six months later. These six repeated measures for the same participants represent three scores across **time** (pretest, posttest, and follow-up) from each of two **sources** (the juvenile and the staff member).

Assume, too, that there is an additional (between-groups) variable, namely age group, whereby the young offenders are categorised into younger (15-years-old and below) and older (over 15 years) age groups.

Note that in studies of this type researchers need to take steps to ensure the important assumption of **independence across individuals** is met, that is, that none of the young offenders can influence any of the others in the study. In reality this can be easier said than done, and is all too often overlooked—but it is a very important assumption upon which the accuracy of statistical tests depends.

This A x (B x C x S), or more specifically, 2 x (3 x 2) mixed repeated measures design is represented schematically as follows:

	Repeated measures Factors (IVs) B: Time and C: Source					
Between-Groups Factor (IV) A: Age-group	Pretest juvenile	Pretest staff	Posttest juvenile	Posttest staff	Follow-up juvenile	Follow-up staff
Younger						
Older						

Age group is the between-groups IV, and **time** and **source** are the two repeated measures IVs.
Note that lines are **not** drawn between the levels of the repeated measures factors, thereby indicating that these scores are from the same group of participants, not different (independent) groups.

Step 1: Enter data in SPSS and save data file

Enter the data listed below into SPSS and save them in a file called ***MixedANOVA.sav***.

ID	Participant number	*Set **Measure** column to **Nominal***
Agegrp	Age group 1=Younger 2 =Older	*Set **Measure** column to **Nominal***
Jpre	Juvenile pretest	
Spre	Staff pretest	
Jpost	Juvenile posttest	*Set **Measure** column to **Scale***
Spost	Staff posttest	
Jfollow	Juvenile follow-up	
Sfollow	Staff follow-up	

ID	Agegrp	Jpre	Spre	Jpost	Spost	Jfollow	Sfollow
1	1	5	7	8	10	7	8
2	2	6	5	14	15	5	4
3	2	9	8	13	10	7	7
4	1	5	5	7	8	6	8
5	1	4	5	13	11	9	7
6	1	5	4	12	14	7	5
7	1	9	9	12	10	11	9
8	2	3	7	10	13	4	5
9	1	8	7	15	16	12	11
10	2	5	6	11	12	4	4
11	2	7	5	13	14	6	5
12	1	7	8	10	8	8	9

Step 2: Screen data and check assumptions

Screen the data, check for outliers, and perform normality assumption checks (given the small sample size) for each of the six variables, using the **Split File...** option to screen the data separately for each age group.

Step 3: Request the analysis

31 How to: Conduct a factorial, mixed repeated measures ANOVA

Analyze
General Linear Model
Repeated Measures...

In the box for **Within-Subject Factor Name** type a name for the first repeated measures factor: **Time**

In the box for **Number of Levels** type: **3**

Click on the **Add** button

In the box for **Within-Subject Factor Name** type a name for the second repeated measures factor: **Source**

In the box for **Number of Levels** type: **2**

Click on the **Add** button

Click on the **Define** button

[Select the **Within-Subjects Variables** in the appropriate **Time** then **Source** order, as follows:]

Juvenile pretest [Jpre] ▸ **_?_ [1,1]**
Staff pretest [Spre] ▸ **_?_ [1,2]**
Juvenile postttest [Jpost] ▸ **_?_ [2,1]**
Staff postttest [Spost] ▸ **_?_ [2,2]**
Juvenile follow-up [Jfollow] ▸ **_?_ [3,1]**
Staff follow-up [Sfollow] ▸ **_?_ [3,2]**

Age group[agegrp] ▸ **Between-Subjects Factor(s):**

If you do not have a between-groups factor this last option is left blank.

Continued.../

Options...

[From **Estimated Marginal Means, Factor(s) and Factor Interactions** select:]

Agegrp }
Time }
Source }
Agegrp*Time }
Agegrp*Source }
Time*Source }
Agegrp*Time*Source } ▸ **Display Means for:**

☑ **Compare main effects**
[From **Confidence interval adjustment:** select] **Sidak** ▾

[From **Display** select:]
☑ **Descriptive Statistics**
☑ **Estimates of Effect Size**
☑ **Observed power**
☑ **Homogeneity tests** ← This option is only available when there is a between-groups factor.

Continue

Plots...

Time ▸ **Horizontal Axis**
Source ▸ **Separate Lines**
Agegrp ▸ **Separate Plots**

Plots: Click on the **Add** button

Continue

OK

Note also that the **Contrasts...** option can be used to identify linear and other trends in the main effects, or to conduct specific *k* - 1 comparisons (to be demonstrated later).

Step 4: Examine and interpret the output

Descriptive Statistics

	Age group	Mean	Std. Deviation	N
Juvenile pretest	Younger	6.14	1.864	7
	Older	6.00	2.236	5
	Total	6.08	1.929	12
Staff pretest	Younger	6.43	1.813	7
	Older	6.20	1.304	5
	Total	6.33	1.557	12
Juvenile posttest	Younger	11.00	2.828	7
	Older	12.20	1.643	5
	Total	11.50	2.393	12
Staff posttest	Younger	11.00	3.000	7
	Older	12.80	1.924	5
	Total	11.75	2.667	12
Juvenile follow-up	Younger	8.57	2.225	7
	Older	5.20	1.304	5
	Total	7.17	2.517	12
Staff follow-up	Younger	8.14	1.864	7
	Older	5.00	1.225	5
	Total	6.83	2.250	12

1

Inspect the descriptive statistics. Note that the age group sizes are slightly unequal.

Homogeneity of covariance:
In this particular example, note also that a warning appears first in the output to advise that "Box's Test of Equality of Covariance Matrices is not computed because there are fewer than two nonsingular cell covariance matrices." The assumption is met in these circumstances. (See p. 192 for information on how to interpret Box's Test when it is shown).

Mauchly's Test of Sphericity[b]

Measure: MEASURE_1

					Epsilon[a]		
Within Subjects Effect	Mauchly's W	Approx. Chi-Square	df	Sig.	Greenhouse-Geisser	Huynh-Feldt	Lower-bound
Time	.265	11.938	2	.003	.577	.669	.500
Source	1.000	.000	0	.	1.000	1.000	1.000
Time * Source	.856	1.395	2	.498	.874	1.000	.500

Tests the null hypothesis that the error covariance matrix of the orthonormalized transformed dependent variables is proportional to an identity matrix.

a. May be used to adjust the degrees of freedom for the averaged tests of significance. Corrected tests are displayed in the Tests of Within-Subjects Effects table.

b.
Design: Intercept+Agegrp
Within Subjects Design: Time+Source+Time*Source

Mauchly's test of sphericity **is** significant for **Time**; therefore, the assumption of sphericity is **not** met for **Time**, nor therefore, for any between-group IV interactions with Time (i.e., **Time x Agegrp**).

There is nothing for **Source**, because with only two levels the assumption of sphericity does not apply.

Most of the epsilon values are quite high (and two exceed .75), so the Huynh-Feldt correction is preferred.

Below, in the **Within-Subjects Effects** table, is the full set of effects involving the repeated measures variables. Note the level of complexity. The Huynh-Feldt results need to be used for the **Time** effect and any between-groups variable interaction with **Time**, given the violation of the sphericity assumption.

The only significant effects are the **main effect for Time**, and more importantly the **Time*Agegrp interaction**, which indicates that the effects over time are different for the two age groups.

Alternative:
You will recall that when Mauchly's Test is significant some authorities recommend using the multivariate results. They are not reproduced here, but if you conduct this analysis yourself you will see they yield the same pattern of significant and nonsignificant effects.

Tests of Within-Subjects Effects

Measure: MEASURE_1

Source		Type III Sum of Squares	df	Mean Square	F	Sig.	Partial Eta Squared	Noncent. Parameter	Observed Power[a]
Time	Sphericity Assumed	438.539	2	219.269	47.189	.000	.825	94.377	1.000
	Greenhouse-Geisser	438.539	1.153	380.342	47.189	.000	.825	54.409	1.000
	Huynh-Feldt	438.539	1.338	327.788	47.189	.000	.825	63.132	1.000
	Lower-bound	438.539	1.000	438.539	47.189	.000	.825	47.189	1.000
Time * Agegrp	Sphericity Assumed	67.872	2	33.936	7.303	.004	.422	14.607	.896
	Greenhouse-Geisser	67.872	1.153	58.865	7.303	.017	.422	8.421	.731
	Huynh-Feldt	67.872	1.338	50.731	7.303	.013	.422	9.771	.780
	Lower-bound	67.872	1.000	67.872	7.303	.022	.422	7.303	.684
Error(Time)	Sphericity Assumed	92.933	20	4.647					
	Greenhouse-Geisser	92.933	11.530	8.060					
	Huynh-Feldt	92.933	13.379	6.946					
	Lower-bound	92.933	10.000	9.293					
Source	Sphericity Assumed	.102	1	.102	.041	.843	.004	.041	.054
	Greenhouse-Geisser	.102	1.000	.102	.041	.843	.004	.041	.054
	Huynh-Feldt	.102	1.000	.102	.041	.843	.004	.041	.054
	Lower-bound	.102	1.000	.102	.041	.843	.004	.041	.054
Source * Agegrp	Sphericity Assumed	.268	1	.268	.109	.748	.011	.109	.060
	Greenhouse-Geisser	.268	1.000	.268	.109	.748	.011	.109	.060
	Huynh-Feldt	.268	1.000	.268	.109	.748	.011	.109	.060
	Lower-bound	.268	1.000	.268	.109	.748	.011	.109	.060
Error(Source)	Sphericity Assumed	24.676	10	2.468					
	Greenhouse-Geisser	24.676	10.000	2.468					
	Huynh-Feldt	24.676	10.000	2.468					
	Lower-bound	24.676	10.000	2.468					
Time * Source	Sphericity Assumed	1.344	2	.672	.631	.542	.059	1.262	.141
	Greenhouse-Geisser	1.344	1.749	.768	.631	.523	.059	1.103	.134
	Huynh-Feldt	1.344	2.000	.672	.631	.542	.059	1.262	.141
	Lower-bound	1.344	1.000	1.344	.631	.445	.059	.631	.111
Time * Source * Agegrp	Sphericity Assumed	.344	2	.172	.161	.852	.016	.323	.072
	Greenhouse-Geisser	.344	1.749	.196	.161	.825	.016	.282	.070
	Huynh-Feldt	.344	2.000	.172	.161	.852	.016	.323	.072
	Lower-bound	.344	1.000	.344	.161	.696	.016	.161	.065
Error(Time*Source)	Sphericity Assumed	21.295	20	1.065					
	Greenhouse-Geisser	21.295	17.489	1.218					
	Huynh-Feldt	21.295	20.000	1.065					
	Lower-bound	21.295	10.000	2.130					

a. Computed using alpha = .05

Levene's Test of Equality of Error Variances[a]

	F	df1	df2	Sig.
Juvenile pretest	.000	1	10	.989
Staff pretest	1.254	1	10	.289
Juvenile posttest	1.926	1	10	.195
Staff posttest	.954	1	10	.352
Juvenile follow-up	1.954	1	10	.192
Staff follow-up	.638	1	10	.443

Tests the null hypothesis that the error variance of the dependent variable is equal across groups.

a.
Design: Intercept+Agegrp
Within Subjects Design: Time+Source+Time*Source

4

These Levene test results are **not** significant, indicating that the assumption of homogeneity of variance **is** met for the two age groups on each of the six measures.

Tests of Between-Subjects Effects

Measure: MEASURE_1
Transformed Variable: Average

Source	Type III Sum of Squares	df	Mean Square	F	Sig.	Partial Eta Squared	Noncent. Parameter	Observed Power[a]
Intercept	4734.173	1	4734.173	396.370	.000	.975	396.370	1.000
Agegrp	7.340	1	7.340	.615	.451	.058	.615	.110
Error	119.438	10	11.944					

a. Computed using alpha = .05

5

This is the table of **Between-Subjects Effects**. It shows that the **main effect** of **Agegrp** was not significant.

Estimated Marginal Means

2. Time

Estimates

Measure: MEASURE_1

Time	Mean	Std. Error	95% Confidence Interval	
			Lower Bound	Upper Bound
Pre	6.193	.473	5.138	7.247
Post	11.750	.677	10.241	13.259
Follow	6.729	.477	5.666	7.792

Pairwise Comparisons

Measure: MEASURE_1

(I) Time	(J) Time	Mean Difference (I-J)	Std. Error	Sig.[a]	95% Confidence Interval for Difference[a]	
					Lower Bound	Upper Bound
Pre	Post	-5.557*	.818	.000	-7.898	-3.217
	Follow	-.536	.289	.254	-1.361	.290
Post	Pre	5.557*	.818	.000	3.217	7.898
	Follow	5.021*	.665	.000	3.120	6.923
Follow	Pre	.536	.289	.254	-.290	1.361
	Post	-5.021*	.665	.000	-6.923	-3.120

Based on estimated marginal means

*. The mean difference is significant at the .05 level.

a. Adjustment for multiple comparisons: Sidak.

6

The **Estimates** table above gives the **unweighted means** for the three levels of time (combined for juvenile and staff scores and across the two age groups). They will be a little different to normal **weighted means**, because of the unequal numbers in the two age groups. Recall that unweighted means remove any bias towards the larger group.

These are, of course, the **Time main effect** means. Given the main effect of Time was significant, we can refer to the **Pairwise Comparisons** table that uses a Sidak post hoc comparison to establish which levels are different from which. The comparisons tell us that overall, the post-test mean was significantly higher than both the pretest mean and the follow-up mean (overall it appears that the young offenders improved at post-test, but had declined again at follow-up). All this is of limited interest, however, in view of the significant time by age group interaction.

Note that you need to add the Time **labels** manually in SPSS output, as described on page 115.

Note also that the main effect means for Source and Agegrp have not been included, as both only involved two levels; but certainly examine them in your own output, as well as the marginal means for the nonsignificant interactions, and the plots of the means.

4. Age group * Time

Measure: MEASURE_1

Age group	Time	Mean	Std. Error	95% Confidence Interval	
				Lower Bound	Upper Bound
Younger	Pre	6.286	.611	4.924	7.647
	Post	11.000	.874	9.052	12.948
	Follow	8.357	.616	6.985	9.730
Older	Pre	6.100	.723	4.489	7.711
	Post	12.500	1.034	10.195	14.805
	Follow	5.100	.729	3.476	6.724

7

Given the significant **Time*Agegrp interaction**, these are the important marginal means that represent the interaction.

Step 5: Analysing a significant interaction

At this stage it is necessary to analyse the significant interaction between age group and time. We follow the same approach as outlined in the chapter on factorial between-groups ANOVA. In general terms interactions are analysed by breaking the design down into smaller components: to two-way interactions (if dealing with a three-way interaction); to simple effects; then ultimately to the relevant analytical comparisons; with decisions made in advance about how to address the problem of family-wise error.

The difference with repeated measures designs, though, is that error terms are derived from the specific conditions being analysed (not the overall ANOVA, as for between-groups), thus simple effects are just one-way ANOVAs and comparisons are *t* tests, using Bonferroni adjusted alpha levels, as appropriate. Having made this simple statement, however, it is necessary to point out that there are alternative approaches that favour the use of pooled error terms under certain circumstances; moreover, there are many ways in which interactions can be broken down into smaller components. The topic is complex and beyond the scope of this book, where a relatively straightforward approach is taken. Interested readers, or researchers facing challenges in their design, need to refer to more specialised statistics text books, of which Keppel and Wickens (2004, especially pp. 408-416) is recommended. Howell (2002, especially pp. 493, and 498-500) is also useful, as is Field (2005, chapters 11 and 12).

Returning to our example, the research focus in this hypothetical study was on whether the rehabilitation program resulted in improved social behaviour, both at its conclusion, and six months later. Given a significant interaction with age group, we can use split file to perform the repeated measures analysis (i.e., simple effects) separately for the younger and older boys. We could then rely on the **Compare main effects** option to pinpoint the differences among the three levels of time, but this is a little wasteful, given our research question. Instead, we will use an automatic SPSS option of a **Simple** contrast that compares every level of a repeated measures factor to a reference level (either the first or the last), with the first level (the pretest) being the appropriate reference level in this case.

32 How to: Conduct a repeated measures simple effects analysis

Data
Split File...
⊙ **Compare Groups**
[Select variable for which you want separate analyses]
Age group [Agegrp] ▸ Groups Based on:
OK

Now repeat the analysis on pages 160-161, modifying as follows:

Omit any reference to age group

Under **Options...** only ask for relevant marginal means, in this case only **Time**, as we know there are no Source effects.

Only ask for **Compare main effects** if you want to compare all Time levels, and ask for **LSD** (not Sidak or Bonferroni) if you want to apply a *manual* Bonferroni adjustment, based on any adjusted alpha already applied to the simple effects analysis.

From **Display** you might only want to select **Estimates of Effect Size**.

Plots... selections may not be required.

Because we want simple contrasts for Time, add the following:

Contrasts...
[From **Factors:** click on] **Time(Polynomial)**
[From **Change Contrast**, then **Contrast...** select] **Simple** ▾
[From **Reference Category:** select] ⊙ **First**
Click on the **Change** button [Time should now read **Time(Simple(first))**]
Continue

OK

> ***Note:***
> The results from **Contrasts** are exactly the same as the results from **Compare main effects**, but the output for Contrasts, although complicated to look at, gives additional information, including effect sizes.

Given no significant difference in Source scores; SPSS **Transform**, **Compute...** could have been used to create three new variables (Pretest, Posttest, and Follow) by averaging, as for example: Pretest = (Jpre + Spre) / 2, etc. This analysis could then have been conducted using these scores for the **one** repeated measures IV of Time. You need to do this anyway, asking for **Descriptive Statistics** as well under **Options...** in order to obtain the combined means and standard deviations that are reported in the sample Results section.

Step 6: Examine and interpret the simple effects analysis output

You will find that **all** the repeated measures output is generated by SPSS, separately for the younger and older age groups. By now you should be getting adept at finding what you need in SPSS output, so only the essential sections are reproduced here.

Mauchly's Test of Sphericity[b]

Measure: MEASURE_1

						Epsilon[a]		
Age group	Within Subjects Effect	Mauchly's W	Approx. Chi-Square	df	Sig.	Greenhouse-Geisser	Huynh-Feldt	Lower-bound
Younger	Time	.299	6.034	2	.049	.588	.646	.500
	Source	1.000	.000	0	.	1.000	1.000	1.000
	Time * Source	.804	1.092	2	.579	.836	1.000	.500
Older	Time	.122	6.316	2	.043	.532	.566	.500
	Source	1.000	.000	0	.	1.000	1.000	1.000
	Time * Source	.914	.271	2	.873	.920	1.000	.500

Tests the null hypothesis that the error covariance matrix of the orthonormalized transformed dependent variables is proportional to an identity matrix.

a. May be used to adjust the degrees of freedom for the averaged tests of significance. Corrected tests are displayed in the Tests of Within-Subjects Effects table.

b.
Design: Intercept
Within Subjects Design: Time+Source+Time*Source

1

Mauchly's Test of Sphericity confirms that the assumption of sphericity is **not** met for **Time** in either age group. In view of the *relatively* high epsilon values, we are advised to use the Huynh-Feldt correction.

Tests of Within-Subjects Effects

Measure: MEASURE_1

Age group	Source		Type III Sum of Squares	df	Mean Square	F	Sig.	Partial Eta Squared
Younger	Time	Sphericity Assumed	156.333	2	78.167	13.660	.001	.695
		Greenhouse-Geisser	156.333	1.176	132.951	13.660	.006	.695
		Huynh-Feldt	156.333	1.292	121.033	13.660	.005	.695
		Lower-bound	156.333	1.000	156.333	13.660	.010	.695
	Error(Time)	Sphericity Assumed	68.667	12	5.722			
		Greenhouse-Geisser	68.667	7.055	9.733			
		Huynh-Feldt	68.667	7.750	8.860			
		Lower-bound	68.667	6.000	11.444			
Older	Time	Sphericity Assumed	322.400	2	161.200	53.143	.000	.930
		Greenhouse-Geisser	322.400	1.065	302.763	53.143	.001	.930
		Huynh-Feldt	322.400	1.133	284.659	53.143	.001	.930
		Lower-bound	322.400	1.000	322.400	53.143	.002	.930
	Error(Time)	Sphericity Assumed	24.267	8	3.033			
		Greenhouse-Geisser	24.267	4.259	5.697			
		Huynh-Feldt	24.267	4.530	5.356			
		Lower-bound	24.267	4.000	6.067			

2

SPSS produces the full, very large, **Table of Within-Subjects Effects**, but because the interaction we are analysing is only between age group and time, we are only interested in the **Time** main effects for the two age groups. Everything else has been suppressed here by double-clicking on the table and editing out unnecessary rows, leaving the two simple effects of time for younger and older boys.

Using a Bonferroni adjusted alpha for two tests (.05/2) of .025, both simple effects are significant, even using the Huynh-Feldt correction.

Tests of Within-Subjects Contrasts

Measure: MEASURE_1

Age group	Source	Time	Source	Type III Sum of Squares	df	Mean Square	F	Sig.	Partial Eta Squared
Younger	Time	Level 2 vs. Level 1		311.143	1	311.143	15.976	.007	.727
		Level 3 vs. Level 1		60.071	1	60.071	20.680	.004	.775
	Error(Time)	Level 2 vs. Level 1		116.857	6	19.476			
		Level 3 vs. Level 1		17.429	6	2.905			
	Source		Linear	.008	1	.008	.014	.909	.002
	Error(Source)		Linear	3.381	6	.563			
	Time * Source	Level 2 vs. Level 1	Linear	.286	1	.286	.097	.766	.016
		Level 3 vs. Level 1	Linear	1.786	1	1.786	1.389	.283	.188
	Error(Time*Source)	Level 2 vs. Level 1	Linear	17.714	6	2.952			
		Level 3 vs. Level 1	Linear	7.714	6	1.286			
Older	Time	Level 2 vs. Level 1		409.600	1	409.600	41.584	.003	.912
		Level 3 vs. Level 1		10.000	1	10.000	20.000	.011	.833
	Error(Time)	Level 2 vs. Level 1		39.400	4	9.850			
		Level 3 vs. Level 1		2.000	4	.500			
	Source		Linear	.100	1	.100	.083	.788	.020
	Error(Source)		Linear	4.844	4	1.211			
	Time * Source	Level 2 vs. Level 1	Linear	.400	1	.400	.186	.688	.044
		Level 3 vs. Level 1	Linear	.400	1	.400	.286	.621	.067
	Error(Time*Source)	Level 2 vs. Level 1	Linear	8.600	4	2.150			
		Level 3 vs. Level 1	Linear	5.600	4	1.400			

3

The **Tests of Within-Subjects Contrasts** table is always provided in repeated measures, but is not routinely used. Here, we are interested in the two contrasts specially selected to enable us to interpret the interaction: Level 2 (posttest) versus Level 1 (pretest), and also Level 3 (follow-up) versus Level 1, for both younger and older boys.
The grey boxes have been added to clearly identify these contrasts (or comparisons).

Assume we had decided to apply strict control over familywise error to maintain it at .05 throughout.
This would mean that each comparison is evaluated using a further Bonferroni adjusted alpha of the adjusted alpha used for the simple effect (.025) divided by the number of comparisons (i.e., 2), therefore the adjusted Bonferroni alpha for the comparisons is .025/2 = .0125.

Reading these results in conjunction with the marginal means below, we conclude that:

For younger boys:

The Level 2 (posttest) mean of 11.00 is significantly higher than the Level 1 (pretest) mean of 6.29, p = .007.
The Level 3 (follow-up) mean of 8.36 is significantly higher than the Level 1 (pretest) mean of 6.29, p = .004.

For older boys:

The Level 2 (posttest) mean of 12.50 is significantly higher than the Level 1 (pretest) mean of 6.10, p = .003.
The Level 3 (follow-up) mean of 5.10 is significantly **lower** than the Level 1 (pretest) mean of 6.10, p = .011.

Estimated Marginal Means

Time

Estimates

Measure: MEASURE_1

Age group	Time	Mean	Std. Error	95% Confidence Interval Lower Bound	95% Confidence Interval Upper Bound
Younger	1	6.286	.662	4.665	7.907
	2	11.000	1.041	8.453	13.547
	3	8.357	.705	6.633	10.081
Older	1	6.100	.620	4.377	7.823
	2	12.500	.632	10.744	14.256
	3	5.100	.534	3.618	6.582

Mixed repeated measures ANOVA: Interpretation and research report sample Results section

Results

Scores from the two different sources (self-report and staff assessment) on socially desirable behaviour for the 12 young offenders across the three time conditions, pretest, post-test, and at six months follow-up, were analysed using a mixed repeated measures analysis of variance (ANOVA), with age group as the between-groups factor with two levels: younger (15-years-old and below, $n = 7$), and older (over 15 years, $n = 5$). The ANOVA test assumptions were satisfactory with the exception of Mauchly's test of sphericity, which indicated that the sphericity assumption was not met for time.

The main effect for age group was not significant, $F(1, 10) = 0.62$, $p = .45$, $\eta_p^2 = .06$, nor was the main effect for source, $F(1, 10) = 0.04$, $p = .84$, $\eta_p^2 = .004$. However, the main effect for time was significant with a Huynh-Feldt adjustment to the degrees of freedom, $F(1.34, 13.38) = 47.19$, $p < .001$, $\eta_p^2 = .83$. Means (with standard errors in parentheses) and 95% confidence intervals for the pretest, post-test, and follow-up were 6.19 (0.47) [5.14, 7.25], 11.75 (0.68) [10.24, 13.26], and 6.73 (0.48) [5.67, 7.79] respectively.

The time by age group interaction, $F(1.34, 13.38) = 7.30$, $p = .013$, $\eta_p^2 = .42$ was significant, but none of the other interactions achieved significance ($p > .50$). Descriptive statistics for the scores averaged across the two sources are shown in Table 1.

The age group by time interaction was analysed by simple effect analyses performed separately for the younger and older boys using separate error terms and a Bonferroni adjusted alpha of .025 in view of the two simple effects. Each simple effect analysis was followed by simple contrasts comparing each level of time to the pretest, using a Bonferroni adjusted alpha of .0125 in view of the two further tests conducted.

Table 1

Mean Scores on Socially Desirable Behaviour, Averaged Across Self-Report and Staff Assessment, for Younger and Older Boys[1]

	Younger Boys (n = 7)		Older Boys (n = 5)	
Stage	*M (SD)*	95% CI	*M (SD)*	95% CI
Pretest	6.29 (1.75)	[4.92, 7.65]	6.10 (1.39)	[4.49, 7.71]
Post-test	11.00 (2.75)	[9.05, 12.95]	12.50 (1.41)	[10.20, 14.81]
Follow-up	8.36 (1.86)	[6.99, 9.73]	5.10 (1.19)	[3.48, 6.72]

Note. The possible range of scores was 0 -20, with higher scores indicative of more socially desirable behaviour. CI = confidence interval.

The simple effect for the younger boys, using the Huynh-Feldt correction, was significant, $F(1.29, 7.75) = 13.66$, $p = .005$, $\eta_p^2 = .70$; with the simple comparisons showing that compared to pretest they had significantly higher means at post-test, $F(1, 6) = 15.98$, $p = .007$, $\eta_p^2 = .73$; and also at follow-up, $F(1, 6) = 20.68$, $p = .004$, $\eta_p^2 = .78$; despite an apparent downward trend from post-test to follow-up, as evident in the graph of the interaction in Figure 1.

Although the simple effect for the older boys, using the Huynh-Feldt correction, was also significant, $F(1.13, 4.53) = 53.14$, $p = .001$, $\eta_p^2 = .93$; the simple comparisons revealed a different pattern of results. The older boys also showed a significant improvement at post-test compared to pretest, $F(1, 4) = 41.58$, $p = .003$, $\eta_p^2 = .91$; but by follow-up they had declined to have a significantly *lower* mean than at pretest, $F(1, 4) = 20.00$, $p = .011$, $\eta_p^2 = .83$.

---

[1] To obtain the averaged means you need to create new variables that combine juvenile and staff scores (i.e., Pretest, Posttest, and Follow) as explained at the bottom of page 167. The CIs are obtained from Output 7 on page 166 as they relate to the full factorial analysis..

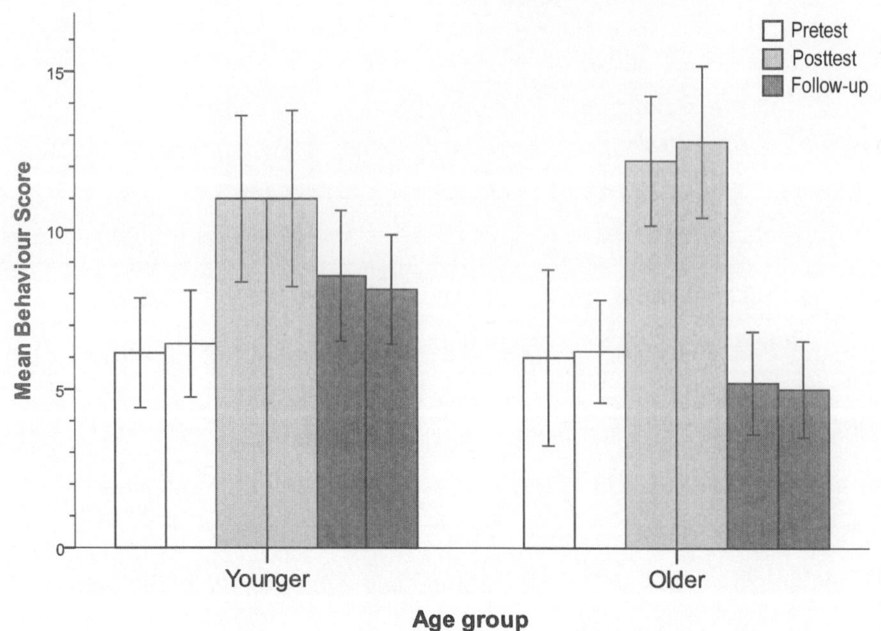

Figure 1. Mean behaviour scores at pretest, post-test, and follow-up for the younger and older offenders (error bars are 95% CI). The juvenile and staff sources are represented by the first and second bars respectively in each pair.

Obtaining a repeated measures bar chart using syntax

Creating a bar chart using **Graphs** options is becoming *more*, not less, cumbersome in successive versions of SPSS. I would argue it is actually becoming easier to use syntax, and being able to use syntax puts you much more "in charge" of things. The simple syntax below was used to create the above chart, which was then edited to suit in **Chart Editor**. (Make sure **Split File** is turned off or **Reset** to include both age groups). Alternatively, See Appendix 2 for how to create this graph using the **Legacy Dialogs** option of **Graphs** that is still relatively easy to use.

33 How to: Use syntax to create a repeated measures bar graph

Open a syntax window, type in and run the following syntax (remembering the stop at the end).

```
GRAPH
  /BAR(GROUPED)=MEAN(Jpre) MEAN(Spre) MEAN(Jpost) MEAN(Spost)
     MEAN(Jfollow) MEAN(Sfollow) BY agegrp
  /MISSING=LISTWISE
  /INTERVAL CI (95.0).
```

To obtain bars for repeated measures variables, it is necessary to list each variable you want, specifying the mean,

Had you created new variables of combined juvenile and staff scores (i.e., Pretest, Posttest, and Follow) as explained at the bottom of page 167 the combined means could then have been graphed instead.

If you have trouble positioning the legend using **Chart Editor** in the version of SPSS you are using, you might need to select **Hide Legend** and create it manually, as I have had to do here.

15 Analysis of Covariance (Ancova)

With Analysis of Covariance (ANCOVA) we are getting into an advanced level of understanding that brings ANOVA and correlation together, so you need to have already studied the earlier ANOVA chapters as well as the chapters on bivariate correlation and regression, and ideally multiple regression and reliability analysis.

When do you use ANCOVA?

ANCOVA can be used as an extension to any type of ANOVA design, including MANOVA in the next chapter. It **statistically controls** for a **covariate** (or covariates), that is, another variable that is **related to the DV**. By control, we mean it removes the effect of the covariate, or removes it as a source of background error variance. Scores on the DV are actually changed! They are adjusted, to what we predict they would have been if everyone had scored the same on the covariate. Now recall what ANOVA, the *F* ratio, actually is. It is the ratio of between-groups variance divided by error variance, so reducing the error variance (the denominator) gives a larger *F* ratio; one that has a greater likelihood of achieving significance. In other words, the great advantage of ANCOVA is that it is a **more powerful** version of ANOVA.

That's the good news, now for the bad news, or at least the news that brings us back down to earth. ANCOVA needs to be used with discretion by researchers, and above all they need to have a thorough understanding of the assumptions of ANCOVA, and of what is actually happening with the analysis. It is all too easy to throw in lots of covariates, all of which might adjust the DV, but what the final adjusted DV score actually means is anyone's guess.

ANCOVA is based on the correlation between the covariate(s) and the DV, and regression is used to predict DV scores from the covariate(s). That part of the DV score that can be predicted by the covariate(s) is, literally, removed from the observed DV score for each participant to produce an adjusted DV score. In the case of a positive correlation this would mean that participants with high scores on a covariate would have their DV score reduced somewhat, and those with low scores on the covariate would have their DV scores increased somewhat—to remove the portion of the DV that can be accounted for by the covariate.

Requirements and assumptions in ANCOVA

In addition to the normal ANOVA assumptions, ANCOVA has some of its own:

1. The covariate(s) needs to be at the interval or ratio level of measurement (or may be dichotomous in some circumstances).
2. To ensure the reliability of ANCOVA results, the correlation between covariates and the DV should be significant.
3. Multiple covariates need to be chosen judiciously and parsimoniously, based on a sound theoretical understanding of their relevance. They should not be too highly correlated with one another.
4. The covariate(s) should be normally distributed (in each group).
5. There needs to be a **linear** relationship between the covariate(s) and DV (in each group).
6. The covariate(s) should be reliable; measures with poor reliability should be avoided.
7. There must be **homogeneity of regression**, that is, relationships between the covariates(s) and the DV must be the same across groups. A moment's reflection makes clear why this needs to be so. Adjustments are made to the DV scores according to the overall correlation between the co-

variate(s) and the DV. Such adjustments would be nonsensical if the correlations were actually different across groups.

Additional requirements in true experiments

In **true experimental** designs the covariate(s) must not be related to the IV, otherwise there is a confound, and even if covariates are **statistically** removed using ANCOVA there is no way of knowing what else may have been associated with them; hence causal inference is compromised. Participants also need to be measured on the covariate(s) **before** being exposed to the experimental manipulations, to avoid the possibility of the manipulation also affecting the covariate. Were this to happen, removing the effect of the covariate might also remove part of the effect of the IV.

In quasi experiments and natural group designs

In **quasi experiments** that use intact groups, covariates must again be measured prior to any experimental manipulation. However, in both **quasi experiments** and **nonexperimental natural group designs**, the covariate(s) may well be related to the IV. This can occur in quasi experiments when groups differ on the pretest (and also the post-test) and the pretest is used as a covariate to statistically remove the effect of the pre-existing differences, thereby "equating" the groups. Correlations between the covariate(s) and DV are quite common in natural group designs. In both these cases, especially the latter, "interpretation is fraught with difficulty" (Tabachnick & Fidell, 2007, p. 196). Researchers really need to know what they are doing to use ANCOVA appropriately, and need to consult more specialised texts before contemplating its use, especially in nonexperimental research. Texts such as Tabachnick and Fidell (2007), Stevens (2002), and Keppel and Wickens (2004) are a good place to start.

Populations and individuals: A caution

Another sobering thought when using ANCOVA is the realisation that adjustments to DV scores are *predicted* adjustments, and we know from regression that there is always a degree or error in such predictions. While ANCOVA is a powerful technique to help us establish if population groups differ **overall** (i.e., in their means), it tells us very little about what to expect with particular individuals, especially when effect sizes are only small to moderate. This caution applies to all ANOVA designs (in fact, to all statistical tests and their results), but is even more stark in ANCOVA. Real understanding of any area to the point of knowing much about individual behaviour and experience is something that is accumulated slowly over time, as we piece together the results of many studies employing many different variables, with the aim of building a comprehensive theoretical framework. Arguably, the headlong rush to churn out quantities of research papers, together with a rather cavalier approach to the use of statistical tests, hinders rather than helps this aim.

Example of ANCOVA

Let us take a very simple one-way between-groups ANOVA design and extend it to be a one-way between-groups ANCOVA. Dr G. Whizz, an educational psychologist, has developed a brilliant new reading program that she believes can dramatically improve the reading ability of children. She decides to conduct an evaluation of its effectiveness, but realises that there is quite an intelligence range in the children she works with. Before conducting the evaluation at the start of a new year she obtains IQ scores from fifteen children who will be taking part in the evaluation, then randomly assigns them to one of three groups: A control group that receives no reading instruction over the one-month duration of the study; a normal group that receives the normal program of reading instruction; and the Whizz group that receives the new Whizz Reading Program.

At the end of the evaluation period, reading scores are obtained from each of the children on a measure of reading ability with a possible range of 0 (very poor) to 25 (highly competent). Because of the very small sample sizes, and consequent reduced power, she believes a more liberal alpha level of .10 is justified for her analysis of the results.

Using SPSS to conduct ANCOVA

Step 1: Enter data in SPSS and save data file

Here are the data for the 15 children. Enter them into SPSS and save in a file called ***ANCOVA.sav***.

ID	Participant number	*Set **Measure** column to **Nominal***
Group	1=Control 2=Normal Program 3=Whizz Program	*Set **Measure** column to **Nominal***
IQ		*Set **Measure** column to **Scale***
Reading	Reading score	*Set **Measure** column to **Scale***

Control Group			Normal Group			Whizz Group		
ID	**IQ**	**Reading Score**	**ID**	**IQ**	**Reading Score**	**ID**	**IQ**	**Reading Score**
1	110	16	6	105	15	11	105	17
2	125	18	7	100	13	12	120	19
3	95	11	8	95	12	13	90	17
4	101	13	9	90	11	14	95	14
5	110	15	10	101	15	15	105	15

Step 2: Screen the data and check assumptions

Use **Split File** to check the usual ANOVA assumptions for each group separately, and to also check that the covariate is normally distributed in each group.

You can request a scatterplot for each group showing the relationship between the covariate and DV. Not only will this give an indication of whether there is a similar relationship in each group, it will also indicate any obvious departures from linearity. As part of a study such as this you would check the reliability of covariates using **reliability analysis** if they were scores from measurement intruments.

The SPSS commands (and output) on the next page are required to test the assumption of homogeneity of regression.

To illustrate how ANCOVA is more powerful than ANOVA, perform a simple one-way ANOVA on these data, with group as the IV and reading score as the DV, without using IQ as the covariate. You will find that it is **not significant** even using an alpha of .10, $F(2,12) = 2.70$, $p = .11$. Dr Whizz would have to conclude that there is no significant effect of either the normal program or the Whizz program on reading ability.

Also perform a one-way ANOVA with group as the IV and IQ as the DV. You will find this is also not significant, $F(2,12) = 1.27$, $p = .32$; satisfying the requirement in a true experiment that the covariate be **unrelated** to the IV. The groups do not differ significantly on IQ.

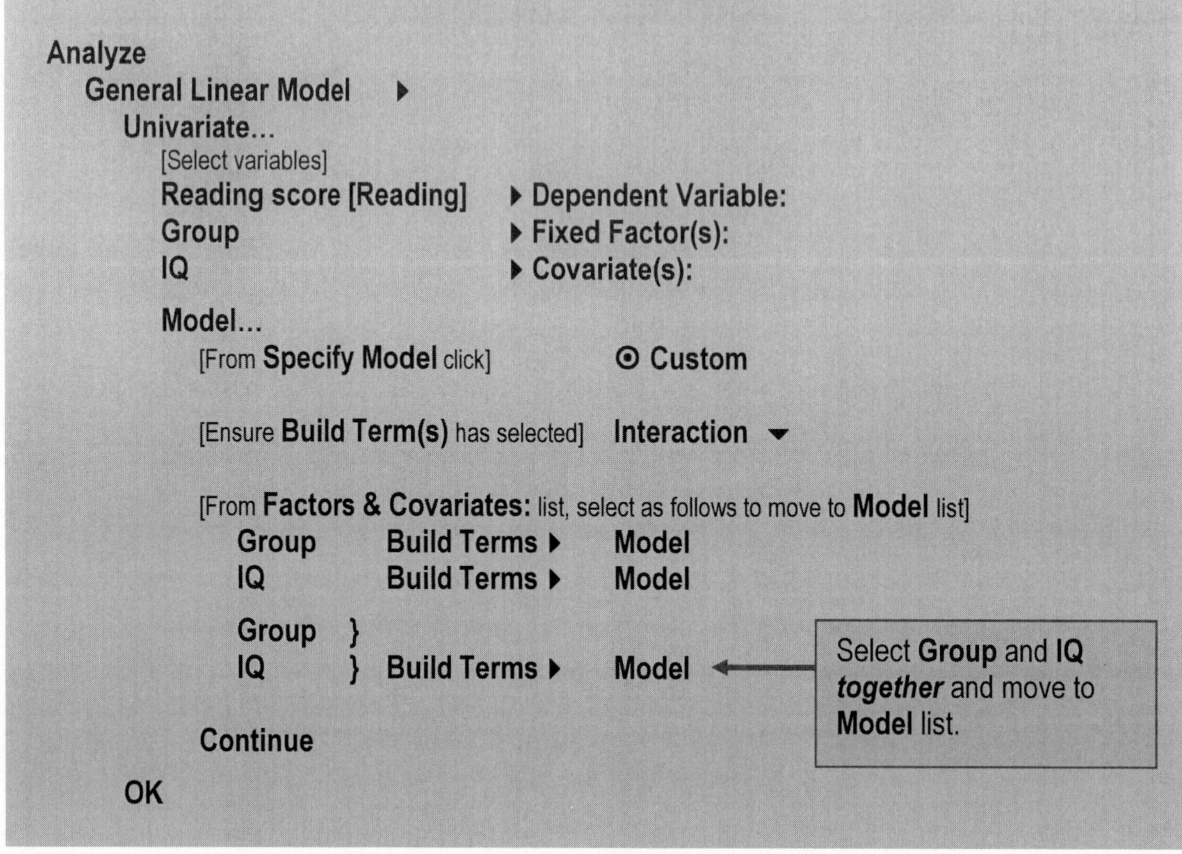

34 **How to: Test homogeneity of regression in ANCOVA**

Analyze
General Linear Model ▶
Univariate...
[Select variables]
Reading score [Reading] ▶ **Dependent Variable:**
Group ▶ **Fixed Factor(s):**
IQ ▶ **Covariate(s):**

Model...
[From **Specify Model** click] ⊙ **Custom**

[Ensure **Build Term(s)** has selected] **Interaction** ▼

[From **Factors & Covariates:** list, select as follows to move to **Model** list]
Group **Build Terms** ▶ **Model**
IQ **Build Terms** ▶ **Model**

Group }
IQ } **Build Terms** ▶ **Model** ←

Select **Group** and **IQ** ***together*** and move to **Model** list.

Continue

OK

Output: **Homogeneity of regression**

Tests of Between-Subjects Effects

Dependent Variable: Reading score

Source	Type III Sum of Squares	df	Mean Square	F	Sig.
Corrected Model	70.096[a]	5	14.019	9.829	.002
Intercept	3.362	1	3.362	2.357	.159
Group	8.373	2	4.187	2.935	.104
IQ	34.359	1	34.359	24.088	.001
Group * IQ	6.190	2	3.095	2.170	.170
Error	12.837	9	1.426		
Total	3339.000	15			
Corrected Total	82.933	14			

a. R Squared = .845 (Adjusted R Squared = .759)

The only thing relevant in this output is the **interaction term** of the **IV with the covariate** (Group*IQ). This tests the assumption of homogeneity of regression, that the relationship between the covariate and the DV is the same across groups. For the assumption to be met it must be **not** significant at alpha = .05.

At Sig. = .170 (> .05) the assumption **is met** in this example.

Step 3: Request the ANCOVA analysis

35 **How to:** Conduct a one-way ANCOVA using GLM

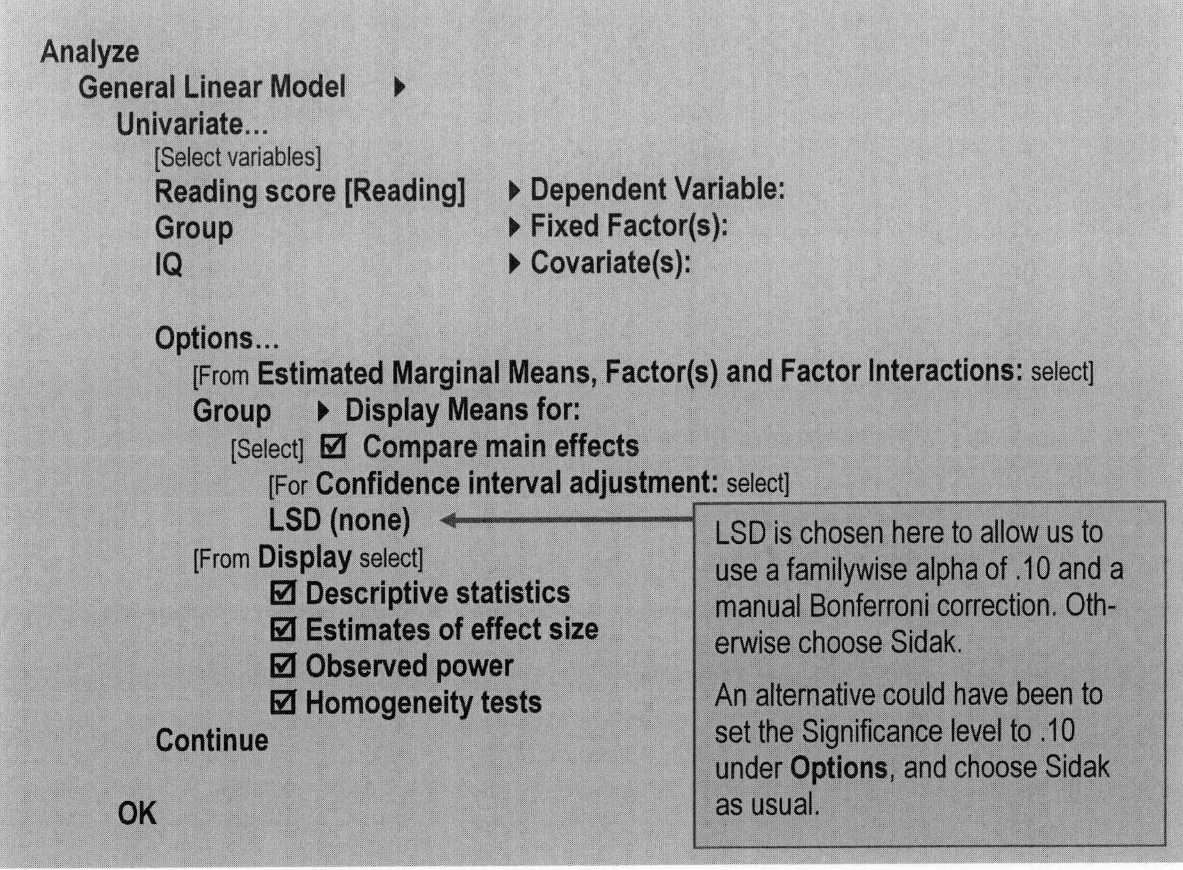

Analyze
  General Linear Model ▸
    Univariate...
      [Select variables]
      Reading score [Reading] ▸ **Dependent Variable:**
      Group ▸ **Fixed Factor(s):**
      IQ ▸ **Covariate(s):**

      Options...
        [From **Estimated Marginal Means, Factor(s) and Factor Interactions:** select]
        Group ▸ **Display Means for:**
          [Select] ☑ **Compare main effects**
            [For **Confidence interval adjustment:** select]
            LSD (none)
        [From **Display** select]
            ☑ **Descriptive statistics**
            ☑ **Estimates of effect size**
            ☑ **Observed power**
            ☑ **Homogeneity tests**
      Continue

    OK

LSD is chosen here to allow us to use a familywise alpha of .10 and a manual Bonferroni correction. Otherwise choose Sidak.

An alternative could have been to set the Significance level to .10 under **Options**, and choose Sidak as usual.

Step 4: Examine and interpret the output

Descriptive Statistics

Dependent Variable: Reading score

Group	Mean	Std. Deviation	N
Control	14.60	2.702	5
Normal Program	13.20	1.789	5
Whizz Program	16.40	1.949	5
Total	14.73	2.434	15

1 Here are the **observed** means. Had we just performed a normal ANOVA on them, without using the covariate, it would have indicated there are no significant differences among them. Note the normal program even has a lower mean than the control group receiving no reading instruction.

Levene's Test of Equality of Error Variances[a]

Dependent Variable: Reading score

F	df1	df2	Sig.
1.160	2	12	.346

Tests the null hypothesis that the error variance of the dependent variable is equal across groups.

a. Design: Intercept+IQ+Group

2 The Levene's test result is **not** significant, indicating that the ANOVA assumption of homogeneity of variance is met.

Tests of Between-Subjects Effects

Dependent Variable: Reading score

Source	Type III Sum of Squares	df	Mean Square	F	Sig.	Partial Eta Squared	Noncent. Parameter	Observed Power[a]
Corrected Model	63.906[b]	3	21.302	12.315	.001	.771	36.946	.994
Intercept	1.606	1	1.606	.928	.356	.078	.928	.143
IQ	38.173	1	38.173	22.069	.001	.667	22.069	.989
Group	21.740	2	10.870	6.284	.015	.533	12.568	.792
Error	19.027	11	1.730					
Total	3339.000	15						
Corrected Total	82.933	14						

a. Computed using alpha = .05

b. R Squared = .771 (Adjusted R Squared = .708)

3

The first thing to note in the ANCOVA summary table is the figure for IQ. It has an *F* value of 22.07 (Sig. = .001). This confirms that the **covariate (IQ) is significantly related to the DV**, and is making a significant adjustment to DV scores—that is, the adjustment is not a spurious one based on a chance relationship with the DV in this particular sample.

Then, after controlling for the covariate **there is now a significant effect of group**, *F* = 6.28, Sig. = .015, with a partial Eta squared of .53 and adequate power at .79.

Estimated Marginal Means

Group

Estimates

Dependent Variable: Reading score

Group	Mean	Std. Error	95% Confidence Interval Lower Bound	Upper Bound
Control	13.689[a]	.619	12.325	15.052
Normal Program	14.087[a]	.618	12.728	15.447
Whizz Program	16.424[a]	.588	15.129	17.719

a. Covariates appearing in the model are evaluated at the following values: IQ = 103.13.

4

When there is a covariate the marginal means are the adjusted means; the DV means after they have been adjusted to what they would be predicted to be if everyone had the overall mean score on the covariate (103.13).

In comparing them with the observed means on the previous page, note how the normal program group now has a **higher** mean than the control group. This is because it had a **lower** IQ mean (see below), so to "compensate" for this in terms of reading score, it was adjusted upwards. The control group had the highest IQ, so its reading scores have been adjusted downwards. The Whizz Group with an IQ in between the other two, and close to the overall mean of 103.00, has been adjusted very little.

Group means on IQ

If you performed the one-way ANOVA on IQ you would have discovered IQ means as follow:

Control group 108.20
Normal Program 98.20
Whizz Program 103.00

Pairwise Comparisons

Dependent Variable: Reading score

(I) Group	(J) Group	Mean Difference (I-J)	Std. Error	Sig.[a]	95% Confidence Interval for Difference[a]	
					Lower Bound	Upper Bound
Control	Normal Program	-.399	.916	.672	-2.414	1.617
	Whizz Program	-2.735*	.855	.008	-4.618	-.853
Normal Program	Control	.399	.916	.672	-1.617	2.414
	Whizz Program	-2.337*	.852	.019	-4.211	-.462
Whizz Program	Control	2.735*	.855	.008	.853	4.618
	Normal Program	2.337*	.852	.019	.462	4.211

Based on estimated marginal means

*. The mean difference is significant at the .05 level.

a. Adjustment for multiple comparisons: Least Significant Difference (equivalent to no adjustments).

The three main effect comparisons with no automatic adjustment to alpha (i.e., using the Least Significant Difference option), but a manual Bonferroni adjustment of .033 based on a familywise alpha of .10 (i.e., .10 /3), indicate that the Whizz group did indeed perform significantly better on reading than either the control or normal program groups after we controlled for IQ (for each comparison Sig. < .033).

There is no significant difference between the control and normal program groups.

ANCOVA analysis: Research report sample Results section

Results

The reading test scores for the control group and the two reading program groups were analysed with a one-way analysis of covariance (ANCOVA), using IQ scores as the covariate, and with alpha set at .10 in view of the loss of power with very small samples. The IQ scores had high reliability, and assumptions of normality, homogeneity of variance, linearity, and homogeneity of regression were all met. There was no significant difference among the groups on IQ, but it did make a significant adjustment to scores on the dependent variable ($p = .001$). After adjusting for IQ the mean reading scores (see Table 1) were significantly different overall: $F(2, 11) = 6.28$, $p = .02$, $\eta_p^2 = .53$. Pairwise comparisons among the adjusted group means using a Bonferroni correction for multiple tests of .033 (.10/3) revealed that the Whizz program group had a significantly higher mean reading score than either the control group (M_{Diff} = 2.74, p = .01, 95% CI [0.85, 4.62]) or the normal program group (M_{Diff} = 2.34, p = .02, 95% CI [0.46, 4.21]). There was no significant difference between the normal program and control group (M_{Diff} = 0.40, p = .67, 95% CI [-1.62, 2.41]).

Table 1

Adjusted and Unadjusted Reading Means for the Groups

	Adjusted Means		Unadjusted Means	
Group	*M* (*SE*)	95% CI	*M*	*SD*
Control	13.69 (0.62)	[12.33, 15.05]	14.60	2.70
Normal Program	14.09 (0.62)	[12.73, 15.45]	13.20	1.79
Whizz Program	16.42 (0.59)	[15.13, 17.72]	16.40	1.95

Note: n = 5 for each group. CI = confidence interval.

More on ANCOVA

In the next chapter on multivariate ANOVA, or MANOVA, there is another use of ANCOVA in what is known as stepdown analysis. In the context of stepdown analysis the chapter covers more complex aspects of ANCOVA that arise with factorial designs.

16 Multivariate Analysis of Variance (MANOVA)

Now we are really into advanced territory. Please do not even look at this chapter without a sound grasp of all the earlier ANOVA chapters, especially ANCOVA, as well as those on correlation, multiple regression, and factor analysis.

Again, to assist in navigating the complexity of MANOVA, a chapter table of contents is provided, not only as a "find-quick", but also as a review facility, whereby you can tick off each section once you are satisfied you understand it. You need to work through the concepts, and then follow the analysis systematically to see them illustrated. You should then review the concepts in light of what you learn from the analysis.

Chapter contents

What is MANOVA, and when is it used?

MANOVA is an extension of univariate analysis of variance (ANOVA) to the situation where there are two or more **dependent variables**. The main reason it is used in preference to simply performing separate ANOVAs on each DV is that it controls familywise error across the multiple tests needed, just as ANOVA is preferred to a series of *t* tests. If MANOVA is **not** significant no further tests are conducted, but if it **is** significant, researchers typically proceed to the univariate ANOVAs and then to analytical comparisons where necessary, using the procedures covered in the previous ANOVA chapters.

MANOVA is a controversial technique, and given it tends to be followed by ANOVAs anyway (when significant) many argue that there is little point to it for routine control of familywise error with multiple DVs. In their paper on the issue of MANOVA versus multiple ANOVAs Huberty and Morris (1989) argued it can be appropriate to use the Bonferroni test to control familywise error for multiple ANOVAs, even when the DVs are related, and certainly in the rare situation where they are unrelated (see also Tabachnick & Fidell, 2007, p. 268).

The real advantage of MANOVA is that it can take account of correlations among DVs, and can identify group differences on the **linear combination** of DVs; differences that might not be apparent when they are considered in isolation. Under certain circumstances (and this will demonstrated in the example to follow) MANOVA can be more powerful than a series of ANOVAs when small differences on individual DVs combine to produce an overall significant effect.

Choice of variables for MANOVA

MANOVA should **not** be regarded as mandatory just because there are multiple DVs in a study. Tabachnick and Fidell (2007, p. 268) have argued that it has lower power than multiple ANOVAs using a Bonferroni correction when the DVs are uncorrelated. They have also argued against using highly correlated DVs (unless negative) because the redundancy makes interpretation of the role of individual variables ambiguous. MANOVA is most suited to moderately correlated variables (around .60), although Field (2005, pp. 573-574) has noted the importance of both the degree of intercorrelation **and** effect sizes. The question of power (and everything else) is anything but simple so before embarking on MANOVA specialised references need to be consulted (see also Stevens, 2002).

A **general guide** for the use of MANOVA is that it should have a sound theoretical or empirical basis, with a limited number of DVs carefully chosen so as to make conceptual sense when considered in combination; ideally with moderate intercorrelations. If there are a large number of variables it may well be more appropriate to perform separate MANOVAs on meaningful subsets, making adjustments for familywise error accordingly. It is not advisable to include all sorts of DVs simply because you think they might be "interesting".

Approaches to analysis in MANOVA

These issues aside, a common approach is to use MANOVA with multiple DVs and, if significant, to follow it with univariate ANOVAs (with a Bonferroni correction for familywise error) on the grounds that MANOVA protects against Type I error at the first step. If the MANOVA is not significant no further analysis is undertaken even where one or more of the univariate ANOVAs might be significant.

An alternative approach is to use **stepdown** analysis (or more specifically Roy-Bargmann stepdown analysis). This depends on being able to order the DVs in some meaningful order of priority, and allows their unique contributions to differences among the IVs to be determined. The first DV is tested in a univariate ANOVA, with successive DVs tested with ANCOVA, using prior DVs as covariates. Although known as stepdown analysis this is identical to a series of ANCOVAs. The intercorrelations among the DVs are taken into account, so group differences on the unique component of each DV can be determined. Stepdown analysis does, however, depend upon being **able to order** the DVs in some defensible manner (just as in hierarchical multiple regression). The order may be causal, theoretical, logical, or practical. In the absence of any of these it is also possible to order the variables according to the size of the univariate F values, but because these may reflect chance differences in the sample, any interpretation based on this ordering may be unreliable. The example that follows takes a stepdown approach to MANOVA.

Another approach to MANOVA is somewhat different, in that rather than looking at individual DVs, linear composite functions (similar to factor analysis) are identified that discriminate the groups, and one looks at how the individual DVs are related to the linear composites. This is the **discriminant function** approach. It will not be illustrated here, although discriminant function analysis is addressed separately in the next chapter. For more details on its use as an approach in MANOVA see Field (2005) or Stevens (2002).

A hypothetical example of MANOVA

Imagine you are an environmental psychologist working for a large international insurance company. A new office tower has been built in a large capital city and just so happens to have a unique orientation. One side has a splendid view of the ocean; another side overlooks a green and tranquil park; while the remaining two sides overlook the rest of the city. Being aware of the literature on health benefits associated with nature views, you arrange to conduct an experiment. You are permitted to randomly select 180 employees, and then to randomly allocate a third of them to offices overlooking the ocean; another third to offices overlooking the park; and the last third to offices that only have views of the city. To maintain as much independence as possible the participants are randomly interspersed on various floors, such that half in each group are located on upper floors (floors 8-10) and half on lower floors (floors 2-4). The building opens at the beginning of September, and at the end of December, you obtain measures from each participant on three interval-level scales that will serve as your dependent variables:

> **Satisfaction with work (WorkSat):** This is a multi-item instrument of items that refer to aspects of employee satisfaction with their overall working environment. It has a possible range of 0 (totally dissatisfied) to 20 (totally satisfied).
>
> **Wellbeing:** This instrument provides an overall measure of employees' general level of physical and psychological wellbeing, including general levels of stress and mood. It is structured in such a way that each person obtains a score in the range 0 -12 (with 0 being the lowest possible level of wellbeing, and 12 being the highest possible level).
>
> **Work performance (Perform)**: The final measure is a company instrument that assesses each employee's overall level of work performance at the end of each quarter. Scores in the range 1-5 indicate unsatisfactory performance; those from 6-8 are satisfactory; 9-11 is highly satisfactory, 12-16 is the high performance range; and 17-20 is indicative of outstanding performance.

A logical ordering is proposed for the DVs whereby satisfaction with work is regarded as a precursor of wellbeing, and both are expected to contribute to work performance as the final variable in the sequence. It is hypothesised that there will be higher work satisfaction, wellbeing, and work performance scores for participants with natural views compared to those with city views.

It is always a good idea in planning any ANOVA (including MANOVA) to draw a schematic representation of your design. This is two-way, fully crossed, between-groups factorial MANOVA; that is, an A x B or 3 x 2 MANOVA design, as there are three levels of factor A fully crossed with two levels of factor B, producing six cells in the design. Note that the A x B, or 3 x 2 refers to the IVs **not** the DVs. The two IVs are window view with three levels (city, ocean, and park), and floors with two levels (upper and lower), and there are three DVs (work satisfaction, wellbeing, and work performance).

	Factor (IV) B: Floors of building (Floors)					
Factor (IV) A: Window view (View)	Upper			Lower		
DVs:	WorkSat	Wellbeing	Perform	WorkSat	Wellbeing	Perform
City						
Ocean						
Park						

It is probably more accurate to describe this design as a **field experiment**, where there is a "natural" **manipulation** of the IV, not one directly implemented by the researcher. There is **random assignment**, but because the study takes place in the field, **control** over other relevant variables would be very limited. As such, the internal validity of the study is compromised. A researcher would need to be careful about assuming any effects of the IVs are causal ones, so in causal terms the study would have more in common with a natural groups design.

All ANOVA designs (including MANOVA) are less complicated if the design is **balanced**, that is, with equal numbers of participants in each cell. Aim for a balanced design wherever possible, and certainly strive to avoid smallest to largest ratios exceeding 1:1.5 (see Stevens, 2002; Tabachnick & Fidell, 2007). In this example, we have a balanced design with 30 participants in each cell.

Unbalanced designs, in which there are unequal numbers in each cell, occur frequently, and computer programs are quite able to handle them (provided they are not too unbalanced). You just need to be aware that **marginal** means in **unbalanced** designs are **unweighted** means (the "average" of contributing cells), and will be different to the observed means, which are weighted in favour of larger cells.

A step-by-step MANOVA analysis using SPSS

Step 1: Enter data in SPSS and save data file

With more complex analyses, it is often useful to keep a workbook in which to record notes about the study, including a register of the files you use and what they contain; and to use in planning code names and value labels for variables.

Because there are 180 cases the data have already been entered and saved in a file called ***16MANOVA.sav***. The file can be downloaded from the link 'Student downloads' on this website: www.pearson.com.au/9781442549821

The data in this example are as follows:

ID	Participant number	***Nominal*** *variable*
View	1=City 2=Ocean 3=Park	***Nominal*** *variable*
Floors	1=Upper 2=Lower	***Nominal*** *variable*
Cell	1=City-Upper 2=City-Lower 3=Ocean-Upper 4=Ocean-Lower 5=Park-Upper 6=Park-Lower	***Nominal*** *variable*
WorkSat	Satisfaction with work	***Scale*** *variable*
Wellbeing		***Scale*** *variable*
Perform	Performance	***Scale*** *variable*

Step 2: Screen data and check assumptions

Screen the data in view of the assumptions of MANOVA. Refer also to Tabachnick and Fidell (2007) as this guide only contains brief notes. You are also advised to consult the excellent chapter on *Assumptions in MANOVA* in Stevens (2002).

Data screening for the usual ANOVA assumptions needs to be conducted separately for each DV for **each cell** of the design, using the SPSS **Split File** option. In addition to the normal ANOVA assumptions the following assumptions and issues need to be addressed.

Cell size

There **must** be more participants in each cell than the number of DVs. Ideally have cell n = 30+ (or at least 20-30), as assumptions of normality and equal variances are then of little concern. You need not have equal cells, but avoid ratios of smallest to largest greater than 1:1.5. Assumptions become very important with **small and unequal** cells. Cell size is satisfactory in this example, but you should still check normality and homogeneity of variance as part of routine data screening.

Normality and outliers

Univariate and multivariate outliers, **identified separately for each cell**, are a serious problem for MANOVA. Multivariate outliers among the DVs are identified using **Mahalanobis distance**, as explained in the chapter on multiple regression (i.e., perform a multiple regression analysis with the DVs as predictor variables and ID or any other continuous variable as the criterion, requesting the Mahalanobis distance for each case, so that it is added to the datafile).

Tabachnick and Fidell (2007) identify multivariate outliers as those with a Mahalanobis distance value greater than or equal to a critical value on the chi-square distribution (recall that chi-square is another type of probability distribution, and has the symbol χ^2). The critical value is obtained from their table C.4 (p. 949) with *df* equal to the number of DVs and α = .001. The critical values are as follows for up to ten DVs (in brackets): 13.816(2), **16.266(3)**, 18.467(4), 20.515(5), 22.458(6), 24.322(7), 26.125(8), 27.877(9), and 29.588(10).

You should find that none of the cases equal or exceed the critical Mahalanobis distance of 16.266 that applies to this example with three DVs, so we have no multivariate outliers.

There is little that can be done with multivariate outliers, other than to delete them from the study, although if ever presented with this as a serious problem you need to read specialised texts before deciding how to proceed (e.g., Tabachnick & Fidell, 2007, pp. 72-77).

Linearity

The relationships among the DVs need to be linear. This can be assessed by an inspection of scatterplots among the DVs for each cell of the design.

Multicollinearity and singularity

From multiple regression you will recall these are unacceptably high (or perfect) correlations among the DVs. This can be assessed from the pooled within-cells correlation matrix to be demonstrated shortly. Recommendations can differ, but generally correlations above .80 or .90 suggest redundant variables, which either need to be combined, or one of them needs to be removed from the analysis.

Stepdown analysis and the assumption of homogeneity of regression

When performing stepdown analysis, as is the case here, the assumption of homogeneity of regression needs to be tested, given that stepdown analysis is simply ANCOVA.

The first stepdown analysis is the ANCOVA on wellbeing as the DV, using satisfaction with work as the covariate. The second one is for performance as the DV with combined covariates of satisfaction with work and wellbeing.

Step 2.1: Checking within–cell correlations (for multicollinearity)

The easiest way to check this is with MANOVA syntax, as follows on the next page.

36 How to: Use SPSS command syntax to obtain pooled within-cell correlations for MANOVA

```
MANOVA
    WorkSat Wellbeing Perform BY View(1 3) Floors (1 2)
    /PRINT CELLINFO(CORR) ERROR(CORR)
    /DESIGN .
```

Do **not** forget the full stop at the end.

Save the syntax commands as *MANOVA.sps*

To run this analysis, position the cursor at the beginning, or somewhere within the commands; or highlight the commands with the mouse:

Then click on the run button ▶ located in the icon bar.
Alternatively, select from the **Run** option in the menu bar at the top

Output: **MANOVA pooled within-cel correlations**

```
- - - - - - - - - - - - - - - - - - - - - - - - - - - - - - - - - - - - - - - -
WITHIN+RESIDUAL Correlations with Std. Devs. on Diagonal

              WorkSat   Wellbein    Perform
WorkSat         1.670
Wellbein         .426       .941
Perform          .483       .464      2.135

- - - - - - - - - - - - - - - - - - - - - - - - - - - - - - - - - - - - - - - -
Statistics for WITHIN+RESIDUAL correlations

Log(Determinant) =                  -.57826
Bartlett test of sphericity =    99.55777 with 3 D. F.
Significance =                         .000

F(max) criterion =                5.15099 with (3,174) D. F.
- - - - - - - - - - - - - - - - - - - - - - - - - - - - - - - - - - - - - - - -
```

As of SPSS version 15 MANOVA syntax still results in the old style of output.
To be able to scroll through **all** this output you need to double-click on it so it opens in its own window.

The intercorrelations for each cell of the design are provided for you to check, but if you scroll down you reach the table of pooled within-cell correlations, as above.

There is no problem here with multicollinearity, as there are only moderate correlations among the three DVs, and there is certainly no **statistical** problem for the analysis, which only occurs when the determinant "is near zero (say, less than .0001)" (Tabachnick & Fidell, 2007, p. 253), that is, when the Log(Determinant) is less than -4.

If **Bartlett's Test of Sphericity** is significant at α = .05 (as it certainly is here) it indicates that the DVs are significantly related, so MANOVA with stepdown analysis **is** appropriate. If not significant, it indicates that the DVs are unrelated and univariate ANOVAs with adjustment for familywise error would be more appropriate.

Step 2.2: Testing the assumption of homogeneity of regression

The easiest way to check the assumption of homogeneity of regression for these two stepdown analyses is to proceed as shown in the previous chapter on ANCOVA, but instead of clicking OK at the end, click **Paste** to paste the commands into a syntax window. Then it is simply a matter of modifying the commands as shown on the following pages to check the assumption for combined effects across the factorial design.

Be sure to save the syntax (perhaps name the file ***MANOVA.sps***) because another great advantage of syntax is that you can rerun it as often as you like without having to repeat all the tedious selecting and clicking (e.g., you might notice an outlier you missed and have to rerun the analysis after modifying it).

Anyone who feels really confident with syntax can dispense with the following procedure and simply type all the syntax commands in from scratch in a new syntax window.

37 How to: Test homogeneity of regression in a factorial design

Analyze
General Linear Model ▸
Univariate...
[Select variables]
Wellbeing ▸ **Dependent Variable:**

Window view [View] }
Floors } ▸ **Fixed Factor(s):**

Satisfaction with work [WorkSat] ▸ **Covariate(s):**

Model...
[From **Specify Model** click] ⊙ **Custom**
[Ensure **Build Term(s)** has selected] **Interaction** ▾

[From **Factors & Covariates:** list, select as follows to move to **Model** list]

WorkSat **Build Terms ▸** **Model**
View **Build Terms ▸** **Model**
Floors **Build Terms ▸** **Model**

View }
Floors } **Build Terms ▸** **Model**

View }
WorkSat } **Build Terms ▸** **Model**

Floors }
WorkSat } **Build Terms ▸** **Model**

View }
Floors }
WorkSat } **Build Terms ▸** **Model**

Remember to select each of these ***together*** and move to **Model** list.

Continue

Paste ◂—— Click **Paste** to paste commands to a syntax window, **not OK**.

The following syntax will be pasted into a syntax window by SPSS.

```
UNIANOVA
 Wellbeing  BY View Floors  WITH WorkSat
 /METHOD = SSTYPE(3)
 /INTERCEPT = INCLUDE
 /CRITERIA = ALPHA(.05)
 /DESIGN = WorkSat View Floors Floors*View View*WorkSat Floors*WorkSat
 Floors*View*WorkSat .
```

The syntax is now modified, as shown below. Neither bold text nor wide-spacing is necessary; it has only been included to make it easier for you to read. The commas are not essential either, but they help to separate the terms. As with all syntax it must be perfect, with no mistakes or omissions.

1. Replace all but the last two lines with the first four lines below.
2. Then add the two + signs where indicated by the arrows. This creates a combined homogeneity of regression test for all combinations of the covariate with the main effect groups, then with the interaction cells.

   You do not have to add the commas, or change * to BY, or change the order of Floors*View to View BY Floors. (I have done this to maintain my own sense of order, and to *invite* you to think about what it means.)

   The design line is simply a list of the covariates, main effects, and interactions, **with each of these repeated but combined in interaction with the covariate(s)**.
3. Then copy the last three lines (/ANALYSIS and modified /DESIGN lines) and paste them at the end, removing the first full stop, so there is **only one full stop at the very end**.
4. Then carefully modify the final three lines as shown, so that now the analysis is on Perform, **combining** or **pooling** both WorkSat and Wellbeing as covariates.

38 How to: Use SPSS command syntax to test homogeneity of regression in factorial designs

```
MANOVA
WorkSat, Wellbeing, Perform BY View(1 3), Floors(1 2)
/PRINT=SIGNIF(BRIEF)
/ANALYSIS Wellbeing                                                    ↓                    ↓
/DESIGN = WorkSat, View, Floors, Floors BY View, View BY WorkSat+Floors BY WorkSat+
   Floors BY View BY WorkSat
/ANALYSIS Perform
/DESIGN = WorkSat, Wellbeing, View, Floors, Floors BY View, View BY pool(WorkSat,
   Wellbeing)+Floors BY pool(WorkSat, Wellbeing)+ Floors BY View BY pool(WorkSat,
   Wellbeing) .
```

Add these syntax commands to the file *MANOVA.sps*, and save.

To run this analysis, position the cursor at the beginning, or somewhere within the commands (or highlight the commands with the mouse if you only want these ones executed):

Then click on the run button ▶ located in the icon bar.
Alternatively, select from the **Run** option in the menu bar at the top

```
* * * * * A n a l y s i s   o f   V a r i a n c e -- design  1 * * * * *

Tests of Significance for Wellbeing using UNIQUE sums of squares
 Source of Variation          SS      DF        MS         F  Sig of F

 WITHIN+RESIDUAL          123.37     168       .73
 WORKSAT                   27.18       1     27.18     37.01      .000
 VIEW                       1.35       2       .68       .92      .400
 FLOORS                     1.72       1      1.72      2.35      .127
 VIEW BY FLOORS              .20       2       .10       .13      .875
 VIEW BY WORKSAT + FL       2.67       5       .53       .73      .603
 OORS BY WORKSAT + VI
 EW BY FLOORS BY WORK
 SAT

 (Model)                   37.86      11      3.44      4.69      .000
 (Total)                  161.24     179       .90
 R-Squared =          .235
 Adjusted R-Squared = .185

 * ** * A n a l y s i s  o f   V a r i a n c e -- design  1 * * * * * *

Tests of Significance for Perform using UNIQUE sums of squares
 Source of Variation          SS      DF        MS         F  Sig of F

 WITHIN+RESIDUAL          507.11     162      3.13
 WORKSAT                   80.60       1     80.60     25.75      .000
 WELLBEING                 61.14       1     61.14     19.53      .000
 VIEW                      10.95       2      5.47      1.75      .177
 FLOORS                    10.86       1     10.86      3.47      .064
 VIEW BY FLOORS             2.18       2      1.09       .35      .706
 VIEW BY POOL(WORKSAT      35.94      10      3.59      1.15      .330
  WELLBEING) + FLOORS
  BY POOL(WORKSAT WEL
 LBEING) + VIEW BY FL
 OORS BY POOL(WORKSAT
  WELLBEING)

 (Model)                  338.69      17     19.92      6.36      .000
 (Total)                  845.80     179      4.73
 R-Squared =          .400
 Adjusted R-Squared = .338
```

1

2

This is the output for the test of homogeneity of regression, and the only figures of interest are the two that have been highlighted. These will **not** be significant if the assumption of homogeneity of regression is **met.**

The first one (Sig. of F) = .603 is **not** significant indicating that we **do have homogeneity of regression** for the first stepdown analysis on Wellbeing as the DV, with WorkSat as the covariate.

The second one (Sig. of F) = .303 is also **not** significant indicating that we **also have homogeneity of regression** for the second stepdown analysis on Perform as the DV, with both WorkSat and Wellbeing as covariates.

.

Step 3: Conduct the MANOVA analysis

Note that the MANOVA procedure can be used as an alternative to the General Linear Model. It provides additional information, but is only available using syntax.

39 How to: Conduct a MANOVA analysis using GLM

Analyze
- **General Linear Model**
  - **Multivariate...**
    - [Select variables, as follows:]
    - **Satisfaction with work [WorkSat]** }
    - **Wellbeing** }
    - **Performance [Perform** } ▸ **Dependent Variables:**
    - **Window view [View]** }
    - **Floors** } ▸ **Fixed Factor(s):**
    - **Options...**
      - [From **Estimated Marginal Means, Factor(s) and Factor Interactions** select:]
        - **View** }
        - **Floors** }
        - **View*Floors** } ▸ **Display means for:**
        - [Select] ☑ **Compare main effects**
          - [For **Confidence interval adjustment:** select]
          - **Sidak**
      - [From **Display** select:]
        - ☑ **Descriptive statistics**
        - ☑ **Estimates of effect size**
        - ☑ **Observed power**
        - ☑ **Homogeneity tests**
      - **Continue**
    - **Plots**
      - [From **Factors** select:]
        - **View** ▸ **Horizontal Axis**
        - **Floors** ▸ **Separate Lines**
        - **Plots:** Click on **Add** button
      - **Continue**
  - **OK**

Step 4: Examine and interpret the multivariate output

Descriptive Statistics

	Window view	Floors	Mean	Std. Deviation	N
Satisfaction with work	City	Upper	9.93	1.946	30
		Lower	9.87	1.570	30
		Total	9.90	1.753	60
	Ocean	Upper	9.33	1.446	30
		Lower	9.07	1.721	30
		Total	9.20	1.582	60
	Park	Upper	9.47	1.634	30
		Lower	10.07	1.660	30
		Total	9.77	1.661	60
	Total	Upper	9.58	1.689	90
		Lower	9.67	1.689	90
		Total	9.62	1.685	180
Wellbeing	City	Upper	6.833	.7694	30
		Lower	6.883	1.0642	30
		Total	6.858	.9210	60
	Ocean	Upper	6.917	.8209	30
		Lower	7.433	1.0233	30
		Total	7.175	.9559	60
	Park	Upper	6.967	1.0499	30
		Lower	6.917	.8718	30
		Total	6.942	.9571	60
	Total	Upper	6.906	.8802	90
		Lower	7.078	1.0109	90
		Total	6.992	.9491	180
Performance	City	Upper	11.23	2.417	30
		Lower	11.00	1.875	30
		Total	11.12	2.148	60
	Ocean	Upper	12.47	2.209	30
		Lower	10.80	1.972	30
		Total	11.63	2.240	60
	Park	Upper	11.07	2.490	30
		Lower	11.23	1.736	30
		Total	11.15	2.130	60
	Total	Upper	11.59	2.431	90
		Lower	11.01	1.851	90
		Total	11.30	2.174	180

1

Always look closely at the means and standard deviations as a first step. (Do there appear to be differences in DVs as a function of window view or floors?)

Box's Test of Equality of Covariance Matrices[a]

Box's M	33.653
F	1.072
df1	30
df2	68421.681
Sig.	.360

Tests the null hypothesis that the observed covariance matrices of the dependent variables are equal across groups.

a. Design: Intercept+View+Floors+View * Floors

2

Examine **Box's Test**, which tests **homogeneity of covariance** matrices.

It is very sensitive, and may be assessed using α = .001.

There **is** homogeneity (as here) if this test is **not** significant at α = .001.

Multivariate Tests[d]

Effect		Value	F	Hypothesis df	Error df	Sig.	Partial Eta Squared	Noncent. Parameter	Observed Power[a]
Intercept	Pillai's Trace	.985	3842.636[b]	3.000	172.000	.000	.985	11527.908	1.000
	Wilks' Lambda	.015	3842.636[b]	3.000	172.000	.000	.985	11527.908	1.000
	Hotelling's Trace	67.023	3842.636[b]	3.000	172.000	.000	.985	11527.908	1.000
	Roy's Largest Root	67.023	3842.636[b]	3.000	172.000	.000	.985	11527.908	1.000
View	Pillai's Trace	.108	3.280	6.000	346.000	.004	.054	19.681	.931
	Wilks' Lambda	.892	3.357[b]	6.000	344.000	.003	.055	20.145	.937
	Hotelling's Trace	.120	3.434	6.000	342.000	.003	.057	20.602	.943
	Roy's Largest Root	.120	6.914[c]	3.000	173.000	.000	.107	20.742	.977
Floors	Pillai's Trace	.052	3.115[b]	3.000	172.000	.028	.052	9.345	.718
	Wilks' Lambda	.948	3.115[b]	3.000	172.000	.028	.052	9.345	.718
	Hotelling's Trace	.054	3.115[b]	3.000	172.000	.028	.052	9.345	.718
	Roy's Largest Root	.054	3.115[b]	3.000	172.000	.028	.052	9.345	.718
View * Floors	Pillai's Trace	.098	2.973	6.000	346.000	.008	.049	17.835	.902
	Wilks' Lambda	.902	3.018[b]	6.000	344.000	.007	.050	18.110	.907
	Hotelling's Trace	.107	3.064	6.000	342.000	.006	.051	18.382	.912
	Roy's Largest Root	.102	5.876[c]	3.000	173.000	.001	.092	17.629	.951

a. Computed using alpha = .05

b. Exact statistic

c. The statistic is an upper bound on F that yields a lower bound on the significance level.

d. Design: Intercept+View+Floors+View * Floors

3

This is the MANOVA table for the Multivariate Tests

It lists the main effects and interactions for the IVs, just as in ANOVA, but these are the **multivariate effects**; that is, the IV effects on **linear combinations** of the DVs.

There are several multivariate statistics available, and they often lead to the same conclusion, as here. **Pillai's Trace** is considered to be the more robust (against violations of assumptions) and to have acceptable power.

Wilks' Lambda is most often cited in the literature and is an appropriate choice when assumptions are met, as it can be more powerful.

In this example all three multivariate effects are significant: the main effect of View, the main effect of Floors, and the View by Floors interaction.

Levene's Test of Equality of Error Variances[a]

	F	df1	df2	Sig.
Satisfaction with work	.471	5	174	.797
Wellbeing	.664	5	174	.651
Performance	2.206	5	174	.056

Tests the null hypothesis that the error variance of the dependent variable is equal across groups.

a. Design: Intercept+View+Floors+View * Floors

4

Univariate results appear next in the output. Here are the Levene's Test results for the assumption of homogeneity of variance for **each** of the DVs. The assumption **is met** when these are **not** significant (> .05).

There is homogeneity of variance for all the DVs in this example.

Tests of Between-Subjects Effects

Source	Dependent Variable	Type III Sum of Squares	df	Mean Square	F	Sig.	Partial Eta Squared	Noncent. Parameter	Observed Power[a]
Corrected Model	Satisfaction with work	23.111[b]	5	4.622	1.658	.147	.045	8.288	.566
	Wellbeing	7.312[b]	5	1.462	1.653	.148	.045	8.266	.565
	Performance	52.933[c]	5	10.587	2.323	.045	.063	11.617	.737
Intercept	Satisfaction with work	16665.689	1	16665.689	5976.566	.000	.972	5976.566	1.000
	Wellbeing	8799.013	1	8799.013	9946.586	.000	.983	9946.586	1.000
	Performance	22984.200	1	22984.200	5044.040	.000	.967	5044.040	1.000
View	Satisfaction with work	16.578	2	8.289	2.973	.054	.033	5.945	.572
	Wellbeing	3.233	2	1.617	1.828	.164	.021	3.655	.378
	Performance	10.033	2	5.017	1.101	.335	.012	2.202	.241
Floors	Satisfaction with work	.356	1	.356	.128	.721	.001	.128	.065
	Wellbeing	1.335	1	1.335	1.509	.221	.009	1.509	.231
	Performance	15.022	1	15.022	3.297	.071	.019	3.297	.439
View * Floors	Satisfaction with work	6.178	2	3.089	1.108	.333	.013	2.215	.243
	Wellbeing	2.744	2	1.372	1.551	.215	.018	3.102	.326
	Performance	27.878	2	13.939	3.059	.049	.034	6.118	.585
Error	Satisfaction with work	485.200	174	2.789					
	Wellbeing	153.925	174	.885					
	Performance	792.867	174	4.557					
Total	Satisfaction with work	17174.000	180						
	Wellbeing	8960.250	180						
	Performance	23830.000	180						
Corrected Total	Satisfaction with work	508.311	179						
	Wellbeing	161.237	179						
	Performance	845.800	179						

a. Computed using alpha = .05

b. R Squared = .045 (Adjusted R Squared = .018)

c. R Squared = .063 (Adjusted R Squared = .036)

5

Univariate (*F*) results for **individual** DVs are provided next in the table headed **Tests of Between-Subjects Effects.**

With MANOVA used to control familywise error, univariate results are only considered if the multivariate effect is significant.

All univariate effects can be considered in this example because all three multivariate effects were significant. However, because there are three DVs, each one of the univariate tests needs to be evaluated against a Bonferroni adjusted alpha of .017 (.05/3).

What we find here is a fairly rare example of MANOVA being more powerful than the univariate ANOVAs. The multivariate effects were all significant, but with an adjusted alpha of .017, none of the univariate effects are.

Estimated Marginal Means

1. Window view

Estimates

Dependent Variable	Window view	Mean	Std. Error	95% Confidence Interval	
				Lower Bound	Upper Bound
Satisfaction with work	City	9.900	.216	9.475	10.325
	Ocean	9.200	.216	8.775	9.625
	Park	9.767	.216	9.341	10.192
Wellbeing	City	6.858	.121	6.619	7.098
	Ocean	7.175	.121	6.935	7.415
	Park	6.942	.121	6.702	7.181
Performance	City	11.117	.276	10.573	11.661
	Ocean	11.633	.276	11.089	12.177
	Park	11.150	.276	10.606	11.694

Pairwise Comparisons

Dependent Variable	(I) Window view	(J) Window view	Mean Difference (I-J)	Std. Error	Sig.[a]	95% Confidence Interval for Difference[a]	
						Lower Bound	Upper Bound
Satisfaction with work	City	Ocean	.700	.305	.067	-.035	1.435
		Park	.133	.305	.962	-.602	.868
	Ocean	City	-.700	.305	.067	-1.435	.035
		Park	-.567	.305	.182	-1.302	.168
	Park	City	-.133	.305	.962	-.868	.602
		Ocean	.567	.305	.182	-.168	1.302
Wellbeing	City	Ocean	-.317	.172	.187	-.731	.097
		Park	-.083	.172	.949	-.497	.331
	Ocean	City	.317	.172	.187	-.097	.731
		Park	.233	.172	.440	-.181	.647
	Park	City	.083	.172	.949	-.331	.497
		Ocean	-.233	.172	.440	-.647	.181
Performance	City	Ocean	-.517	.390	.462	-1.456	.423
		Park	-.033	.390	1.000	-.973	.906
	Ocean	City	.517	.390	.462	-.423	1.456
		Park	.483	.390	.519	-.456	1.423
	Park	City	.033	.390	1.000	-.906	.973
		Ocean	-.483	.390	.519	-1.423	.456

Based on estimated marginal means

a. Adjustment for multiple comparisons: Sidak.

6

Estimated Marginal Means are provided next, followed by **Pairwise Comparisons**. The marginal means will be the same as the means in the Descriptive Statistics table at the beginning, except where cells sizes are unequal, in which case they will be the unweighted means (i.e., the "average" of contributing cells).

Pairwise comparisons are only considered if there is a significant univariate main effect (which there was not).

All the marginal means are provided in the SPSS output, together with tables that repeat the multivariate and univariate results for each effect. (Only marginal means and comparisons for Window view are included here, simply to provide an example.)

Profile Plots are also provided at the end of the output if they were requested. They can be useful as an aid to interpretation when univariate interactions are significant.

Step 5: Conducting follow-up analyses (univariate and/or stepdown analysis)

When there are significant MANOVA effects, the most common approach is to proceed to univariate analyses, followed, if necessary, by simple effects analysis or analytical comparisons, as demonstrated in previous ANOVA chapters.

When stepdown analysis is involved, the steps to be taken are listed below.

In either case an adjustment for familywise error should be made. With three DVs in this example, a Bonferroni adjusted alpha of .017 (.05/3) would need to be used to maintain the familywise alpha at .05. With a large number of DVs, you might choose a more liberal familywise alpha of .10 or .15 (rarely is it acceptable for familywise alpha to exceed these values.)

Step 5.1: Analyse the first DV by univariate ANOVA

In this example, the univariate ANOVA for the first DV (WorkSat) requires no further investigation, because none of the effects were significant.

Step 5.2: Analyse subsequent DVs, in order, using ANCOVA, with prior DVs treated as covariates.

In this example two ANCOVAs are needed; the first on wellbeing, with work satisfaction as the covariate; and the second on performance, with both work satisfaction and wellbeing as covariates.

MANOVA syntax can be used to provide stepdown analyses automatically, but we will use two ANCOVAs in this example.

40 How to: Conduct Roy-Bargmann stepdown (ANCOVA) analyses

Analyze
- **General Linear Model**
  - **Univariate...**
    - [Select variables, as follows:]
    - **Wellbeing** } ▸ **Dependent Variables:**
    - **Window view [View]** }
    - **Floors** } ▸ **Fixed Factor(s):**
    - **Satisfaction with work [WorkSat]** } ▸ **Covariates(s)**
    - **Options...**
      - [From **Estimated Marginal Means, Factor(s) and Factor Interactions** select:]
        - **View** }
        - **Floors** }
        - **View*Floors** } ▸ **Display means for:**
        - [Select] ☑ **Compare main effects**
          - [For **Confidence interval adjustment:** select]
          - **LSD**
      - [From **Display** select:]
        - ☑ **Descriptive statistics**
        - ☑ **Estimates of effect size**
        - ☑ **Observed power**
        - ☑ **Homogeneity tests**
      - **Continue**
      - **Plots...**
        - [From **Factors** select:]
        - **View** ▸ **Horizontal axis**
        - **Floors** ▸ **Separate Lines**
        - **Plots:** Click on **Add** button
        - **Continue**
    - **OK**

This whole procedure is then repeated for the second ANCOVA, but this time with **Perform** as the DV, and **both WorkSat and Wellbeing** as covariates.

Step 6: Examine and interpret the ANCOVA (stepdown) output

A full ANOVA output is produced by SPSS, but only the relevant parts are reproduced here.

Tests of Between-Subjects Effects

Dependent Variable: Wellbeing

Source	Type III Sum of Squares	df	Mean Square	F	Sig.	Partial Eta Squared	Noncent. Parameter	Observed Power[a]
Corrected Model	35.189[b]	6	5.865	8.049	.000	.218	48.297	1.000
Intercept	111.783	1	111.783	153.421	.000	.470	153.421	1.000
WorkSat	27.877	1	27.877	38.260	.000	.181	38.260	1.000
View	7.432	2	3.716	5.100	.007	.056	10.200	.816
Floors	1.024	1	1.024	1.406	.237	.008	1.406	.218
View * Floors	4.614	2	2.307	3.167	.045	.035	6.333	.601
Error	126.048	173	.729					
Total	8960.250	180						
Corrected Total	161.237	179						

a. Computed using alpha = .05

b. R Squared = .218 (Adjusted R Squared = .191)

The first stepdown analysis: Wellbeing as the DV with WorkSat as the covariate.

WorkSat, as the covariate, is significant (Sig. < .001), indicating it makes a significant adjustment to DV scores.

Using an adjusted alpha of .017, the only significant (unique) effect on Wellbeing, after adjusting for WorkSat as the covariate is the main effect of **View**.

Because View has more than two levels, we know we need to look at the marginal means for view, and the main effect comparisons to find out which views differ from which.

(Note that while the interaction is just significant at the normal alpha of .05, it is not significant at the adjusted alpha of .017.)

Estimated Marginal Means

1. Window view

Estimates

Dependent Variable: Wellbeing

Window view	Mean	Std. Error	95% Confidence Interval	
			Lower Bound	Upper Bound
City	6.792[a]	.111	6.573	7.010
Ocean	7.276[a]	.111	7.056	7.496
Park	6.907[a]	.110	6.689	7.125

a. Covariates appearing in the model are evaluated at the following values: Satisfaction with work = 9.62.

Pairwise Comparisons

Dependent Variable: Wellbeing

(I) Window view	(J) Window view	Mean Difference (I-J)	Std. Error	Sig.[a]	95% Confidence Interval for Difference[a]	
					Lower Bound	Upper Bound
City	Ocean	-.484*	.158	.003	-.797	-.172
	Park	-.115	.156	.461	-.423	.192
Ocean	City	.484*	.158	.003	.172	.797
	Park	.369*	.157	.020	.059	.680
Park	City	.115	.156	.461	-.192	.423
	Ocean	-.369*	.157	.020	-.680	-.059

Based on estimated marginal means

*. The mean difference is significant at the .05 level.

a. Adjustment for multiple comparisons: Least Significant Difference (equivalent to no adjustments).

2

From the marginal means (after adjusting for the covariate) there is a trend for higher wellbeing to be associated with the ocean view compared to either the park or city views.

However, to continue to maintain familywise error at .05, we need to apply an adjusted alpha for these comparisons of .017 (that applied to the "parent" ANCOVA test) divided by 3 further comparisons, that is, .017/3, which equals .006.

With an adjusted alpha of .006 the only significant difference is that favouring the ocean view over the city view.

They are not reproduced here, but you should also look at the other marginal means (for Floors and the interaction), even though these effects were not significant. Also look at the Profile Plot to help you understand what is happening.

Tests of Between-Subjects Effects

Dependent Variable: Performance

Source	Type III Sum of Squares	df	Mean Square	F	Sig.	Partial Eta Squared	Noncent. Parameter	Observed Power[a]
Corrected Model	302.756[b]	7	43.251	13.699	.000	.358	95.893	1.000
Intercept	10.805	1	10.805	3.422	.066	.020	3.422	.452
WorkSat	79.149	1	79.149	25.069	.000	.127	25.069	.999
Wellbeing	64.580	1	64.580	20.454	.000	.106	20.454	.994
View	12.598	2	6.299	1.995	.139	.023	3.990	.408
Floors	24.473	1	24.473	7.752	.006	.043	7.752	.791
View * Floors	30.135	2	15.068	4.772	.010	.053	9.545	.789
Error	543.044	172	3.157					
Total	23830.000	180						
Corrected Total	845.800	179						

a. Computed using alpha = .05

b. R Squared = .358 (Adjusted R Squared = .332)

3

The second stepdown analysis: Performance as the DV with WorkSat and Wellbeing as covariates.

WorkSat and Wellbeing, as covariates are both significant (Sig. < .001), indicating they both make a significant adjustment to DV scores.

Using an adjusted alpha of .017, there is a significant (unique) effect on Performance (after adjusting for the covariates) for the main effect of **Floors** (Sig. =.006), and also for the **View by Floors** interaction (Sig. = .01). There is no significant main effect of View.

Because Floors has only two levels, we only need to look at the marginal means to see how they differ.

Estimated Marginal Means

1. Window view

Estimates

Dependent Variable: Performance

Window view	Mean	Std. Error	95% Confidence Interval	
			Lower Bound	Upper Bound
City	11.088[a]	.233	10.629	11.547
Ocean	11.691[a]	.236	11.224	12.157
Park	11.121[a]	.230	10.667	11.575

a. Covariates appearing in the model are evaluated at the following values: Satisfaction with work = 9.62, Wellbeing = 6.992.

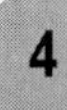

There is still a trend favouring ocean, park, then city in these means, but the main effect for view was not significant.

2. Floors

Estimates

Dependent Variable: Performance

Floors	Mean	Std. Error	95% Confidence Interval	
			Lower Bound	Upper Bound
Upper	11.670[a]	.188	11.300	12.041
Lower	10.930[a]	.188	10.559	11.300

a. Covariates appearing in the model are evaluated at the following values: Satisfaction with work = 9.62, Wellbeing = 6.992.

5

With a significant main effect of Floors, we see here that performance was better on the upper floors than on the lower floors.

3. Window view * Floors

Dependent Variable: Performance

Window view	Floors	Mean	Std. Error	95% Confidence Interval	
				Lower Bound	Upper Bound
City	Upper	11.208[a]	.327	10.561	11.854
	Lower	10.968[a]	.326	10.325	11.612
Ocean	Upper	12.649[a]	.325	12.007	13.291
	Lower	10.732[a]	.340	10.061	11.403
Park	Upper	11.154[a]	.325	10.513	11.795
	Lower	11.089[a]	.328	10.442	11.735

a. Covariates appearing in the model are evaluated at the following values: Satisfaction with work = 9.62, Wellbeing = 6.992.

6

These are the marginal means for the interaction. It is fairly apparent what is driving the interaction, especially when we examine the profile plot on the next page; there is a much higher level of performance for the ocean upper floors.

Interactions are not always this apparent, so let us proceed to interpret it, using simple effect analyses followed by simple comparisons.

Be careful with the profile plot on the next page. It appears much bigger than it actually is, because the vertical axis (*y*-axis) does not begin at zero.

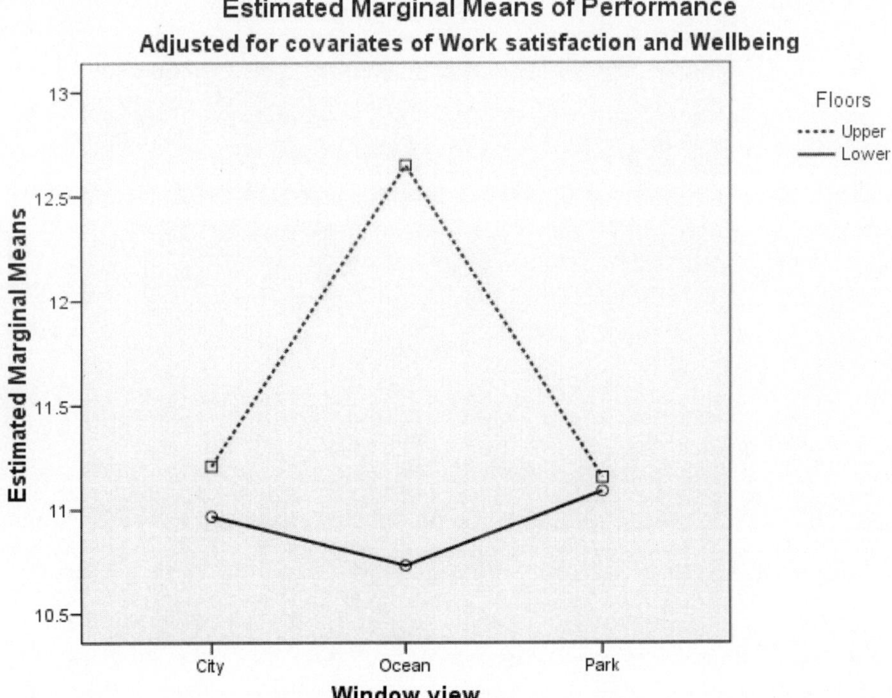

Step 7: Performing simple effects analysis on marginal means adjusted for covariates, following a significant stepdown (ANCOVA) interaction

An interaction tells us that differences across views for the upper floors will be different for the lower floors, and this demands two simple effect analyses (on View for upper and lower floors separately). We might also be interested in the three other possible simple effects: between floors for each level of view.

To continue to maintain familywise error at .05, we would need to apply an adjusted alpha for these simple effects of .017 (that applied to the parent ANCOVA test) divided by 5 further tests, that is, .017/5, which equals .003. However, this is getting **very small** such that subsequent tests have very little power. To maintain a reasonable level of power for subsequent tests, we could at this stage set a familywise alpha of .10, and so use .006 as the adjusted alpha level for the simple effects.

To conduct simple effect analyses and simple comparisons on **adjusted** means, there is no option but to use syntax. The time has come to begin to master it.

41 How to: Using SPSS command syntax to obtain ANCOVA simple effects analyses

```
MANOVA
   Perform BY View(1 3) Floors(1 2) WITH WorkSat Wellbeing
  /ERROR WITHIN
  /DESIGN = View WITHIN Floors(1)
  /DESIGN = View WITHIN Floors(2)
  /DESIGN = Floors WITHIN View(1)
  /DESIGN = Floors WITHIN View(2)
  /DESIGN = Floors WITHIN View(3).
```

Save these additional syntax commands in *MANOVA.sps*

To run this analysis, position the cursor at the beginning, or somewhere within the commands (or highlight the commands with the mouse if you only want these ones executed):

Then click on the run button ▶ located in the icon bar.
Alternatively, select from the **Run** option in the menu bar at the top

Output: **ANCOVA simple effects**

```
* ** * * A n a l y s i s   o f   V a r i a n c e -- design  1 *  * * * *

 Tests of Significance for Perform using UNIQUE sums of squares
 Source of Variation           SS      DF        MS         F  Sig of F

 WITHIN CELLS              543.04     172      3.16
 REGRESSION                249.82       2    124.91     39.56      .000
 VIEW WITHIN FLOORS(1)      42.88       2     21.44      6.79      .001

* * * * * A n a l y s i s   o f   V a r i a n c e -- design  2 * * * * *
 Tests of Significance for Perform using UNIQUE sums of squares

 Source of Variation           SS      DF        MS         F  Sig of F

 WITHIN CELLS              543.04     172      3.16
 REGRESSION                249.82       2    124.91     39.56      .000
 VIEW WITHIN FLOORS(2)       1.78       2       .89       .28      .755
```

1

The first analysis VIEW WITHIN FLOORS (1) refers to view levels across the **upper floors** (because we coded the upper floors as 1 in the dataset). This **is** significant (Sig. of F = .001) even with an adjusted α = .006, and tells us that there is a difference **somewhere** across the adjusted means for the upper floor views.

The second analysis VIEW WITHIN FLOORS(2) is **not** significant (Sig. of F = .755), indicating there is no significant difference anywhere across the different views when we are only looking at the **lower floors**.

(Note how the WITHIN CELLS error term is the same as the error term for the ANCOVA on Perform.)

```
* * * * * A n a l y s i s   o f   V a r i a n c e -- design   3 * * * * *

 Tests of Significance for Perform using UNIQUE sums of squares
 Source of Variation              SS       DF        MS         F  Sig of F

 WITHIN CELLS                 543.04      172      3.16
 REGRESSION                   249.82        2    124.91     39.56      .000
 FLOORS WITHIN VIEW(1)           .86        1       .86       .27      .603


* * * * * A n a l y s i s   o f   V a r i a n c e -- design   4 * * * * *

 Tests of Significance for Perform using UNIQUE sums of squares
 Source of Variation              SS       DF        MS         F  Sig of F

 WITHIN CELLS                 543.04      172      3.16
 REGRESSION                   249.82        2    124.91     39.56      .000
 FLOORS WITHIN VIEW(2)         52.91        1     52.91     16.76      .000


* * * * * A n a l y s i s   o f   V a r i a n c e -- design   5 * * * * *

 Tests of Significance for Perform using UNIQUE sums of squares
 Source of Variation              SS       DF        MS         F  Sig of F

 WITHIN CELLS                 543.04      172      3.16
 REGRESSION                   249.82        2    124.91     39.56      .000
 FLOORS WITHIN VIEW(3)           .06        1       .06       .02      .888
```

2

These are the simple effects for floors.

The first one, FLOORS WITHIN VIEW (1), refers to the **city view** (which was coded 1) and tells us there is no significant difference between upper and lower floors for the city view (Sig. of F = .603).

The second one, FLOORS WITHIN VIEW (2), refers to the **ocean view** (which was coded 2) and tells us there **is** a significant difference between upper and lower floors for the ocean view (Sig. of F = < .001, which is < adjusted α of .006). Referring back to the marginal means for the ocean view we see performance was higher on the upper floors.

The third one, FLOORS WITHIN VIEW (3), refers to the **park view** (which was coded 3) and tells us there is no significant difference between upper and lower floor for the park view (Sig. of F = .888).

Step 8: Performing simple comparisons following a significant stepdown (ANCOVA) simple effects analysis involving more than two levels

In this example, we now need simple comparisons to ascertain which view differed from which for the upper floors which had the significant simple effect for View. To continue to maintain familywise error at .10, we would need to apply a further adjusted alpha for these comparisons of .006 (that applied to the simple effect analysis) divided by 3 further tests, that is, .006/3, which equals .002.

To do all this, let us master comparison syntax in GLM. The numbers to the left below are **not** part of the syntax, but are used to enable you to cross-reference to the lines from the key that follows.

42 How to: Use SPSS command syntax to obtain ANCOVA comparisons

```
1   GLM Perform  BY Floors View  WITH WorkSat Wellbeing
2       /METHOD=SSTYPE(3)
3       /LMATRIX "City vs Park for Upper Floor"
4           View*Floors 1 0 -1 0 0 0 View 1 0 -1
5       /LMATRIX "Ocean vs Park for Upper Floor"
6           View*Floors 0 1 -1 0 0 0 View 0 1 -1
7       /LMATRIX "Ocean vs City for Upper Floor"
8           View*Floors -1 1 0 0 0 0 View -1 1 0
9       /LMATRIX "City-Upper vs City-Lower"
10              View*Floors 1 0 0 -1 0 0  Floors 1 -1
11      /LMATRIX "Ocean-Upper vs Ocean-Lower"
12              View*Floors 0 1 0 0 -1  0  Floors 1 -1
13      /LMATRIX "Park-Upper vs Park-Lower"
14              View*Floors 0 0 1 0 0 -1 Floors 1 -1.
```

Save these additional syntax commands in *MANOVA.sps*

To run this analysis, position the cursor at the beginning, or somewhere within the commands (or highlight the commands with the mouse if you only want these ones executed):

Then click on the run button ▶ located in the icon bar.

Alternatively, select from the **Run** option in the menu bar at the top

Key to control lines:

Line 1:

This simply specifies the variables, and has the form DV (Perform) **BY** IVs (Floors View) **WITH** covariate(s) (WorkSat Wellbeing). Note that it is easier to keep track of the coefficients if the IV with the **smallest number of levels** (i.e., Floors) goes **first** in the IV list. This will mean the first three coefficients code for the lower floor, and the second three code for the upper floor.

Lines 3, 5, 7, 9, 11, and 13:

These /LMATRIX lines are followed by a description of what is being compared enclosed within quotation marks. The last three are not the comparisons we want at this stage, but are another way to obtain the simple effect comparisons for floors, given it has only two levels.

Lines 4, 6, 8, 10, 12, and 14:

These are the **critical** lines; the ones you **must get right**; the ones that specify the coefficients to run the comparisons we want.

Line 4:

This is a **View** comparison between city and park.

In GLM with specification of a **View*Floors set of coefficients**, the first three coefficients refer to the three **views** for the **upper floors** (that were coded 1) and the second three refer to views for the **lower floors** (which were coded 2).

The views were coded 1=city, 2=ocean, 3=park, so to contrast the **city with the park for the upper floor only** we need coefficients as follows:

1	0	**-1**	0	0	0
Upper City	Upper Ocean	**Upper Park**	Lower City	Lower Ocean	Lower Park

Because it is a **View** comparison, GLM requires the View only coefficients to follow. View only coefficients for city versus park are: 1 0 -1.

Lines 6 and 8 compare upper floor View levels as follows:

Ocean vs Park	0	**1**	**-1**	0	0	0
Ocean vs City	**-1**	**1**	0	0	0	0
	Upper City	Upper Ocean	Upper Park	Lower City	Lower Ocean	Lower Park

Each line is followed by **View** only coefficients of: 0 1 -1 and -1 1 0 respectively.

Lines 10, 12, and 14

These follow the same logic, but they are not contrasting views within the upper floors, but rather, they are the simple effect contrasts between floors for each level of View.

Upper vs lower for **city view**	**1**	0	0	**-1**	0	0
Upper vs lower for **ocean view**	0	**1**	0	0	**-1**	0
Upper vs lower for **park view**	0	0	**1**	0	0	**-1**
	Upper City	Upper Ocean	Upper Park	Lower City	Lower Ocean	Lower Park

Because these are **Floors** only contrasts they are followed by the floors only coefficients: 1 -1

If you cannot cope with contrast coefficients there is a very simple way to obtain all the comparisons between pairs of means, and that is to conduct the ANCOVA on performance, with work satisfaction and wellbeing as covariates, just as before, but to use **Cell** as the IV. Then, when you select ☑ **Compare main effects, using LSD confidence interval adjustment** (i.e., no adjustment), everything will be compared with everything. Then it is just a matter of ignoring the irrelevant comparisons, and only attending to those of interest, applying a Bonferroni adjusted alpha to determine significance.

A reason this is **not** ever a good idea is that it can tempt you into post hoc decision-making, that is, “deciding” a comparison is of theoretical interest **after** you see it is significant.

Custom Hypothesis Tests #1

Contrast Results (K Matrix)[a]

Contrast			Dependent Variable Performance
L1	Contrast Estimate		.054
	Hypothesized Value		0
	Difference (Estimate - Hypothesized)		.054
	Std. Error		.462
	Sig.		.907
	95% Confidence Interval for Difference	Lower Bound	-.858
		Upper Bound	.966

a. Based on the user-specified contrast coefficients (L') matrix: City vs Park for Upper Floor

Test Results

Dependent Variable: Performance

Source	Sum of Squares	df	Mean Square	F	Sig.
Contrast	.043	1	.043	.014	.907
Error	543.044	172	3.157		

1

This contrast is not significant (Sig. = .907).

There is no significant difference between the **city and park upper floor** performance means after controlling for the covariates of work satisfaction and wellbeing.

If you are not sure it is the correct contrast, check by subtracting the relevant marginal means from the earlier output: City (11.208) – park (11.154) = 0.054. This agrees with the **Contrast Estimate**.

Custom Hypothesis Tests #2

Contrast Results (K Matrix)[a]

Contrast			Dependent Variable Performance
L1	Contrast Estimate		1.495
	Hypothesized Value		0
	Difference (Estimate - Hypothesized)		1.495
	Std. Error		.459
	Sig.		.001
	95% Confidence Interval for Difference	Lower Bound	.589
		Upper Bound	2.401

a. Based on the user-specified contrast coefficients (L') matrix: Ocean vs Park for Upper Floor

Test Results

Dependent Variable: Performance

Source	Sum of Squares	df	Mean Square	F	Sig.
Contrast	33.519	1	33.519	10.617	.001
Error	543.044	172	3.157		

2

This contrast **is** significant (Sig. = .001), using our adjusted alpha of .002.

There **is** a significant difference between the **ocean and park upper floor** performance means after controlling for the covariates of work satisfaction and wellbeing.

Custom Hypothesis Tests #3

Contrast Results (K Matrix)[a]

Contrast			Dependent Variable Performance
L1	Contrast Estimate		1.441
	Hypothesized Value		0
	Difference (Estimate - Hypothesized)		1.441
	Std. Error		.463
	Sig.		.002
	95% Confidence Interval for Difference	Lower Bound	.528
		Upper Bound	2.355

a. Based on the user-specified contrast coefficients (L') matrix: Ocean vs City for Upper Floor

Test Results

Dependent Variable: Performance

Source	Sum of Squares	df	Mean Square	F	Sig.
Contrast	30.639	1	30.639	9.704	.002
Error	543.044	172	3.157		

3

This contrast **is also** significant (Sig. = .002) using our adjusted alpha of .002.

There **is** a significant difference between the **ocean and city upper floor** performance means after controlling for the covariates of work satisfaction and wellbeing.

Custom Hypothesis Tests #4

Contrast Results (K Matrix)[a]

Contrast			Dependent Variable Performance
L1	Contrast Estimate		.239
	Hypothesized Value		0
	Difference (Estimate - Hypothesized)		.239
	Std. Error		.459
	Sig.		.603
	95% Confidence Interval for Difference	Lower Bound	-.666
		Upper Bound	1.145

a. Based on the user-specified contrast coefficients (L') matrix: City-Upper vs City-Lower

Test Results

Dependent Variable: Performance

Source	Sum of Squares	df	Mean Square	F	Sig.
Contrast	.859	1	.859	.272	.603
Error	543.044	172	3.157		

Custom Hypothesis Tests 4, 5, and 6 are the simple effects for Floors at each level of View.

Check back to see how they agree with those obtained with MANOVA syntax previously at Step 7.

This is city upper versus city lower.

Custom Hypothesis Tests #5

Contrast Results (K Matrix)[a]

Contrast			Dependent Variable Performance
L1	Contrast Estimate		1.917
	Hypothesized Value		0
	Difference (Estimate - Hypothesized)		1.917
	Std. Error		.468
	Sig.		.000
	95% Confidence Interval for Difference	Lower Bound	.993
		Upper Bound	2.842

a. Based on the user-specified contrast coefficients (L') matrix: Ocean-Upper vs Ocean-Lower

Test Results

Dependent Variable: Performance

Source	Sum of Squares	df	Mean Square	F	Sig.
Contrast	52.911	1	52.911	16.759	.000
Error	543.044	172	3.157		

Custom Hypothesis Tests 4, 5, and 6 are the simple effects for Floors at each level of View.

Check back to see how they agree with those obtained with MANOVA syntax previously at Step 7.

This is ocean upper versus ocean lower.

Custom Hypothesis Tests #6

Contrast Results (K Matrix)[a]

Contrast			Dependent Variable Performance
L1	Contrast Estimate		.065
	Hypothesized Value		0
	Difference (Estimate - Hypothesized)		.065
	Std. Error		.462
	Sig.		.888
	95% Confidence Interval for Difference	Lower Bound	-.847
		Upper Bound	.978

a. Based on the user-specified contrast coefficients (L') matrix: Park-Upper vs Park-Lower

Test Results

Dependent Variable: Performance

Source	Sum of Squares	df	Mean Square	F	Sig.
Contrast	.063	1	.063	.020	.888
Error	543.044	172	3.157		

Custom Hypothesis Tests 4, 5, and 6 are the simple effects for Floors at each level of View.

Check back to see how they agree with those obtained with MANOVA syntax previously at Step 7.

This is park upper versus park lower.

Summarising approaches to MANOVA

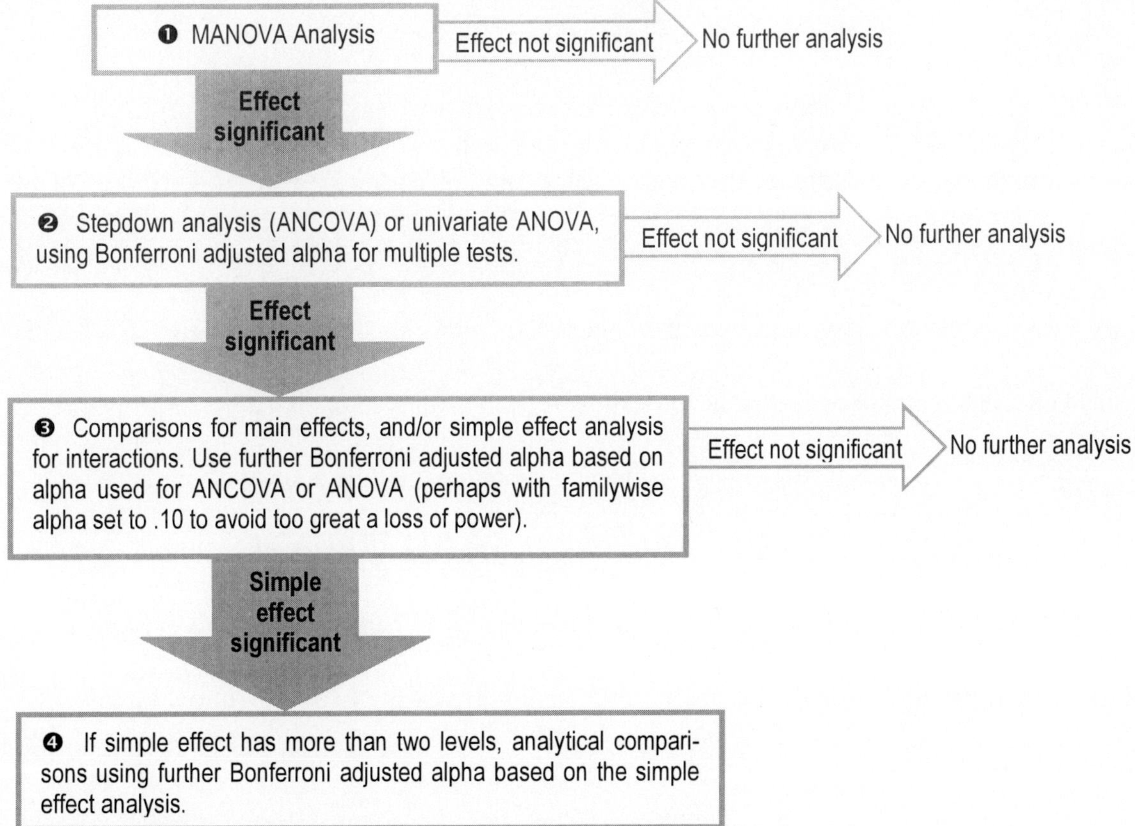

An alternative approach altogether from stage 2 is to use **discriminant analysis**; or another alternative is to proceed using **multivariate comparisons**. At stages 3 and 4 in some research contexts **trend analysis** might be more appropriate than simple comparisons among means. Refer to Stevens (2002) or Tabachnick and Fidell (2007) for all the rich detail about the options available.

MANOVA, repeated measures, and profile analysis

There are circumstances in which a repeated measures analysis is more appropriate than MANOVA, or indeed, as we saw in the repeated–measures chapter, when MANOVA is a preferred alternative to repeated measures ANOVA. When DVs are measured on the same scale, as they are in repeated measures, a special form of MANOVA known as **profile analysis** can be used. Refer to Tabachnick and Fidell (2007, Ch. 8).

For those addicted to complexity, it is also possible to have **doubly-multivariate designs** where different DVs are measured at different times, but do not embark on one of these unless you truly are prepared to deal with statistical complexity—big time!

MANOVA analysis: Research report sample Results section, and opening Discussion

Results

The three dependent variables (DVs) associated with the working environment were analysed using a 3 x 2 multivariate analysis of variance, with independent variables of view (city, ocean, and park) and floors (upper and lower). Roy-Bargmann stepdown analysis was used to assess unique differences as a function of the IVs for each DV, with prior DVs treated as covariates. The ordering of the variables was based on a causal logic whereby work satisfaction was expected to be associated with overall wellbeing, which in turn was expected to be associated with work performance. Consistent with this expectation the DVs were moderately intercorrelated (see Table 1), with no evidence of multicollinearity, and were sufficiently reliable to warrant the use of stepdown analysis. With an *N* of 180, there were no univariate or multivariate outliers, and assumptions of normality, homogeneity of variance-covariance matrices, linearity, and homogeneity of regression for the stepdown analyses were satisfactory.

Table 1

Pooled Within-cell Correlations among the Dependent Variables (with Standard Deviations on the Diagonal)

	Work satisfaction	Wellbeing	Performance
Work satisfaction	1.67		
Wellbeing	.43	0.94	
Performance	.48	.46	2.14

Using Wilks' criterion[1] there were significant multivariate main effects for both view, Λ (6,346) = 3.36, p = .003, η_p^2 = .06; and floors, Λ (3,172) = 3.12, p = .03, η_p^2 = .05. There was also a significant multivariate view by floors interaction, Λ (6,346) = 3.02, p = .007, η_p^2 = .05. Using an adjusted alpha of .017 to maintain familywise error at .05 across the three tests performed, there were no significant main effects or interactions on any of the univariate ANOVAs.

Descriptive statistics for the first DV, work satisfaction are shown in Table 2.

Table 2

Descriptive Statistics for Work Satisfaction

Group	*M* (*SD*)	95% CI
City view	9.90 (1.75)	[9.48, 10.33]
Upper floors	9.93 (1.95)	[9.33, 10.54]
Lower floors	9.87 (1.57)	[9.27. 10.47]
Ocean view	9.20 (1.58)	[8.78, 9.63]
Upper floors	9.33 (1.45)	[8.73, 9.94]
Lower floors	9.07 (1.72)	[8.47, 9.67]
Park view	9.77 (1.66)	[9.34, 10.19]
Upper floors	9.47 (1.63)	[8.87, 10.07]
Lower floors	10.07 (1.66)	[9.47, 10.67]

Note: CI = Confidence interval. The unweighted mean and standard error is shown for main effects.[2]

[1] Λis the symbol for Wilks's lambda multivariate criterion.
[2] The note about unweighted means is **only** necessary when cell sizes are unequal, so it would not actually be used in this table where we have equal cell sizes. The confidence intervals (and unweighted means and standard errors) are obtained from the relevant outputs of *Estimated Marginal Means.*

As the first variable in the stepdown analysis, it was tested in a univariate ANOVA. Neither the main effects nor the interaction were significant: views, $F(2,174) = 2.97$, $p = .054$, $\eta_p^2 = .03$; floors, $F(1,174) = 0.13$, $p = .72$, $\eta_p^2 = .001$; view by floors interaction, $F(2,174) = 1.11$, $p = .33$, $\eta_p^2 = .013$.

Results for the next two DVs using a Bonferroni adjusted alpha of .017 are shown in Table 3, with descriptive statistics in Table 4. For the wellbeing stepdown analysis there was a significant main effect of view after adjusting for the covariate, work satisfaction (see Table 3). Three pairwise comparisons among the view means using a Bonferroni adjusted alpha level of .006 (.017/3) showed that wellbeing was higher with the ocean view compared to the city view ($M_{Diff} = 0.49$, $p = .003$, 95% CI [0.17, 0.80]), but not compared to the park view ($M_{Diff} = 0.37$, $p = .02$, 95% CI [0.06, 0.68]). There was no significant difference between the city and park views ($M_{Diff} = 0.12$, $p = .46$, 95% CI [-0.42, 0.19]).

Table 3

Univariate and Stepdown Tests for Wellbeing and Work Performance

DV	Univariate *F*	*df*	*p*	Stepdown *F*	*df*	*p*	η_p^2
View main effect							
Wellbeing	1.83	2,174	.16	5.10	2,173	.007*	.056
Performance	1.10	2,174	.34	2.00	2,172	.14	.023
Floors main effect							
Wellbeing	1.51	1,174	.22	1.41	1,173	.237	.008
Performance	3.30	1,174	.07	7.75	1,172	.006*	.043
Interaction							
Wellbeing	1.55	2,174	.22	3.17	2,173	.045	.035
Performance	3.06	2,174	.05	4.77	2,172	.010*	.053

* $p < .017$.

As can be seen in Table 3, a different pattern of stepdown results occurred for performance after controlling for work satisfaction and wellbeing as covariates. There was a significant main effect of floors, and a significant view by floors interaction, but no significant main effect of view. Performance was better on the upper floors (adjusted M = 11.67, SE = 0.19, 95% CI [11.30, 12.04]) than the lower floors (adjusted M = 10.93, SE = 0.19, 95% CI [10.56, 11.30]).

Table 4

Means for Wellbeing, Adjusted for Work Satisfaction as the Covariate, and Unadjusted

	Adjusted Means		Unadjusted Means	
Group	*M* (*SE*)	95% CI	*M*	*SD*
City view	6.79 (0.11)	[6.57, 7.01]	6.86	0.92
Upper floors	6.76 (0.16)	[6.45, 7.07]	6.83	0.77
Lower floors	6.83 (0.16)	[6.52, 7.13]	6.88	1.06
Ocean view	7.28 (0.11)	[7.06, 7.50]	7.18	0.96
Upper floors	6.99 (0.16)	[6.68, 7.29]	6.92	0.82
Lower floors	7.57 (0.16)	[7.26, 7.88]	7.43	1.02
Park view	6.91 (0.11)	[6.69, 7.13]	6.94	0.96
Upper floors	7.00 (0.16)	[6.70, 7.31]	6.97	1.05
Lower floors	6.81 (0.16)	[6.50, 7.12]	6.92	0.88

Note: CI = Confidence interval.

The significant interaction on performance was investigated using two simple effect analyses across view conditions separately for the upper and lower floors. Three further simple effect analyses investigated differences between floors separately for each of the three views. To minimise the loss of power in view of the overall number of tests being performed all of these were evaluated at an adjusted alpha of .006, which maintained familywise error for these and subsequent tests at .10. The simple effect of view was not significant for the lower floors, $F(2,172) = 0.28, p = .76$; but it was for the upper floors, $F(2,172) = 6.79, p = .001$.

Simple comparisons among the view means for the upper floors, using a Bonferroni adjusted alpha of .002 (.006/3), showed no significant difference in performance between the city and park views, $F(1,172) = 0.01, p = .91$, η_p^2 effectively zero[1]; but the ocean view had significantly higher performance than the both the city view, $F(1,172) = 9.70, p = .002$, $\eta_p^2 = .05$; and the park view, $F(1,172) = 10.62, p = .001$, $\eta_p^2 = .06$ (see Table 5 for the relevant descriptive statistics).

The simple effect of floors on performance was not significant for the city view, $F(1,172) = 0.27, p = .60$, $\eta_p^2 = .002$; nor for the park view, $F(1,172) = 0.02, p = .89$, $\eta_p^2 < .001$; but it was for the ocean view, $F(1,172) = 16.76, p < .001$, $\eta_p^2 = .09$. As can be seen in Table 5, performance was higher for the ocean view on the upper floors than it was on the lower floors.

Discussion

Contrary to expectations there were no significant univariate effects for any of the dependent variables. However, both main effects and the interaction were significant when the moderately and positively intercorrelated dependent variables were tested in combination using MANOVA. The stepdown analysis then revealed a moderate association of view with wellbeing. After controlling for work satisfaction (where means were slightly lower for those with an ocean view compared to a city view) wellbeing was significantly *higher* for participants with an ocean view than it was for those with a city view, in partial support of the hypothesis for this dependent variable.

---

[1] The effect size for each comparison is calculated using the partial η^2 effect size formula of $SS_{contrast}/(SS_{contrast}+SS_{error})$.

Table 5

Means for Work Performance, Adjusted for Work Satisfaction and Wellbeing as Covariates, and Unadjusted

	Adjusted Means		Unadjusted Means	
Group	*M* (*SE*)	95% CI	*M*	*SD*
City view	11.09 (0.23)	[10.63, 11.55]	11.12	2.15
Upper floors	11.21 (0.33)	[10.56, 11.85]	11.23	2.42
Lower floors	10.97 (0.33)	[10.33, 11.61]	11.00	1.88
Ocean view	11.69 (0.24)	[11.22, 12.16]	11.63	2.24
Upper floors	12.65 (0.33)	[12.01, 13.29]	12.47	2.21
Lower floors	10.73 (0.34)	[10.06, 11.40]	10.80	1.97
Park view	11.12 (0.23)	[10.67, 11.58]	11.15	2.13
Upper floors	11.15 (0.33)	[10.51, 11.80]	11.07	2.49
Lower floors	11.09 (0.33)	[10.44, 11.74]	11.23	1.74

Note: CI = Confidence interval.

For work performance, after controlling for both work satisfaction and wellbeing there was a significant main effect favouring the upper floors over the lower floors, however, there was also a significant view by floors interaction, driven primarily by a moderately strong effect of floors for employees with an ocean view. Wellbeing tended to be slightly lower on the upper floors than the lower floors for those with an ocean view (with very little difference on work satisfaction), leading to a similar expectation for work performance. However, after controlling for the two prior variables, the mean work performance for the ocean view upper floors was *higher* than anticipated, in the high performance range, and significantly higher than for the lower floors. There were no significant differences between floors for either the park or city views, and no significant differences among the views for participants working on lower floors. For participants working on upper floors, the mean for those with an ocean view was significantly higher than for those with either a park or city view.

Concluding comments

The opening paragraphs for the discussion section of a research report have been provided on the previous page to give you an idea of how to integrate the complex results from a MANOVA analysis to tell a coherent "story"; and also to provide a general idea of how any opening Discussion needs to present the results "in English" for the all-too-typical reader who may not have a good grasp of statistics.

You should re-read the previous chapter on ANCOVA, then come back to this analysis to reflect on how the covariates have operated to increase the power of the MANOVA.

17 DISCRIMINANT FUNCTION ANALYSIS

Before tackling this chapter you need to be thoroughly familiar with the chapters on MANOVA, multiple regression, and factor analysis. Discriminant analysis (or discriminant function analysis) is MANOVA in reverse. Instead of having groups (representing the IV) and asking if the groups differ on a set of DVs, the DVs become IVs or **predictors** in **discriminant analysis** and one asks if they are able to predict group membership (which becomes the DV, not IV).

Discriminant analysis can be used for MANOVA to describe how the groups differ on combinations of DVs. It is limited to one-way designs, but factorial designs can always be converted to one-way designs by using cells as groups. This is only necessary when a MANOVA analysis reveals a significant interaction; otherwise main effect groups are sufficient.

More commonly, discriminant analysis is used when the research question is about predicting group membership or classifying cases into groups on the basis of a set of predictor variables. In this sense, it is similar to multiple regression except that the criterion is a **categorical**, not continuous, variable. In fact, a form of regression, **logistic regression**, can often be used instead of discriminant analysis. Each logistic regression equation requires a dichotomous criterion, but more than two outcome categories can be accommodated by taking a dummy variable approach using separate regression equations. Logistic regression is often preferable to discriminant analysis because it has less restrictive assumptions and can be used with predictors that are discrete or continuous.

Assumptions

Discriminant analysis has basically the same assumptions as MANOVA; has the same sample size requirements, and can accommodate unequal groups.

A hypothetical example of discriminant analysis

Although drawn from a real research area (that of affect intensity) the data in this example are fairly unrealistic, for demonstration purposes, in that they provide extremely accurate classification.

Imagine a clinical psychologist is interested in the individual difference variable of affect intensity, which differentiates between people whose experience of emotions is intense versus more subdued. Suppose a sample of 65 participants has been classified on the basis of clinical assessment into three groups who are high on affect intensity, neuroticism, and extraversion respectively. Seven measures are then used to assess the extent to which they are able to predict group membership. The measures are:

Sensation seeking: Scores on a sensation seeking scale that has a possible range of 0 to 10 with high scores indicating higher sensation seeking.

Emotional responsiveness: Participants' self-reported emotional intensity experienced in response to viewing a set of slides designed to elicit a range of different emotions. Responses are averaged across slides, and the resulting measure has a possible range of 0 to 5, with high scores indicating greater emotional responsiveness.

Sociability: Scores on a sociability scale that has a possible range of 0 to 50 with high scores indicating greater sociability.

Somatic complaints: Scores on a measure of how much participants suffer from a variety of minor somatic complaints (e.g., headaches, muscles pains, gastric upsets etc.). It has a possible range of 0 to 10 with high scores indicating greater somatic problems.

Arousability: Scores on an arousability measure. It is participants' self-reported level of felt arousal averaged across five randomly determined time periods per day for a period of 10 days. It has a possible range of 0 to 10 with high scores indicating higher arousability.

Aggressiveness: Scores on an aggressiveness scale that has a possible range of 0 to 50 with high scores indicating higher aggressiveness.

Wellbeing: Scores on a general wellbeing scale that has a possible range of 0 to 5 with high scores indicating greater wellbeing.

Using SPSS to conduct a discriminant function analysis

Step 1: Enter data in SPSS and save data file

The data in this example are as follows:

ID	Participant number	***Nominal*** *variable*
Group	1=Affectively intense 2=Neurotic 3=Extraverted	***Nominal*** *variable*
Sensat	Sensation seeking	***Scale*** *variables*
Emotion	Emotional responsiveness	
Social	Sociability	
Somat	Somatic complaints	
Arouse	Arousability	
Aggress	Aggressiveness	
Wellbeing		

The data have already been entered and saved in a file called ***17Discrim.sav***. The file can be downloaded from the link 'Student downloads' on this website: www.pearson.com.au/9781442549821

Step 2: Screen data and check assumptions

Screen the data for the assumptions of MANOVA and discriminant function analysis, as discussed in Tabachnick and Fidell (2007).

Screening for the usual ANOVA assumptions needs to be conducted separately for each predictor variable for each of the groups, using the SPSS **Split File** option.

As always, univariate and multivariate outliers must be identified and dealt with.

Step 3: Conduct the discriminant analysis

43 How to: Conduct a discriminant function analysis

Analyze
Classify
Discriminant...

[Select variables, as follows:]
Group [group] ▸ **Grouping Variable:**
[Click **Define Range** button and type code ranges for the Grouping Variable]
Minimum: 1
Maximum: 3
Continue

Sensation seeking [Sensat] }
Emotional responsiveness [Emotion] }
Sociability [Social] }
Somatic complaints [Somatic] }
Arousability [Arouse] }
Aggressiveness [Aggress] }
Wellbeing } ▸ **Independents**

⊙ **Enter independents together**
O **Use stepwise method** ← Choose this option for a **stepwise**, as opposed to **standard** or direct, analysis.

Classify...
[From **Prior Probabilities** select:]
⊙ **Compute from group sizes** ← Only necessary to activate this option when group sizes are unequal.
[From **Display** select:]
☑ **Summary table**
☑ **Leave-one-out classification**
[From **Plots** select:]
☑ **Combined-groups**
Continue

Statistics...
[From **Descriptives** select:]
☑ **Means**
☑ **Univariate ANOVA's**
☑ **Box's M**
[From **Matrices** select:]
☑ **Within-group correlations**
Continue
OK

Step 4: Examine and interpret the discriminant analysis output

Output: **Discriminant function analysis**

Discriminant

Examine these two tables to assess extent and type of any missing data, and examine descriptive statistics.

1

Analysis Case Processing Summary

Unweighted Cases		N	Percent
Valid		65	100.0
Excluded	Missing or out-of-range group codes	0	.0
	At least one missing discriminating variable	0	.0
	Both missing or out-of-range group codes and at least one missing discriminating variable	0	.0
	Total	0	.0
Total		65	100.0

1

Group Statistics

Group		Mean	Std. Deviation	Valid N (listwise)	
				Unweighted	Weighted
Affectively intense	Sensation seeking	4.9474	1.77869	19	19.000
	Emotional responsiveness	3.9342	.49189	19	19.000
	Sociability	24.4737	4.70660	19	19.000
	Somatic complaints	4.3158	1.20428	19	19.000
	Arousability	7.4211	1.26121	19	19.000
	Aggressiveness	20.8947	2.23345	19	19.000
	Wellbeing	3.6368	.25432	19	19.000
Neurotic	Sensation seeking	4.3913	2.27115	23	23.000
	Emotional responsiveness	2.4783	.37623	23	23.000
	Sociability	21.6087	4.32476	23	23.000
	Somatic complaints	7.0870	1.47442	23	23.000
	Arousability	5.4348	1.64665	23	23.000
	Aggressiveness	21.7391	2.41618	23	23.000
	Wellbeing	3.4609	.41750	23	23.000
Extraverted	Sensation seeking	7.1739	1.37021	23	23.000
	Emotional responsiveness	2.5109	.72095	23	23.000
	Sociability	36.8261	5.92876	23	23.000
	Somatic complaints	2.9130	1.59297	23	23.000
	Arousability	4.5217	1.27456	23	23.000
	Aggressiveness	21.6522	2.08040	23	23.000
	Wellbeing	3.3522	.40773	23	23.000
Total	Sensation seeking	5.5385	2.20140	65	65.000
	Emotional responsiveness	2.9154	.85499	65	65.000
	Sociability	27.8308	8.43091	65	65.000
	Somatic complaints	4.8000	2.29265	65	65.000
	Arousability	5.6923	1.82793	65	65.000
	Aggressiveness	21.4615	2.24358	65	65.000
	Wellbeing	3.4738	.38539	65	65.000

Tests of Equality of Group Means

	Wilks' Lambda	F	df1	df2	Sig.
Sensation seeking	.683	14.411	2	62	.000
Emotional responsiveness	.404	45.720	2	62	.000
Sociability	.348	58.057	2	62	.000
Somatic complaints	.386	49.368	2	62	.000
Arousability	.580	22.452	2	62	.000
Aggressiveness	.973	.862	2	62	.427
Wellbeing	.911	3.041	2	62	.055

2

The above table shows the univariate ANOVA results. In this example there are significant group differences on all the predictors except Aggressiveness and Wellbeing.

Pooled Within-Groups Matrices

		Sensation seeking	Emotional responsiveness	Sociability	Somatic complaints	Arousability	Aggressiveness	Wellbeing
Correlation	Sensation seeking	1.000	-.090	.436	-.019	-.189	.239	.063
	Emotional responsiveness	-.090	1.000	-.064	-.026	-.173	-.064	-.292
	Sociability	.436	-.064	1.000	-.173	-.261	.033	-.271
	Somatic complaints	-.019	-.026	-.173	1.000	-.145	-.082	.059
	Arousability	-.189	-.173	-.261	-.145	1.000	-.032	.152
	Aggressiveness	.239	-.064	.033	-.082	-.032	1.000	.015
	Wellbeing	.063	-.292	-.271	.059	.152	.015	1.000

This table of pooled within-groups correlations is useful in assessing multicollinearity. If predictors are too highly intercorrelated (e.g., around .90 or above) a question arises as to whether anything is really to be gained by including both of them.

The information in this table is also useful in understanding differences between coefficients and correlations later in the output (see output sections 7 and 8).

Box's Test of Equality of Covariance Matrices

Log Determinants

Group	Rank	Log Determinant
Affectively intense	7	1.017
Neurotic	7	3.467
Extraverted	7	2.538
Pooled within-groups	7	3.635

The ranks and natural logarithms of determinants printed are those of the group covariance matrices.

4 When the assumption tested by Box's M (below) is **not** met this table of **Log Determinants** may identify groups with very small determinants. If these groups are deleted, the problem may be solved and the assumption may then be met.

Test Results

Box's M		74.963
F	Approx.	1.117
	df1	56
	df2	10269.319
	Sig.	.256

Tests null hypothesis of equal population covariance matrices.

4 Box's M is the very sensitive test of the assumption of homogeneity of variance-covariance matrices, evaluated at α = .001. With adequate sample sizes it is of little importance, **except that violation can affect the accuracy of classification**. In this example, the assumption is met.

Summary of Canonical Discriminant Functions

Eigenvalues

Function	Eigenvalue	% of Variance	Cumulative %	Canonical Correlation
1	3.005[a]	52.1	52.1	.866
2	2.762[a]	47.9	100.0	.857

a. First 2 canonical discriminant functions were used in the analysis.

5

The above table headed **Eigenvalues** shows that in this case two discriminant functions have been extracted (number possible = number of groups – 1). It also shows how much each function contributes to the total variance explained by the **solution**.

The canonical correlation is the **multiple correlation** between the predictors and the discriminant function.

Wilks' Lambda

Test of Function(s)	Wilks' Lambda	Chi-square	df	Sig.
1 through 2	.066	160.044	14	.000
2	.266	78.172	6	.000

6

The **Wilks' Lambda** table takes a step-by-step approach to identifying which functions are significant. In this case, it shows that when all functions are included (1 through 2) the solution is significant (Sig. < .05).

Then, when the first function is removed, the remaining function(s) is also significant. In other words, both functions are significant in this example.

If the second step had **not** been significant it would have meant that the remaining functions (only number 2 in this case) are not significant.

Standardized Canonical Discriminant Function Coefficients

	Function	
	1	2
Sensation seeking	-.041	.098
Emotional responsiveness	.549	.754
Sociability	-.450	.458
Somatic complaints	.448	-.472
Arousability	.449	.387
Aggressiveness	.053	-.103
Wellbeing	.110	.387

7

These **Standardized Canonical Discriminant Function Coefficients** are equivalent to **standardised regression coefficients** in multiple regression. Here, they are the standardised regression coefficients for the predictor variables in predicting the discriminant functions that discriminate between the groups. Their size indicates the relative importance of each predictor.

As in multiple regression, they represent the **unique** relationship between each predictor variable and the discriminant function that discriminates the groups.

Structure Matrix

	Function	
	1	2
Sociability	-.725*	.325
Arousability	.429*	.249
Sensation seeking	-.360*	.166
Wellbeing	.164*	.078
Somatic complaints	.449	-.598*
Emotional responsiveness	.456	.554*
Aggressiveness	-.057	-.081*

Pooled within-groups correlations between discriminating variables and standardized canonical discriminant functions
Variables ordered by absolute size of correlation within function.

*. Largest absolute correlation between each variable and any discriminant function

8

The **Structure Matrix** shows the **correlation** between the predictors and the discriminant functions that discriminate the groups—ordered by size for each function (they are **factor loadings** in effect).

The **Functions at Group Centroids** table shows the average score on each discriminant function for each group, thereby indicating **how the functions discriminate the groups**.

9

Functions at Group Centroids

	Function	
Group	1	2
Affectively intense	1.758	1.881
Neurotic	.773	-2.064
Extraverted	-2.225	.510

Unstandardized canonical discriminant functions evaluated at group means

How is all this interpreted so far?

Output section 8: Generally loadings over .30 in the **Structure Matrix** are used in interpretation.

Output section 9: The group centroids indicate that the **first function** discriminates all three groups, but mainly sets the extraverted (-2.225) apart from the affectively intense (1.758), with the neurotic falling in between (.773), although on the positive side and closer to the affectively intense.

> Referring back to the Structure Matrix for this function we see that high scores are indicative of low sociability (-.73), accompanied by high emotional responsiveness (.46), a high level of somatic complaints (.45), high arousability (.43), and low sensation seeking (-36); and vice versa for low scores.

Output section 9: The second function also discriminates all three groups, but mainly sets the neurotic (-2.064) apart from the affectively intense (1.881), with the extraverted falling between the two but a little closer to the affectively intense.

> Referring back to the Structure Matrix for the second function we see that high scores are indicative of a low level of somatic complaints (-.60) accompanied by high emotional responsiveness (.55) and high sociability (.33); and vice versa for low scores.

Consistent with their nonsignificant univariate results aggressiveness and wellbeing are not related to either discriminant function.

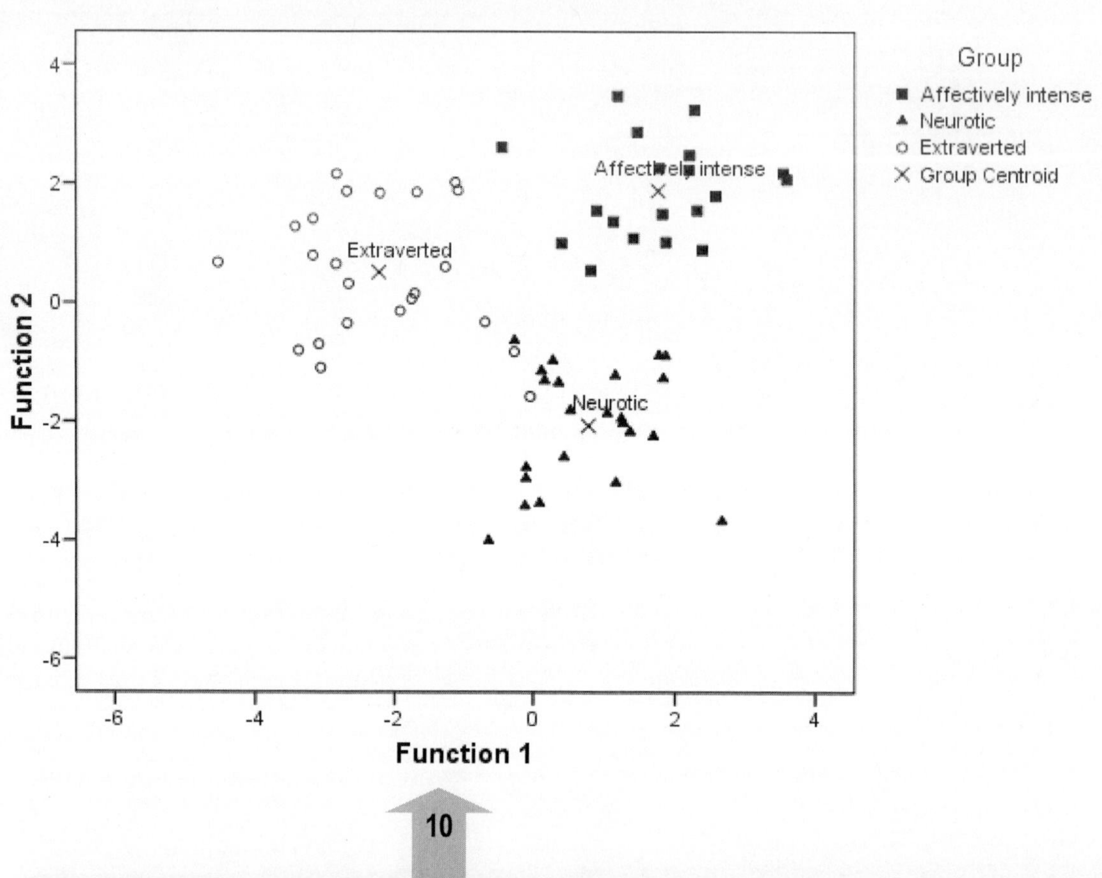

It does not follow immediately in the actual SPSS output, but the **Canonical Discriminant Functions Plot** illustrates the results and interpretation.

It shows how Function 1 tends to set the extraverted apart, especially from the affectively intense. The affectively intense relative to the extraverted have **high** scores on Function 1 (from Group Centroids table at output 9) and tend to be differentiated by relatively **low** scores on sociability and sensation seeking, and relatively **high** scores on emotional responsiveness, somatic complaints, and arousability.

Function 2 tends to set the neurotic apart, especially from the affectively intense. The affectively intense relative to the neurotic have **high** scores on Function 2 (from Group Centroids table at output 9) and tend to be differentiated by relatively **low** scores on somatic complaints, and relatively **high** scores on emotional responsiveness and sociability.

Refer back to the descriptive statistics to examine how the means differ between the groups in accordance with this interpretation.

Classification Statistics

The **Prior Probabilities for Groups** table shows what would be expected by chance (these would be equal at .33 with equal group sizes).

To illustrate, the affectively intense group has 19 of the 65 participants; 19/65 = .292.

11

Prior Probabilities for Groups

Group	Prior	Cases Used in Analysis	
		Unweighted	Weighted
Affectively intense	.292	19	19.000
Neurotic	.354	23	23.000
Extraverted	.354	23	23.000
Total	1.000	65	65.000

Classification Results[b,c]

		Group	Predicted Group Membership			Total
			Affectively intense	Neurotic	Extraverted	
Original	Count	Affectively intense	19	0	0	19
		Neurotic	0	23	0	23
		Extraverted	0	2	21	23
	%	Affectively intense	100.0	.0	.0	100.0
		Neurotic	.0	100.0	.0	100.0
		Extraverted	.0	8.7	91.3	100.0
Cross-validated[a]	Count	Affectively intense	18	0	1	19
		Neurotic	0	23	0	23
		Extraverted	0	2	21	23
	%	Affectively intense	94.7	.0	5.3	100.0
		Neurotic	.0	100.0	.0	100.0
		Extraverted	.0	8.7	91.3	100.0

a. Cross validation is done only for those cases in the analysis. In cross validation, each case is classified by the functions derived from all cases other than that case.

b. 96.9% of original grouped cases correctly classified.

c. 95.4% of cross-validated grouped cases correctly classified.

12

Discriminant analysis also provides information on how well predictors classify participants to groups. This may have practical importance (e.g., in clinical diagnosis). Relevant information is found in the **Classification Results** table.

This is an (unrealistically) excellent solution. From the **Original** classification results table we see that 96.9% of cases are correctly classified for the original groups (note b. beneath table). The **only misclassification** is two participants from the extraverted group incorrectly predicted to be in the neurotic group. They can be detected on the plot on the previous page. (It is possible to edit this plot so that it places the case number or ID next to each data point to identify the participants. The plot may be less useful with large data sets and/or many misclassifications.)

The **Cross-validated** results in at the bottom of the table actually produce what is known as a **"jackknifed classification"**. This helps to eliminate the bias in using cases to classify themselves. Instead, each case is classified according to coefficients derived from the other cases, not including itself. This gives a more realistic classification (refer Tabachnick & Fidell, 2007). In the present example, this produces an additional misclassification from affectively intense into extraverted, so that an overall 95.4% accuracy is obtained (note c. beneath table).

Note that classification accuracy is sensitive to violations of the homogeneity of variance-covariance assumption, "Cases tend to be overclassified into groups with greater dispersion" (Tabachnick & Fidell, 2007, pp. 382-383).

Obtaining additional discriminant output

To obtain a listing of every participant, showing the actual and predicted group membership (useful for identifying large numbers of misclassifications), include the following in the SPSS options:

Classify...
[From **Display** select:]
☑ **Casewise results**

Stepwise analysis

Note also that stepwise discriminant analysis can be used if one wants an analysis based only on the reduced set of variables that discriminates the groups. In this example, stepwise would have eliminated the variables aggressiveness and wellbeing from the analysis.

Discriminant function analysis: Research report sample Results section

If you have got this far you should no longer need me to do things for you, and it is time to wean yourself off the *Foolproof guide* (something students seem to have difficulty doing—but it must be done!). Refer to Tabachnick and Fidell (2007, pp. 426-429) for an example of how to write up the results for discriminant function analysis. Be warned, though, Tabachnick and Fidell can be a little excessive with their results sections, so the example below presents a somewhat simpler approach. It does not include contrasts between the groups, as included in the Tabachnick and Fidell example (pp. 427-428), although these could have been performed.

Of course, you should not blindly copy either if ever this applies to your own research, but rather you should read some actual and recent journal articles where discriminant function analysis has been reported.

Results

A discriminant function analysis was conducted to determine if the seven predictors (aggressiveness, arousability, emotional responsiveness, sensation seeking, sociability, somatic complaints, and wellbeing) could discriminate among the three groups of affectively intense, neurotic, and extroverted individuals (classified according to clinical assessment). With an *N* of 65 (19 in the affectively intense group, and 23 in each of the other groups), there were no univariate or multivariate outliers, and assumptions of linearity, multicollinearity, and homogeneity of variance-covariance matrices, were satisfactory. Normality was generally satisfactory, although there was mild positive skewness in the affectively intense group for arousability and sensation seeking; and in the extraverted group for somatic complaints. Departures from normality by way of rather flat distributions also occurred for the neurotic group on emotional responsiveness; and for the extraverted group on sensation seeking. The pooled within-cell correlations among the predictors are reported in Table 1.

Table 1

Pooled Within-cell Correlations among the Predictors

	1	2	3	4	5	6
1. Aggressiveness						
2. Arousability	-.03					
3. Emotion response	-.06	-.17				
4. Sensation seeking	.24	-.19	-.09			
5. Sociability	.03	-.26	-.06	.44		
6. Somatic complaints	-.08	-.15	-.03	-.02	-.17	
7. Wellbeing	.02	.15	-.29	.06	-.27	.06

Two significant discriminant functions were calculated with a combined Wilks' lambda[1] of $\Lambda = .07$, χ^2 (14, $N = 65$) = 160.04, p < .001. After removal of the first function, the remaining function was also significant, $\Lambda = .27$, χ^2 (6, $N = 65$) = 78.17, p < .001. The functions accounted for 52% and 48% respectively of the between-group variability, and with canonical R^2 of .75 and .73 they accounted for 75% and 73% respectively of the total relationship between the predictors and the groups.

[1] Λ is the symbol for Wilks's lambda multivariate criterion.

As can be seen from Figure 1, the first function mainly discriminated the extraverted from the affectively intense, while function two mainly separated the affectively intense from the neurotic.

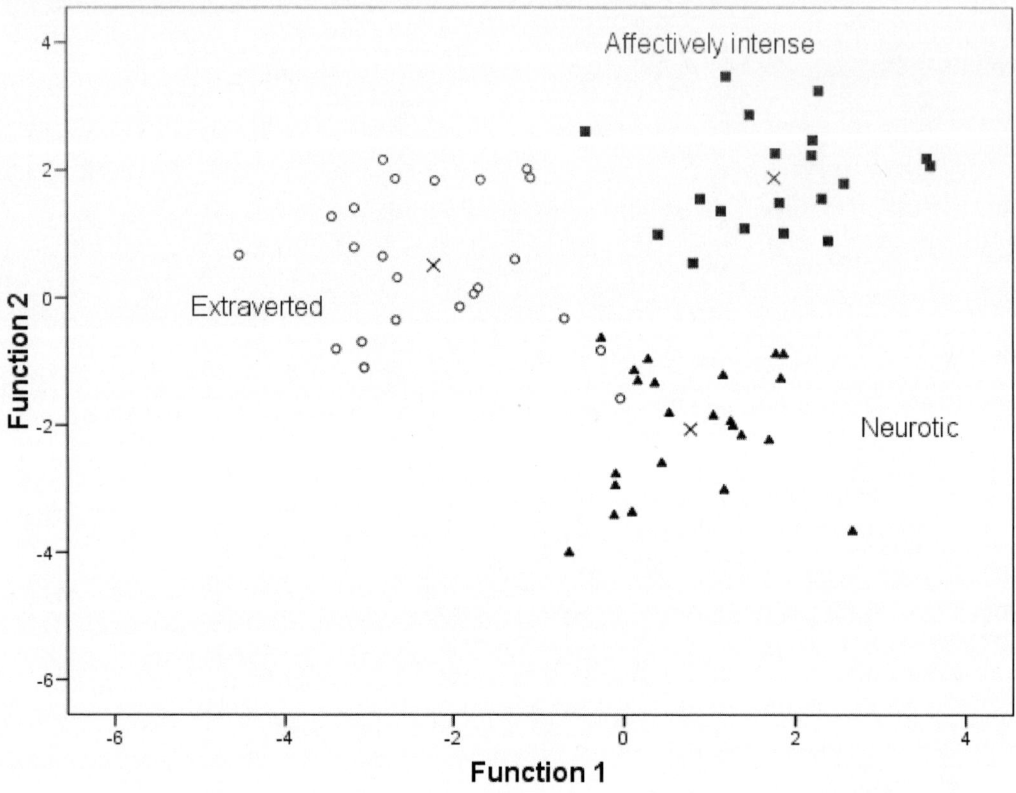

Figure 1. Separation of the three groups on the two discriminant functions.

The standardised coefficients and correlations of the predictor variables with the two discriminant functions are shown in Table 2. From the pattern of correlations we can see that the variables that most differentiated between the extraverted and affectively intense were sociability (r = -.73, extraverted M = 36.83, affectively intense M = 24.47), emotional responsiveness (r = .46, extraverted M = 2.51, affectively intense M = 3.93), somatic complaints (r = .45, extraverted M = 2.91, affectively intense M = 4.32), arousability (r = .43, extraverted M = 4.52, affectively intense M = 7.42), and sensation seeking (r = -.36, extraverted M = 7.17, affectively intense M = 4.95).

The variables that most differentiated between the affectively intense and the neurotic were somatic complaints (r = -.60, affectively intense M = 4.32, neurotic M = 7.09), emotional responsiveness (r = .55, affectively intense M = 3.93, neurotic M = 2.48), and sociability (r = .33, affectively intense M = 24.47, neurotic M = 21.61). The full list of descriptive statistics is shown in Table 3.[1]

Jackknifed classification showed that 62 participants (95.4%) were correctly classified; far in excess of the 33% that could be predicted by chance alone. Of the three who were misclassified, two were from the extraverted group (misclassified as neurotic), and one from the affectively intense group was misclassified as extraverted (see Figure 1).

Table 2

Correlations and Standardised Coefficients of Predictors with Discriminant Functions

	Correlation Coefficients		Standardised Coefficients	
Predictors	Function 1	Function 2	Function 1	Function 2
Aggressiveness	-.06	-.08	.05	-.10
Arousability	.43	.25	.45	.39
Emotion response	.46	.55	.55	.75
Sensation seeking	-.36	.17	-.04	.10
Sociability	-.73	.33	-.45	.46
Somatic complaints	.45	-.60	.44	-.47
Wellbeing	-.16	.08	.11	.39

[1] By now you are perfectly capable of creating these tables in a manner suitable for a Results section, so this one is not produced here. It would be a matter of you suitably adapting the SPSS output table shown in the output section 1.

18 Bivariate Correlation

Correlation is a different type of statistical concept to those considered in previous chapters, which have looked at group means, and differences between group means.

Correlation describes **the relationship between two variables**; consequently its precise name is **bivariate** correlation.

Note, however, that the most common types of correlation only concern **linear** relationships.

Relationships are best shown by a **scattergram** or **scatterplot**.

Types of relationship

A relationship can **be positive** when two variables covary in the **same direction** (e.g., petrol consumption and kilometres travelled).

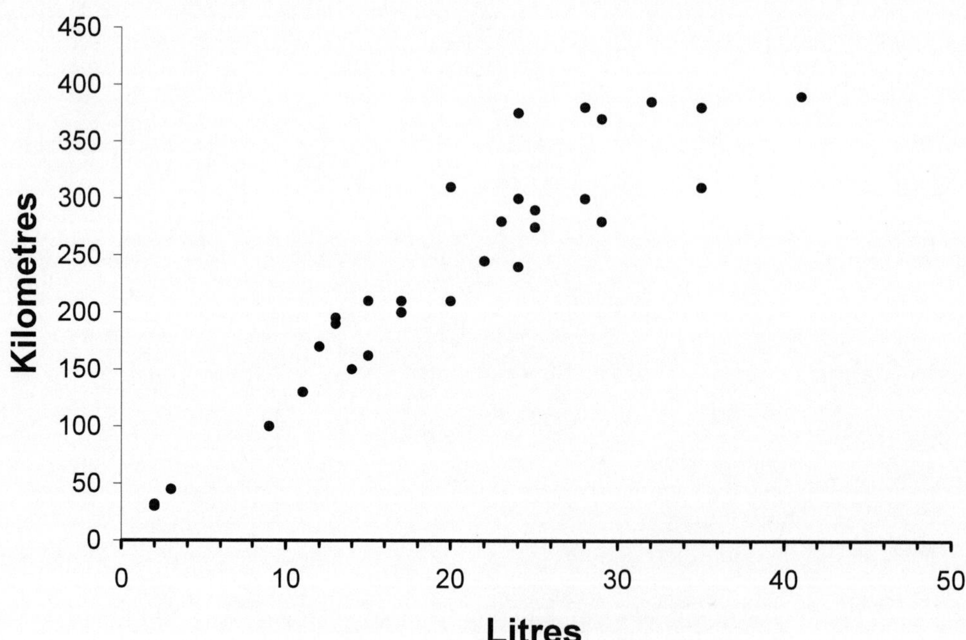

Note that to perform correlation there must be **two** scores from each "participant" (in this case the participants are cars).

A relationship can also be **negative** when two variables covary in **opposite directions** (e.g., fuel gauge reading and kilometres travelled).

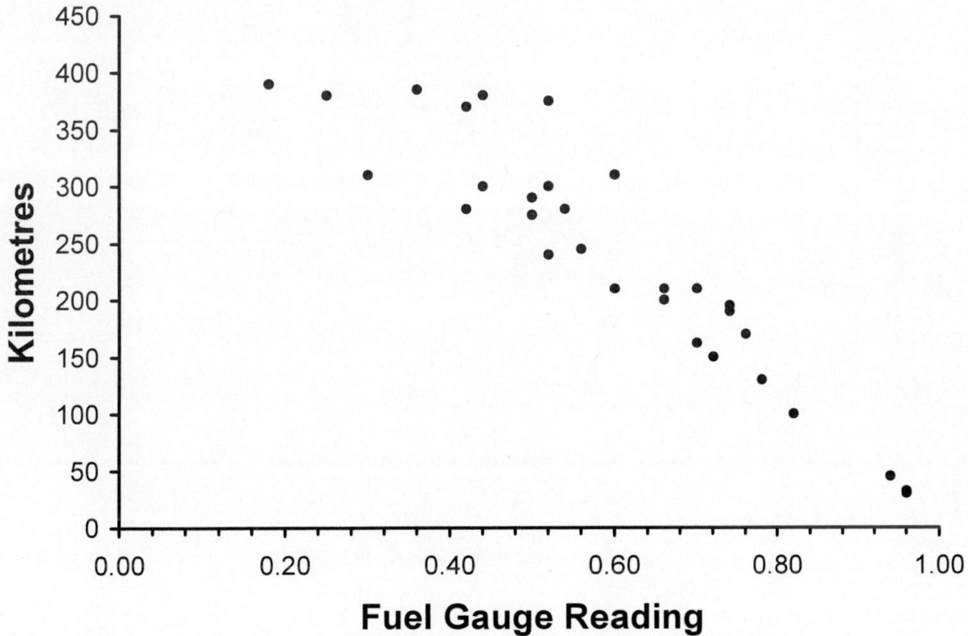

Finally, there is the situation where there is **no** relationship between two variables (e.g., shoe size of driver and kilometres travelled).

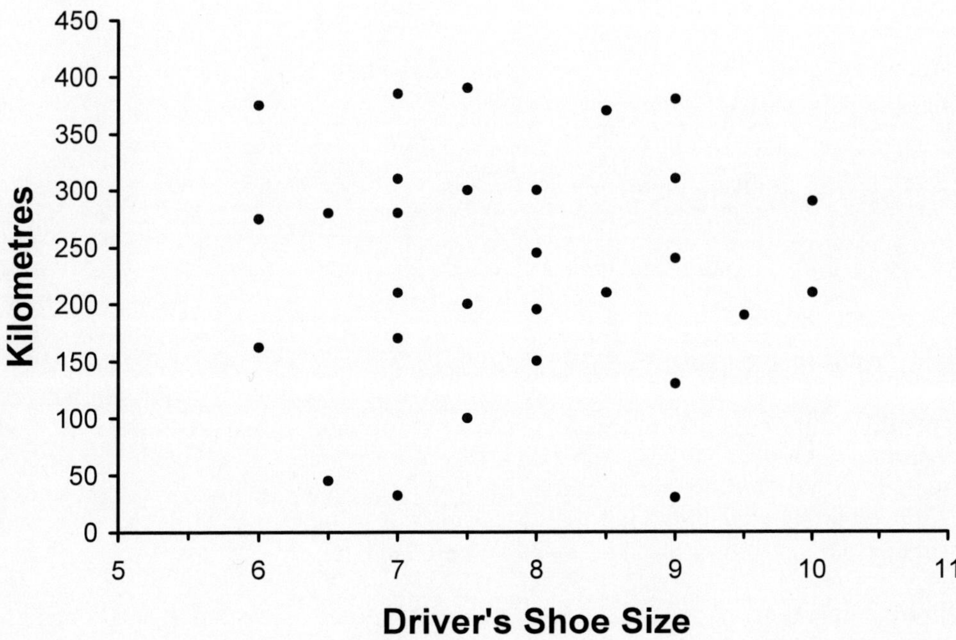

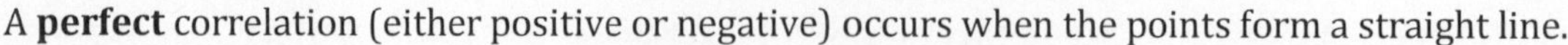

A **perfect** correlation (either positive or negative) occurs when the points form a straight line.

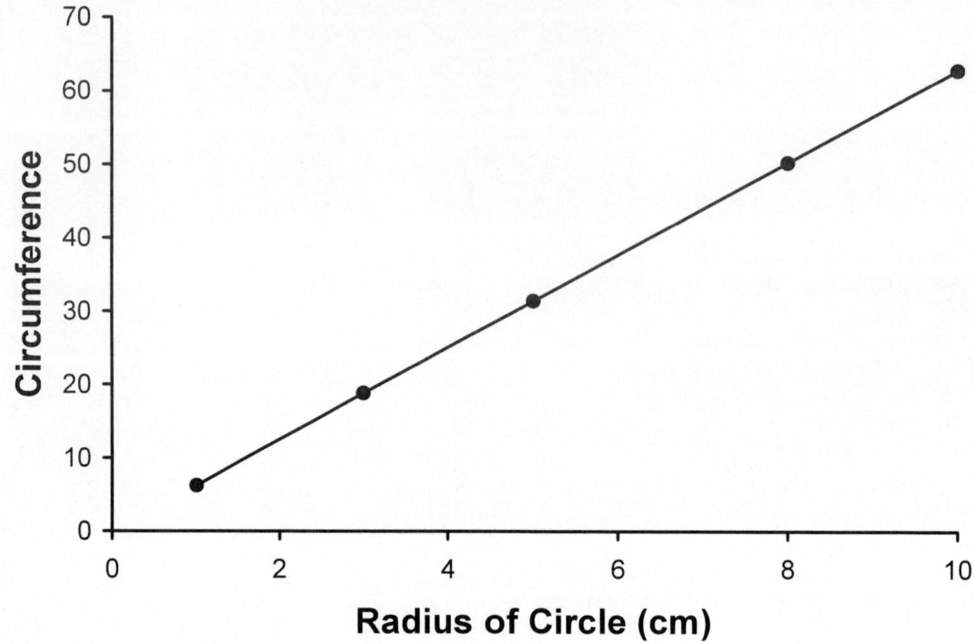

Perfect correlation allows **perfect prediction** of scores on one variable from knowledge of scores on the other.

Correlation and causation

Do not **ever** forget that correlation does **not** mean **causation**. When there is a correlation between two variables (let's call them *X* and *Y*) there are three possibilities:

1. *X* causes *Y*.
2. *Y* causes *X*.
3. Another variable (or variables) causes the observed relationship between *X* and *Y*.

Causation cannot be inferred from the research design and statistical technique, but it may be possible to argue causation on logical or theoretical grounds.

Correlation coefficients

Correlation coefficients are statistical measures of correlation.

Parametric correlation coefficient

Pearson's product-moment correlation (*r*) is the most common measure of **linear** relationship. It is used when both variables have at least interval levels of measurement, although it is possible to perform correlations with **dichotomous** variables (i.e., variables such as gender that only have two possible values). It is preferable to code dichotomous variables as 0 and 1, and the proportions of each value should be similar (e.g., near equal numbers of males and females).

Pearson's product-moment correlation (*r*) is a **parametric linear** correlation coefficient and has the assumption of **normality**, that is, that the variables are normally distributed in the population.

Nonparametric correlation coefficient

The **Spearman rank correlation coefficient** (also known as **Spearman's rho**) is the **nonparametric linear correlation**, used when one or both variables have only an ordinal measurement level. It can also be used when the normality assumption is violated.

Strength and direction of correlation coefficients

The correlation coefficient has a range of possible values from -1.00 through 0 to 1.00. The **number** indicates the **magnitude or strength** of the relationship (from 0 indicating no relationship to 1.00 indicating a perfect relationship), while the **sign** (- or +) indicates the **direction** of the relationship. The sign is negative (-) when the variables covary in opposite directions, and positive (+) when the variables covary in the same direction

By convention correlations are regarded as follows (these are arbitrary and approximate, and different text books can give different cutoffs):

.00 to .09	as very weak, negligible
.10 to .29	as weak, low, small
.30 to .49	as moderate, medium
.50 to .69	as strong, high, large
.70 to 1.0	as very strong, very high

The squared correlation coefficient

Another way to interpret the strength of a correlation is through use of r^2, which indicates the proportion of variance in one variable explained, or more correctly, **predicted**, by the other, and vice versa. It is also known as the **coefficient of determination**.

For example, if $r = .50$, then $r^2 = .25$. This indicates that 25% of the variance in one variable can be explained by the other and vice versa, leaving 75% of the variance in each to be explained by other factors.

The concept of r^2 can be represented diagrammatically, using what is called a **Venn** diagram. Each circle below represents the total variance for each of two variables that are correlated (i.e., variables *X* and *Y*). The shaded overlapping portion represents shared variance, the variance in *Y* that can be explained or accounted for by *X* and vice versa:

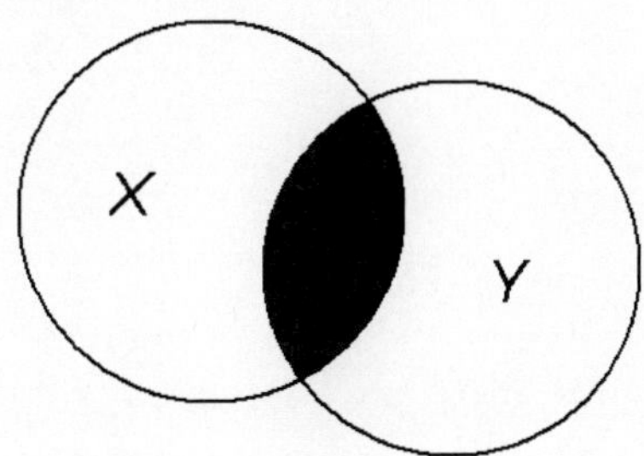

Calculating the correlation coefficient

The formula for calculating a correlation coefficient is given in most statistical text books. Again, there are two formulae: a **deviation score formula** provides the best conceptual understanding of what the correlation coefficient actually does; while a **raw score formula** is most suitable for manual calculation.

The deviation score formula (that best demonstrates what is actually being calculated) for a Pearson correlation coefficient has two steps.

First, the combined variance, or **covariance**, is calculated:

$$\text{Covariance}_{XY} = \frac{\Sigma(X - \overline{X})(Y - \overline{Y})}{N - 1}$$

Second, this covariance is **standardised** by dividing it by the product of the standard deviation of *X* and the standard deviation of *Y*:

$$r_{XY} = \frac{\text{Covariance}_{XY}}{s_X s_Y}$$

The **raw score** formula, which is more convenient for manual calculation, is as follows (note there are several versions of this same formula that are identical even if they may appear different):

$$r_{XY} = \frac{N(\Sigma XY) - (\Sigma X)(\Sigma Y)}{\sqrt{\left[N\Sigma X^2 - (\Sigma X)^2\right]\left[N\Sigma Y^2 - (\Sigma Y)^2\right]}}$$

Degrees of freedom

The *df* for correlation = $N - 2$

Descriptive and inferential correlation

Correlation is a **descriptive** statistic, which **describes** the relationship between two variables.

However, it can also be used as an **inferential** statistic that tests hypotheses about the correlation in the population.

The symbol for the population correlation is ρ (the **population parameter**), while *r* is the symbol for the sample correlation (the **sample statistic**).

Most often the null hypothesis (H_0) is that the relationship between two variables (*X* and *Y*) in the population is 0, as follows:

The null hypothesis H_0: $\rho_{XY} = 0$

The alternative hypothesis may be one of either:

Nondirectional (two-tail) H_1: $\rho_{XY} \neq 0$
Directional negative (one-tail) H_1: $\rho_{XY} < 0$
Directional positive (one-tail) H_1: $\rho_{XY} > 0$

When do you use a correlation analysis?

Correlation is used when the researcher wants a measure of the relationship between two variables.

The accuracy of **linear** correlations (Pearson's and Spearman's) depends on the following assumptions that are evaluated by inspection of the **scatterplot**.

Assumptions and pitfalls

- The relationship between the two variables must be **linear**. There should be no indication of a **curvilinear** relationship of any kind (like the examples below), as this will cause an underestimation of the degree of correlation. Spearman's correlation can cope with **monotonic** curvilinear relationships (i.e., curvilinear relationships where there is no reversal of direction).

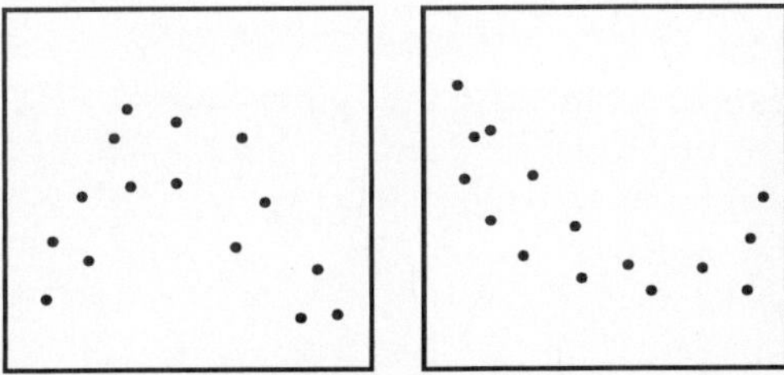

- The distribution should be equal across the range of *X* scores (*X* is plotted on the *x*-axis). In other words, the relationship between two variables should be:

- There should be **no restricted range** on one or both variables, as this will reduce the true correlation between them. For example, a strong positive relationship is evident between the two variables in the scatterplot below, but if we restricted the scores to the upper range only (the box within the scatterplot) there would appear to be no relationship between the variables.

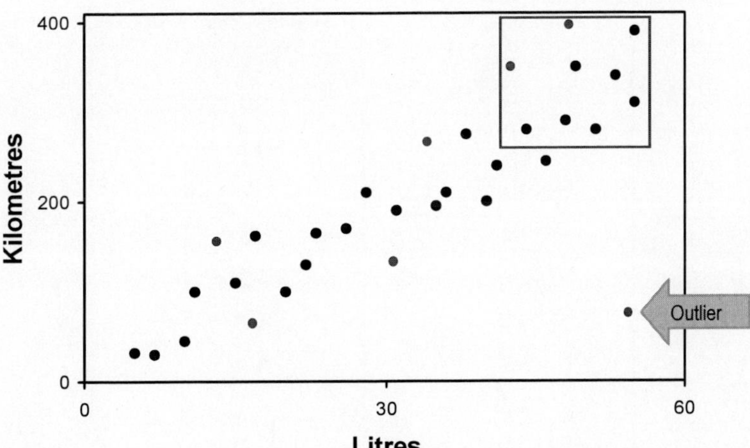

- **Outliers** are a serious problem as they distort correlations. Usually they need to be deleted, although this action must be reported in the Results section and the researcher should attempt to understand and evaluate why a case is an outlier. An outlier is indicated by the arrow in the scatterplot on the previous page. Note that even though it is not extreme on either variable it has a completely different pattern to all the other scores (a car that has used a lot of petrol, but travelled a very short distance; I think someone needs to inspect the petrol tank for leaks!)
- Also beware of using **extreme** groups, or of combining groups with different means, as the correlation may be misleading in these instances.
- For **inferential** tests of correlation the "participants" should be **randomly sampled**, and be **independent** of one another. Sample size is also an important consideration. With small samples the observed correlation is likely to differ considerably from the true population correlation; with large samples the error is normally slight. Try to have an *N* of at least 30 for inferential tests of correlation.

Example of correlation

Imagine you have conducted a pilot survey in which you have randomly sampled from your suburb to make a preliminary assessment of the relationship between age and knowledge of environmental issues. Assume you have developed a knowledge questionnaire comprising 50 items that participants either get right or wrong. The possible range of the scale is 0 (indicating no knowledge at all) to 50 (indicating very high knowledge). This is a nonexperimental or correlational research design, and, as both variables are continuous interval-level, Pearson's correlation is the appropriate statistical test.

You might also be interested in the relationship between gender and knowledge. As gender is a dichotomous variable, correlation would also be appropriate for this analysis provided you have near equal numbers of males and females (alternatively you could perform a *t* test, or Mann-Whitney if the *t* test assumptions were not met).

Dichotomous variables in correlation should be coded as 0 and 1.
Suppose the scores you obtain are as follows:

ID	Gender	Age	Knowledge
1	Female	23	46
2	Female	32	40
3	Male	Missing	30
4	Female	22	42
5	Male	35	33
6	Male	41	30
7	Female	52	19
8	Male	26	31
9	Male	48	20
10	Male	23	35
11	Female	44	25
12	Male	43	24
13	Female	22	42
14	Male	27	31
15	Female	31	38

Using SPSS to conduct correlation

Step 1: Enter data in SPSS and save data file

In the SPSS data window define the variables as follows, then enter the data (leave blank for any missing data):

ID	Participant number	*Set **Measure** column to **Nominal***
Gender	0=Female 1=Male	*Set **Measure** column to **Scale***
Age		*Set **Measure** column to **Scale***
Knowledge	Knowledge of environmental issues	*Set **Measure** column to **Scale***

Remember to save the data in a file called ***Correlation.sav***.

Step 2: Screen data and check assumptions

After checking data accuracy and missing values, data screening for correlation involves an inspection of the scatterplot. Scatterplots are requested as follows.

44 How to: Obtain a scatterplot

Graphs
Legacy Dialogs ▸
Scatter/Dot...
[Click mouse on box for **Simple Scatter**]

Define
[Select variable(s) you want]
Knowledge of environmental issues [Knowledge] ▸ Y Axis
Age ▸ X Axis:

OK

Repeat the above specifying Gender for the X axis.

Note: In earlier versions of SPSS it might just be a matter of selecting **Graphs** then **Scatter...** without needing to go into **Legacy Dialogs**.

Step 3: Request the analysis

45 How to: Conduct a correlation analysis

Analyze
  Correlate ▸
    Bivariate...
    [Select variable(s) you want]
    Age }
    Gender }
    Knowledge of environmental issues [Knowledge] } ▸ **Variables**

    [From **Correlation Coefficients** select] ☑ **Pearson**
    [From **Test of Significance** select] ⊙**Two-tailed**

    [Click on **Options...** button]
      [From **Statistics** select] ☑ **Means and standard deviations**
      Continue

  OK

Note:
If you want a Spearman correlation:
  [From **Correlation Coefficients** select] ☑ **Spearman**
If you want a one-tailed test:
  [From **Test of Significance** select] ⊙ **One-tailed**

Step 4: Examine and interpret the scatterplots and correlation output

Output: **Scatterplots and correlation**

Although there is slight evidence of **heteroscedasticity** (with a greater spread at the low end of age) it is probably not enough to worry about and can often occur with such a small sample (correlations from small samples are, thus, not very reliable estimates of population correlations).

The next scatterplot for gender is more difficult to interpret because gender is dichotomous.

1

There appears to be a weak negative relationship with gender: males with the high code (1) on gender tend to have a little less knowledge than females (coded low, 0).

There is also evidence of **heteroscedasticity** in that there is greater spread among the female scores than the male. This is really only attributable to two low (possibly outlying) female scores.

The scatterplot suggests we need to interpret the correlation between gender and knowledge with caution, especially in view of the small sample size (again, these sorts of anomaly often arise with small samples).

2

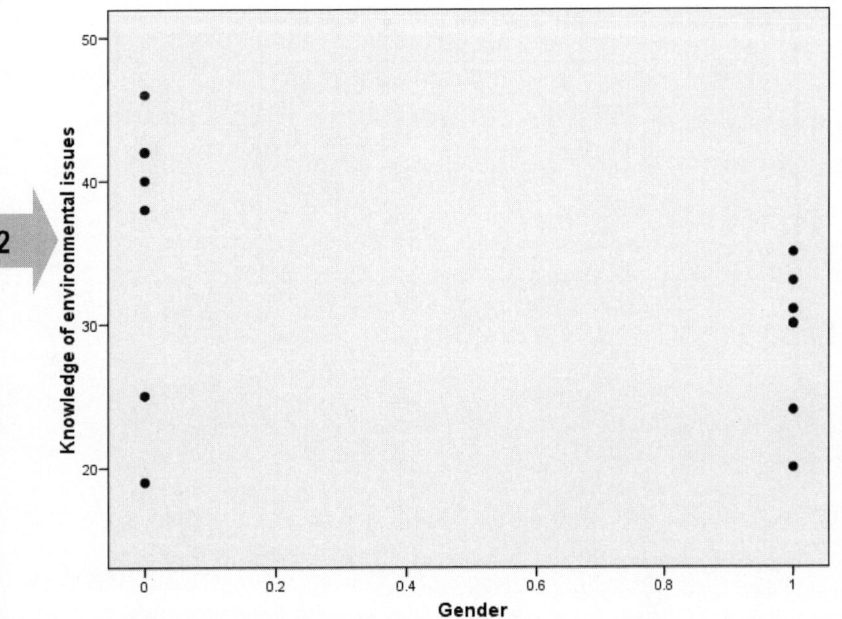

Correlations

Descriptive Statistics

	Mean	Std. Deviation	N
Age	33.50	10.390	14
Knowledge of environmental issues	32.40	8.192	15
Gender	.53	.516	15

3

Note that when you code dichotomous variables as 0 or 1 a mean **does** make sense. It is a **proportion**, showing that 53% of the sample has the code of 1. In this particular example, because 1 is the code for males, it means that 53% of the sample is male.

Correlations

		Age	Knowledge of environmental issues	Gender
Age	Pearson Correlation	1	-.873**	.121
	Sig. (2-tailed)		.000	.680
	N	14	14	14
Knowledge of environmental issues	Pearson Correlation	-.873**	1	-.425
	Sig. (2-tailed)	.000		.114
	N	14	15	15
Gender	Pearson Correlation	.121	-.425	1
	Sig. (2-tailed)	.680	.114	
	N	14	15	15

**. Correlation is significant at the 0.01 level (2-tailed).

4

Correlation tables have three rows; the first reports the correlation; the second, the Sig. (2-tailed), is the *p* value; and the third is *N*.

This table tells us that there is a **significant** (Sig. < .05) and **high negative** correlation of **-.87 between age and knowledge of environmental issues** for people living in your suburb. As the relationship is negative it indicates that the variables covary in **opposite** directions. Low scores on age tend to accompany high scores on knowledge, and vice versa. We must to refer back to the way the knowledge scale was coded to interpret: high knowledge scores indicate a high level of knowledge. Hence, this result tells us that younger people tend to have more knowledge, and older people tend to have less knowledge.

There is also a **moderate negative** correlation between **gender and knowledge**, but it is **not** significant with so small a sample. Again, low "scores" on gender tend to accompany high scores on knowledge, and vice versa. To understand what this means you need to look at how gender was coded. Low scores on gender (0) indicate females; high scores (1) indicate males. Hence, the negative correlation means females had greater knowledge of environmental issues than males in this particular sample.

Correlation: Research report sample Results section

Results

Scores on the 50 environmental knowledge items were summed to provide an overall score for knowledge of environmental issues with a possible range of 0 through to 50. The mean was 32.40 (*SD* = 8.19). Pearson product-moment correlations were performed between age and knowledge, and also gender and knowledge, using an alpha level of .05. One participant had missing data on age, and scatterplots suggested that the assumptions of correlation were satisfactory, although some evidence of heteroscedasticity in the relationship between gender and knowledge indicated that this correlation should be interpreted with caution. A strong, negative relationship between age and knowledge was significant, $r(12) = -.87$, $p < .001$ (younger people had greater knowledge of environmental issues than older people). Although there was a moderate negative relationship between gender and knowledge (suggesting that females have greater knowledge of environmental issues than males), it did not achieve statistical significance, $r(13) = -.43$, $p = .11$.

Notes:

A correlation is reported in the Results section as follows: *r*(*df* of *N* - 2) = correlation coefficient **without leading 0**, *p* = Sig. value (with alpha previously specified).

Means and standard deviations are reported to indicate the range in scores (age and gender means are not given here as they would appear in the Method section under the subheading Participants).

With several correlations it may be more effective to report them in a table as a correlation matrix, as shown on the next page.

Table 1

Correlations Among Age, Gender, and Knowledge of Environmental Issues

	Age	Gender
Gender	.12	
Knowledge of environmental issues	-.87*	-.43

*$p < .05$.

Significance of the difference between two independent correlations

Sometimes we want to know if two correlations are significantly different from one another. It is assumed that the correlations come from random samples, and they **must be independent** (i.e., they cannot come from the same sample of participants).

Suppose we have two correlations of the relationship between age and knowledge of environmental issues. The first comes from your suburb, and as we have discovered is -.873 from a sample size of 14. Suppose a different suburb returns a correlation of -.58 with a sample size of 43. Are these two correlations significantly different? This is determined in two steps (see Pallant, 2007, pp. 138-141 where the calculations are shown in full; or see Howell, 2002, pp. 277-280).

Step 1: Convert both correlations to *z* scores using Fisher's *z* test.

To do this you need to refer to appropriate tables, for example, Howell (2002, p. 746) for the conversion of r to r'(or z_r).

r	z_r (r') from tables
-.873	1.354
-.58	0.662

Step 2: Substitute these values into the following formula and calculate the *z* score of the difference

$$z_{Diff} = \frac{z_{r1} - z_{r2}}{\sqrt{\frac{1}{N_1 - 3} + \frac{1}{N_2 - 3}}}$$

If you perform this calculation you obtain a z_{Diff} of 2.0064. Because it exceeds the critical z of 1.96 that corresponds to a p of .05 (refer back to chapter 5 on the normal distribution if you have forgotten!), the difference is significant. **The two correlations are significantly different**.

Alternatively you can simply enter the values for the two correlations and sample sizes into the calculator on websites that are now available and they will do it all for you, including giving the exact *p* (.02241). For example: http://faculty.vassar.edu/lowry/rdiff.html

This next site also provides confidence intervals as shown below:
http://www.quantitativeskills.com/sisa/statistics/correl.htm

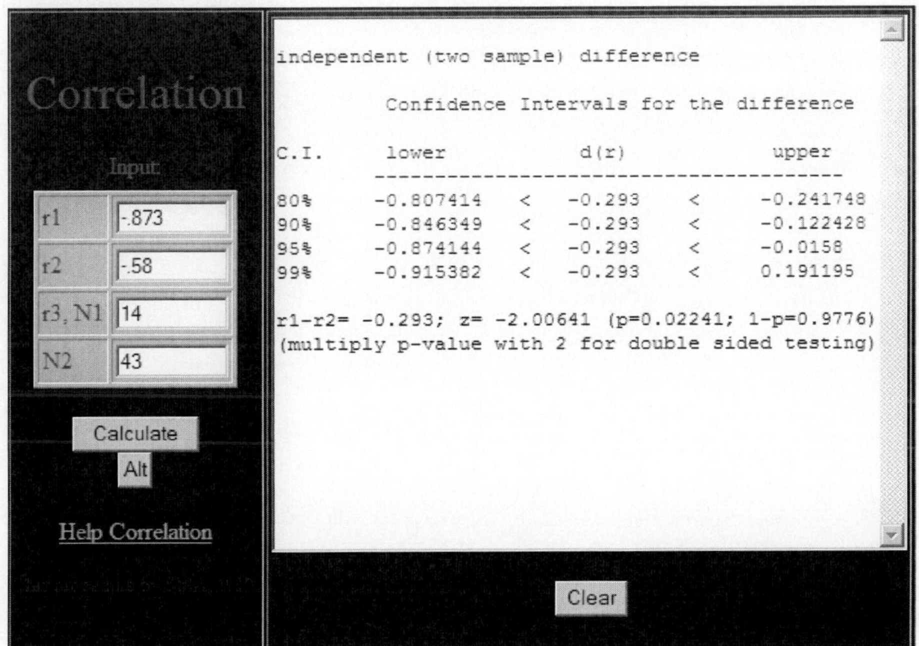

Significance of the difference between two dependent correlations

Howell (2007, pp. 280-281) describes how to test whether two **dependent** correlations (i.e., from the same sample) are significantly different. This does, however, require a correlation matrix of **three** correlations so that the degree of dependence can be calculated.

It can also be calculated automatically from the website calculator on the previous page (and above) by entering the three correlations in the spaces for r1, r2, and r3; then entering the sample size in N2.

19 Bivariate Regression

When there is a relationship between two variables, it is possible to predict scores on one from knowledge of scores on the other—with better than chance accuracy.

The regression line

Prediction is based on calculating a **line of best fit** or **regression line**, a line that is as close as possible to all the points on a scatterplot.

Using the example from the previous chapter, recall that a strong negative correlation of -.87 was found between the variables age and knowledge of environmental issues. This correlation was significant, and indicated that younger people have greater knowledge of environmental issues than older people.

Armed with information about this correlation, we are thus in a position to predict a person's knowledge of environmental issues from their age.

The variable that we want to predict is referred to as the **outcome** or **criterion**. In this example it is knowledge of environmental issues. For convenience, it may also be called the DV, although this is not strictly correct. Age is then referred to as the **predictor** (or the IV).

By convention, the **predictor** is always plotted on the horizontal axis (***x*-axis**), and the **criterion** on the vertical axis (*y*-axis), so the scatterplot now appears as follows:

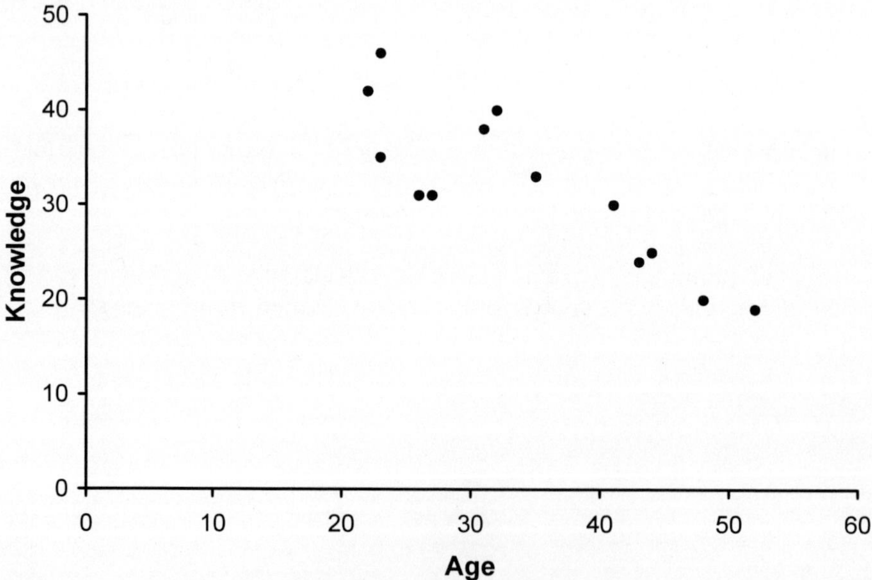

Onto this scatterplot we can now fit a **line of best fit** or **regression line**. This is the line on the scatterplot that is closest on average to all the observation points. It is the line that allows the best possible **prediction** of *Y* scores ($\hat{Y}$) from knowledge of *X* scores.

46 How to: Obtain a regression line on a scatterplot

Graphs
Legacy Dialogs ▸
Scatter/Dot...
[Click mouse on box for **Simple Scatter**]

Define
[Select variable(s) you want]
Knowledge of environmental issues [Knowledge] ▸ **Y Axis**
Age ▸ **X Axis:**

OK

Note: In earlier versions of SPSS it might just be a matter of selecting **Graphs** then **Scatter...** without needing to go into **Legacy Dialogs**.

Now, go to the output window and double-click anywhere on the scatterplot to open **Chart Editor** in its own window.

If you double-click on any of the data points until they are **all** highlighted, you can then change their style from the dialog box that appears.

Lastly, select menu options as follows:

Elements
Fit Line at Total
[In the **Properties** dialog box that appears, select from **Fit Method** as follows]
◉ **Linear**
[Click on **Close** button at bottom]
Exit **Chart Editor** window

As mentioned previously, the Chart Editor was rather different in earlier versions of SPSS. For example, to obtain a regression line in Version 11, you would need to go to the output window and double-click anywhere on the scatterplot to open it in its own window, then select menu options as follows:

Chart
Options...
[Under **Fit Line** select] ☑ **Total**
[Click on **Fit Options...** button]
Click mouse on box for **Linear Regression**
Continue
OK

Below is the scatterplot produced by default by SPSS. Note that in order to provide a good fit in the space SPSS may not draw the axes beginning at zero.

You can modify the scatterplot to suit. For example, you might want to make the data points solid rather than open circles that may be produced by default. To develop your skills, spend some time learning how to modify the appearance of the scatterplot.

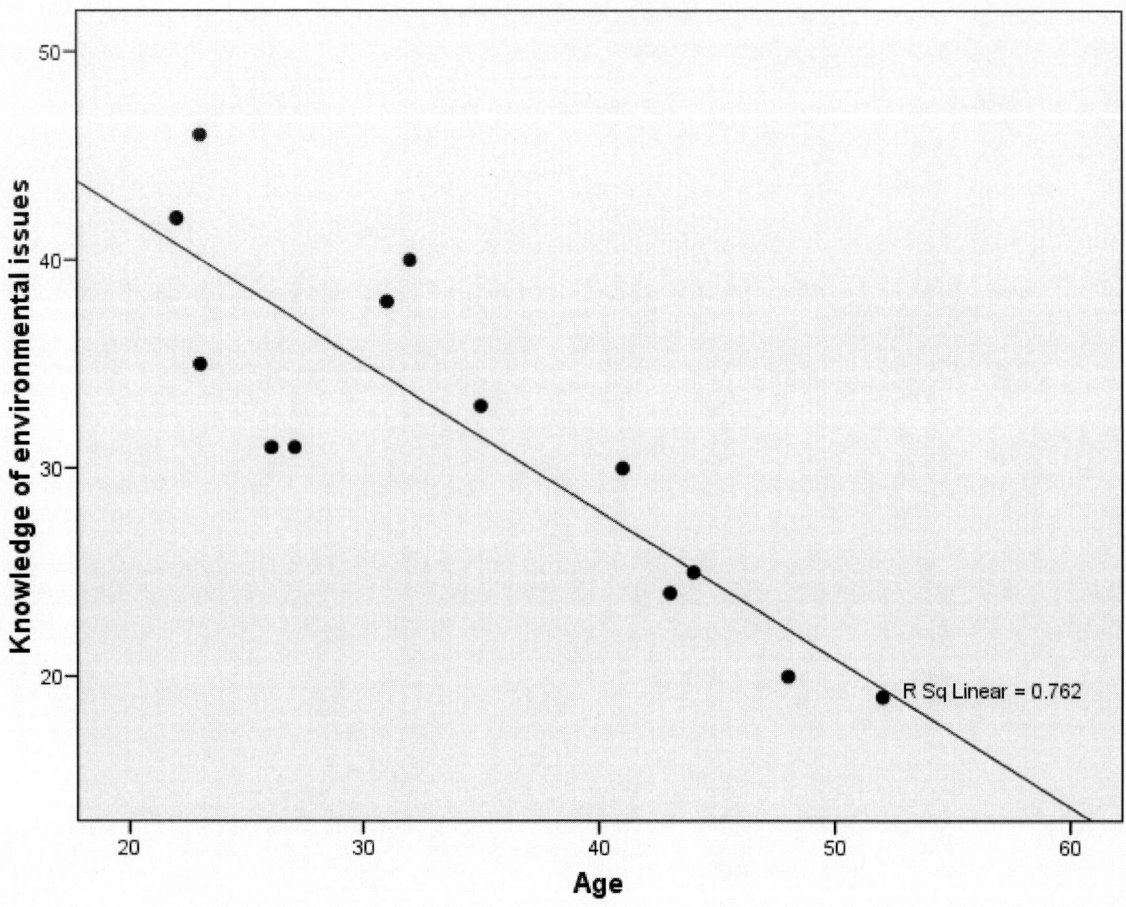

Predicting scores from a regression line

Suppose we want to predict the knowledge score of someone aged 32. We can use the regression line to do so. In the diagram below a line is drawn at age 32 perpendicular to the *x*-axis. At the point where this line meets the regression line another line is drawn perpendicular to the y-axis to give the best possible prediction ($\hat{Y}$) of the person's knowledge score (33 approx.).

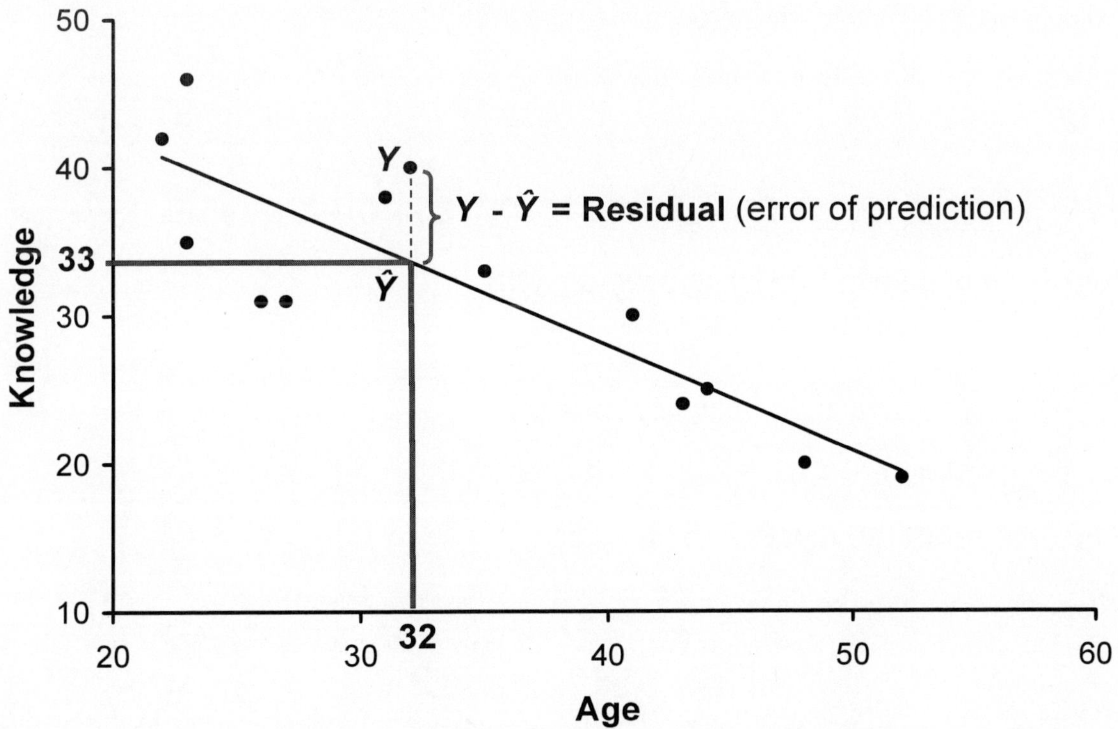

When there is perfect correlation (i.e., ± 1.00), there is perfect prediction as all the points fall along the regression line. However, the smaller the correlation, the more inaccurate the prediction.

In the example above it just so happens that we know the knowledge score of one person aged 32, and it is a score of 40, not 33. There is an error of prediction for this individual of 7.

Errors in prediction are known as **residuals**, and are the difference between actual *Y* scores and predicted *Y* scores ($Y - \hat{Y}$).

To calculate the regression line the **least squares criterion** is used. If all the residuals are added they will sum to 0 because there is "as much" above the regression line as there is below. The least squares regression line, therefore, is the line calculated so that **it minimises the sum** of the **squared** residuals.

The equation for a regression line

The equation for the line of best fit or regression line is:

$$\hat{Y} = a + bX$$

where:

a (known as the **constant** or **intercept**) is the point where the line **intercepts** the y-axis (i.e., the value of Y when X is zero);

b (known as the b weight or regression coefficient) is the **slope** of the line (the amount by which Y increases for every 1 unit increase in X).

Linear regression is a technique for calculating the values of a and b for the least squares regression line.

Using SPSS to conduct a linear regression analysis

Step 1: Enter data in SPSS and save data file

As in previous chapter.

Step 2: Screen data and check assumptions

As in previous chapter.

Step 3: Request the regression analysis

47 How to: Conduct a bivariate linear regression analysis

Analyze
  Regression ▸
    Linear...
      [Select variable(s) you want]
      Knowledge of environmental issues [Knowledge] ▸ **Dependent:**
      Age ▸ **Independent:**

      Statistics. . .
        [From **Regression Coefficients** select]
          ☑ **Confidence intervals**
            [For **Level (%):** type] **95** [assuming it is not already entered]
        Continue
    OK

Step 4: Examine and interpret the output

Regression

Model Summary

Model	R	R Square	Adjusted R Square	Std. Error of the Estimate
1	.873[a]	.762	.742	4.305

a. Predictors: (Constant), Age

1 Value of r ($R = r$ when there are only two variables).

2 Value of r^2

3 Estimate of r^2 in the population.

ANOVA[b]

Model		Sum of Squares	df	Mean Square	F	Sig.
1	Regression	711.080	1	711.080	38.377	.000[a]
	Residual	222.348	12	18.529		
	Total	933.429	13			

a. Predictors: (Constant), Age

b. Dependent Variable: Knowledge of environmental issues

4 Sig. or p of the **regression equation**, which is compared to α to determine if the equation is significant or not. With $\alpha = .05$, this is < .05; therefore it **is** significant.

In the case of **bivariate** regression this p is identical to the p value for r.

Coefficients[a]

Model		Unstandardized Coefficients		Standardized Coefficients	t	Sig.	95.0% Confidence Interval for B	
		B	Std. Error	Beta			Lower Bound	Upper Bound
1	(Constant)	56.416	4.017		14.043	.000	47.663	65.170
	Age	-.712	.115	-.873	-6.195	.000	-.962	-.461

a. Dependent Variable: Knowledge of environmental issues

5 Sig. or **p** of the **regression coefficient (B)**, which in **bivariate** regression will be identical to the p value for the regression equation in the ANOVA table above (i.e., note that $t^2 = F$, in this case, $-6.195^2 = 38.38$).

Interpreting the output: the regression equation

In the SPSS output, the regression equation ($\hat{Y} = a + bX$) appears in the table headed Coefficients:
a is the constant (56.416)
b is referred to as B by SPSS (-.712).

Hence, the regression equation is: $\hat{Y}$ = 56.416 - 0.712X
That is, predicted knowledge score = 56.416 – (0.712 x Age)

So, for an age of 32, the predicted knowledge score is as follows:

Predicted knowledge score = 56.416 - (0.712 x 32)
= 56.416 - 22.784
= **33.63**

Surprise, surprise, this matches the approximate value we were able to read directly off the scatterplot.

About the constant

The constant or intercept in this equation is 56.416, but the regression line crosses the *y*-axis somewhere around age 42 or 43 in the scatterplot that was drawn previously by SPSS. What is going on? Surely something is wrong. ***Hint:*** The explanation has something to do with the way the axes in the scatterplot were constructed.

The problem would be solved if the scatterplot were to be redrawn with both the *x*- and *y*-axes beginning at 0. Watch out for this when interpreting graphs.

Note that **Beta** in the SPSS table is the **standardised *b*** weight, calculated using standard, not raw, scores, with the regression line passing through the origin (0 on the *y*-axis). The standard score formula is:

Predicted standard score value of Y = -.873 standard score value of X
That is: $z_{\hat{Y}}$ = Beta $z\,X$

The standard error of the estimate (Standard error of prediction)

The **Std. Error of the Estimate**, which is the final figure given in the **SPSS Model Summary** table at the beginning, is similar to standard deviation in univariate distributions, and corresponds to the average amount of error in predicted Y scores. When normally distributed we can conclude that 68% of actual Y scores will be within one standard error (i.e., ±4.31 points) of the predicted Y score ($\hat{Y}$).

From the scatterplot it can be seen that our 32-year-old participant with a knowledge score of 40 was quite extreme in having an error of prediction of 6.37 ($Y - \hat{Y}$ = 40 - 33.63).

The higher the correlation, the smaller the standard error of prediction.

Similarly, in the SPSS output table headed **Coefficients** the column for **Unstandardized Coefficients Std. Error** gives the standard error of the *b* weight and constant respectively.

For the time being, you can ignore the other information given in the output. This will become relevant in later chapters on multiple regression.

Things to note about linear regression

Predicting *Y* from *X*, and *X* from *Y*

By convention regression equations are constructed so that *Y* is predicted from *X* (knowledge from age, in this example). However, sometimes we may want to predict *X* from *Y* (age from knowledge, which admittedly makes little sense in this example). To do this a separate regression analysis must be run, and a new scatterplot calculated. The two regressions are not the same.

Assumptions and uses of linear regression

Linear regression is an extension of correlation analyses, and has similar assumptions.

Being able to predict scores on one variable from knowledge of another is useful in many contexts, particularly in education. One might want to use final school examination scores, for example, to predict semester weighted averages in university.

Suppose a lecturer were to lose all the final examination marks for a class of students. One solution to the problem might be to use linear regression to predict final examination marks from scores on a mid-semester test, based on the correlation observed between these variables over the past few years. Were this to happen to your class you should pay particular attention to the size of the correlation and standard error of prediction. Moreover, while a regression equation minimises errors of prediction on average, it does not necessarily minimise the error for any **given individual**—you might be the unlucky individual whose examination score was 98%, but whose predicted score was only 52%! The students in the class might not think this was such a good solution.

Percentage of variance

Suppose everyone on earth had a score on knowledge of environmental issues of 32. The mean would be 32 and there would be no variance to explain. But in reality, of course, scores on any variable vary about the mean. When two variables are correlated we can **explain**, or **account for**, or **(more correctly) predict** part of the variance in one from knowledge of the other, and vice versa.

In fact, a valuable way to interpret the strength of a correlation is through use of r^2, as mentioned in the previous chapter. This indicates the proportion of variance in one variable explained by the other, and vice versa.

The following scatterplot gives an indication of what is meant by variance explained.

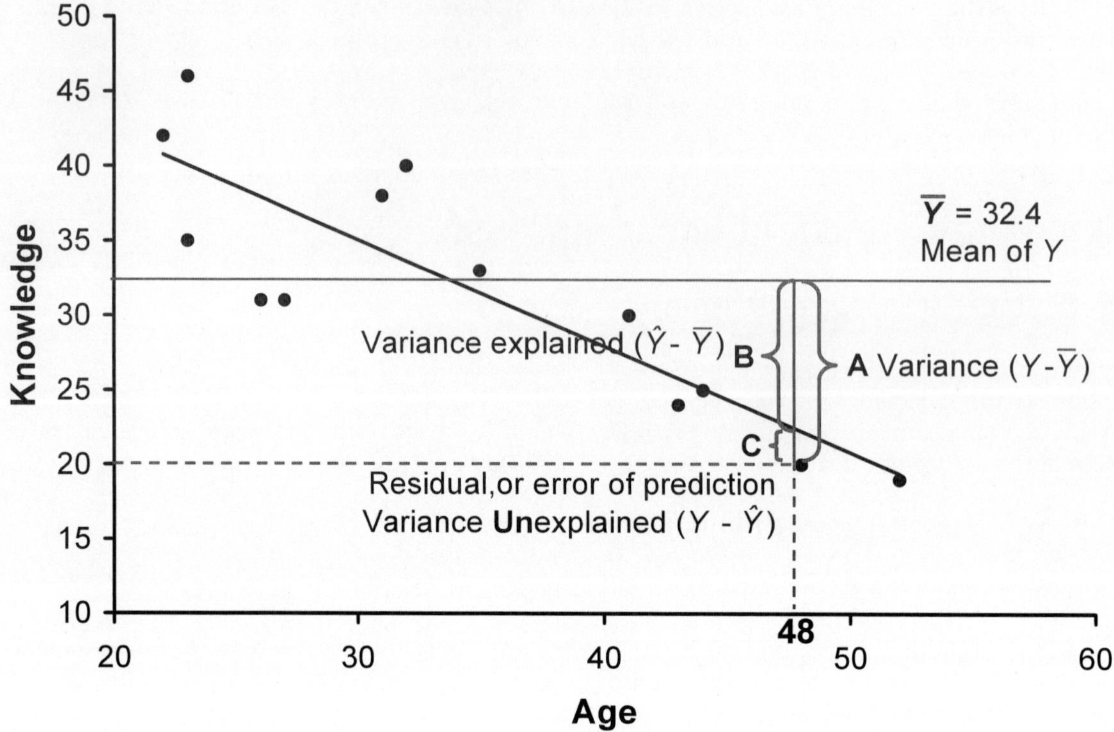

Look at the participant aged 48, whose knowledge score is 20.

The knowledge score of this participant varies from the mean by the amount shown at **A**, but using the regression line we can explain or predict part of this variation (**B**), leaving only the residual (**C**) unexplained or unpredicted.

When these amounts are **squared** then **summed** across **all** participants, the sum of the **B's** as a proportion of the sum of the **A's** corresponds to the **variance that can be explained** by making use of the correlation between the two variables.

The sum of the **C's** as a proportion of the sum of the **A's** corresponds to the variance that is **un**explained by the correlation between the two variables.

20 MULTIPLE REGRESSION

Multiple regression is an extension of bivariate correlation and regression beyond two variables to the situation where the researcher wants to establish the relationship between **one DV** (*Y*) and **several IVs** (*X1, X2, X3, X4* etc.).

Note that the terms DV and IV are only used for convenience. The technically correct terms in multiple regression are **criterion** or **outcome** (instead of DV) and **predictors** (instead of IVs).

Features of multiple regression

The **multiple correlation coefficient** is a measure of the relationship between one criterion (DV) and several predictors (IVs).

The symbol for the bivariate correlation coefficient (Pearson's) is *r*; the symbol for the multiple correlation coefficient is *R*.

While *r* has a range of -1 through 0 to 1, *R* only has a range of 0 to 1 (from no relationship to perfect relationship).

R^2 is the **squared multiple correlation** (**RSQ** or **SMC**) and represents the proportion of variance in the DV that is explained by the **linear combination** of the IVs.

The multiple regression equation is for the line of best fit that allows the best possible prediction of the DV from scores on the IVs. $\hat{Y}$ stands for the predicted DV score, so using the *X*, *Y* notation the formula for the multiple regression equation is as follows:

$$\hat{Y} = a + b_1X1 + b_2X2 + b_3X3 \text{ etc.}$$

The **intercept** or **constant** is represented by *a*; while b_1, b_2, b_3 etc. are the **regression coefficients** (*b* or B weights). When **standardised** the regression coefficients are known as **beta weights** (β) and the constant becomes 0.

Residuals are the difference between the actual *Y* score and the predicted $\hat{Y}$ score, that is, $Y - \hat{Y}$. The greater the *R*, the smaller the combined residuals.

R^2 corresponds to variance explained; residuals correspond to variance **un**explained.

The problem for multiple regression: Correlations among IVs

Uncorrelated IVs

When IVs are **uncorrelated** their individual correlations with the DV are **additive**, and to find R^2 one simply adds the r^2 values for each IV.

This is best represented diagrammatically, using a Venn diagram. Each circle in the diagram below (Figure 20.1) represents the variance in the relevant variable. It shows a DV (*Y*) and three uncorrelated IVs (*X1, X2*, and *X3*).

Overlapping shaded areas represent the variance that is shared between variables, in other words the r^2 between variables. $R^2 = \Sigma\, r^2$ for each IV.

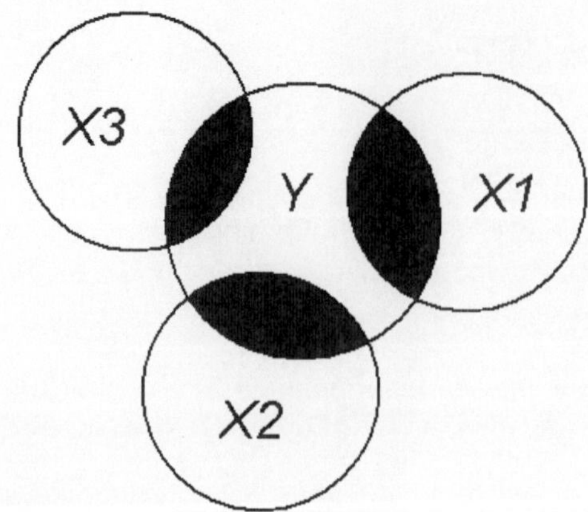

Figure 20.1. Venn diagram of uncorrelated IVs.

Correlated IVs entered together as a set

However, when IVs are **correlated** among themselves their individual correlations with the DV are **not additive**. Correlated IVs are common in the behavioural sciences.

When IVs are correlated the researcher is interested in their **unique** relationship with the DV, that is, that part of their relationship with the DV that cannot be attributed to, or is not associated with, their relationship with other IVs.

The next Venn diagram (Figure 20.2) depicts a DV (*Y*), and three **intercorrelated** IVs (*X1*, *X2*, and *X3*). The shaded area represents R^2: the proportion of variance in *Y* that is explained by the linear combination of IVs. The **dark** shaded areas represent the unique relationship for *X1*, *X2*, and *X3* respectively, that is, the relationship of each of these variables with *Y* after the effects of the other two *X* variables have been removed.

For example, the dark shaded area in the middle represents that portion of R^2 that is uniquely attributable to *X2*. It is known as the **squared semi-partial correlation**, and reflects the semi-partial correlation of *X2* with *Y*.

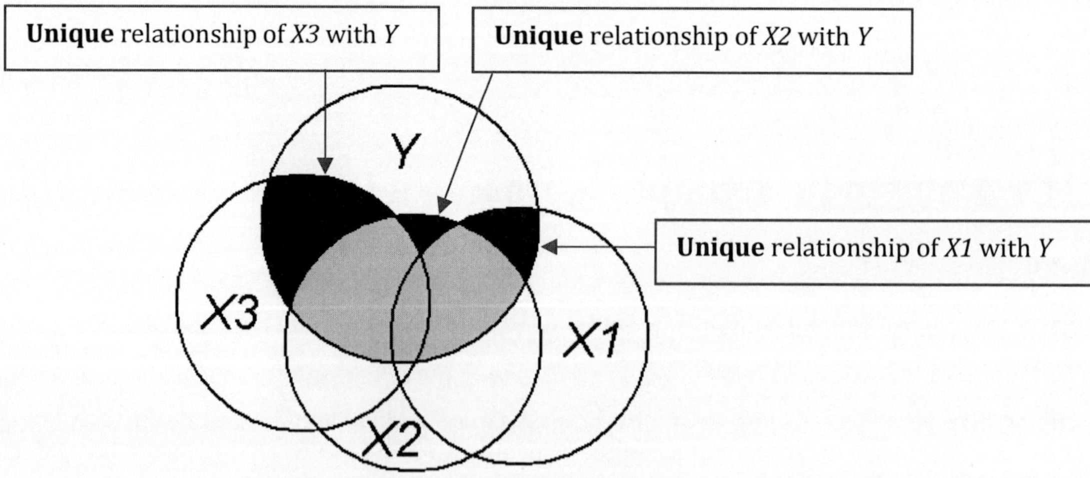

Figure 20.2. Venn diagram of correlated IVs entered together in standard multiple regression.

Note in Figure 20.2 that *X1*, *X2* and *X3* have similar areas of overlap with *Y*, but their unique overlap is rather smaller, especially for *X1* and *X2*. One would not be surprised to find significant correlations with *Y* for all three variables, but nonsignificant semi-partial correlations for *X1* and *X2*.

Figure 20.2 shows how predictor variables are treated in **standard** or **simultaneous** multiple regression. They are all entered together as a set and their **unique contribution** to predicting the criterion is assessed as to whether or not it is **significant**, that is, whether they have a **significant regression coefficient**. *X1* and *X2* probably would not in this example.

Although, strictly speaking, the unique relationships are **semi-partial correlations**, some texts speak in terms of **part correlations**, and **part regression coefficients**. There are also **partial correlations,** which are different.

- **A semi-partial (or part) correlation** is the relationship between *X1* and *Y*, when shared variance with other *X* variables has been removed from *X1* (and so on for other *X* variables).
- **A partial correlation** is the relationship between *X1* and *Y*, when shared variance with other *X* variables has been removed from **both** *X1* **and** *Y* (and so on for other *X* variables).

Correlated IVs entered in priority order

Figure 20.3 on the next page shows how the same three predictor variables might be entered in a **hierarchical** or **sequential** multiple regression. Here, variables are entered according to a priority order (e.g., according to a causal, theoretical, or practical ordering). In this example, *X1* is entered first and is attributed with **all the variance** it shares with *Y*; *X2* is entered second and is attributed with the variance it shares with *Y* after *X1* has taken its share; *X3* is entered third and is attributed with the variance it shares with *Y* after both *X1* and *X2* have taken their share.

This ordering makes a big difference to *X1* and *X2* in this example. They will probably have significant regression coefficients in this ordering.

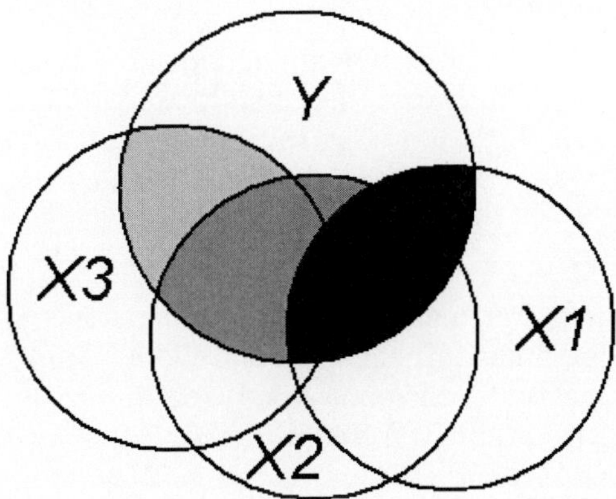

Figure 20.3. Venn diagram of correlated IVs entered according to a priority ordering in hierarchical multiple regression.

Assumptions and pitfalls in multiple regression

When performing multiple regression there are a number of assumptions and potential pitfalls of which one needs to be aware. As it is based on correlation, issues of linearity and restricted range apply. Other issues are addressed briefly as follows. Before conducting a multiple regression appropriate text books should be consulted. Recommended is Chapter 5 on multiple regression in Tabachnick and Fidell (2007), but you might also consult Field (2005), Howell (2002), and Stevens (2002).

Sample size

Stevens (2002, p.143) concludes that around 15 cases per predictor variable are sufficient for a reliable regression equation. However, in their discussion of the issue, Tabachnick and Fidell (2007, pp. 123-124) recommend the following rules-of-thumb for determining sample size (N):

- for testing a multiple correlation: $N \geq 50 + 8m$ (where m is the number of IVs or predictor variables);
- for testing individual predictors: $N \geq 104 + m$.

According to the former, with four IVs the sample size (N) would need to be 50 + (8 x 4), that is, 50+32, or 82. According to the latter, the sample size would need to be 108. Usually, one wants the larger of these two. Adopting the Stevens (2002) recommendation, only 60 cases would be needed (i.e., 4 x 15).

An even larger case-to-variable ratio is preferred if effect sizes are likely to be small, variables are not normally distributed, and/or variables are not very reliable (i.e., they have a high degree of measurement error). A version of multiple regression, known as **stepwise regression**, tends to capitalise on chance and needs even higher case-to-variable ratios of around 40:1 (Tabachnick & Fidell, 2007, p. 123).

Outliers

Researchers **must** always **check for** and **deal with** univariate and multivariate outliers.

Identifying and dealing with univariate outliers has already been described in the data screening chapter. However, multivariate outliers are more complicated and of two kinds, both of which should be deleted from the analysis. They can be a severe problem as there can sometimes be a large number of them, especially since other cases can become outliers when initial outliers are deleted.

Multivariate outliers with large standardised residual

SPSS identifies outliers that have an unusual relationship between the IVs and DV by listing cases with a large standardised residual; greater than three is the default used in SPSS, although Tabachnick and Fidell favour 3.29 which corresponds a little more exactly to a p of .001. The difference is not critical and the SPSS default can be used. The standardised residual is a residual z score.

Multivariate outliers among the IVs

These are cases that have an unusual pattern of scores among the IVs. They are identified by requesting the **Mahalanobis distance** for each case. Tabachnick and Fidell (2007, p. 166) deem multivariate outliers to be those cases with a Mahalanobis distance value greater than or equal to a critical value on the chi-square distribution (chi-square is another type of probability distribution, and has the symbol χ^2). The critical values can be obtained from their Table C.4 (p. 949), with df equal to the number of IVs and α =.001; or directly from the following internet site:
http://www.danielsoper.com/statcalc/calc12.aspx

For example, with four IVs the critical value would be 18.467. Any case with a Mahalanobis distance ≥ 18.467 would be a multivariate outlier.

Normality, linearity, homoscedasticity

These are all assumptions in multiple regression. Be sure to read the description of them in an appropriate text. Violations of normality, linearity, and homoscedasticity can be detected most easily from residual scatterplots as indicated in Tabachnick and Fidell (2007, pp. 125-127). Residuals are plotted along the y-axis for each of the standardised predicted values that are plotted along the x-axis.

Remember, a residual of zero indicates a completely accurate prediction, and if assumptions are met we would expect residuals, which are errors of prediction, to spread out evenly either side of the zero line, as shown in plot A below. Plot B illustrates a failure of normality; plot C a failure of linearity; and plot D a failure of homoscedasticity. A multivariate outlier with a large standardised residual is also illustrated in plot B.

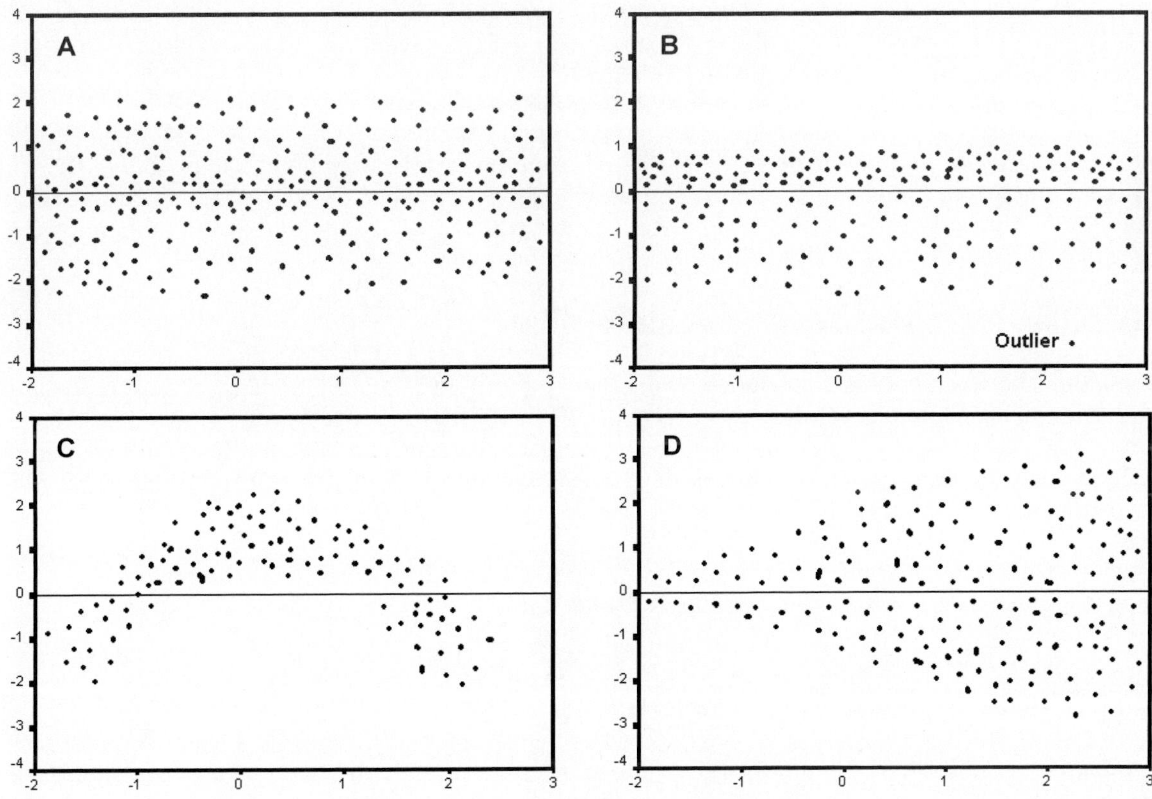

Violations can sometimes be due to outliers, which is why they need to be dealt with as a first step. If violations are present and no action is taken (e.g., no transformation of variables) the researcher should at least interpret results with extreme caution, especially when sample sizes are small.

Multicollinearity and singularity

These are perfect (singularity) or very high (multicollinearity) correlations among IVs. Correlations among IVs >.90 indicate multicollinearity. Most computer programs have default values for multicollinearity and will not admit variables that are too great a problem. The programs issue warnings accordingly. If the problem exists some of the highly intercorrelated IVs need to be deleted or perhaps combined. Because multiple regression gets much harder to interpret with highly correlated predictors, it is advisable to avoid intercorrelations of .80 or above, or even .70 (see Stevens, 2002; or Tabachnick & Fidell, 2007 for more discussion of this issue).

Singularity can occur when researchers make the mistake of including **both** sub-scale scores **and** total scale scores as predictors in a multiple regression equation.

Levels of measurement

The DV must be continuous (interval level), and IVs need to be continuous (interval level) or dichotomous. A categorical IV can be recoded into **dummy dichotomous variables**. To avoid singularity there must be one less dummy variable than categories. For example, suppose religion is a variable with four categories in your particular set of data (Catholic, Protestant, Methodist, Baptist). You would recode into three dummy variables as follows:

Catholic	1=Is a Catholic	0=Is not a Catholic
Protestant	1=Is a Protestant	0=Is not a Protestant
Methodist	1=Is a Methodist	0=Is not a Methodist

The fourth category (Baptist) is perfectly predicted from these three dummy variables. If a participant is not a Catholic, nor a Protestant, nor a Methodist he/she must be a Baptist in this particular set of data. See de Vaus (2002, pp. 328-330) and Field (2005, pp. 212-215) for a discussion of dummy variable analysis.

Interaction terms and curvilinear relationship

Interactions between IVs can be included by using their **product** as a predictor (e.g. *X1* x *X2* represents the interaction of *X1* and *X2*). However, to avoid multicollinearity the variables need to **centred**, that is, converted to deviation scores (score–the mean). Refer to Tabachnick and Fidell (2007, pp. 157-159) or other specialised texts for more information. Note also, that this flexibility of multiple regression means that it can be used as an alternative analysis technique for ANOVA designs.

Curvilinear relationships can be assessed in multiple regression by raising the IVs to powers of 2, 3, or 4 and so on. Specialised texts should also be consulted on this issue.

Standard multiple regression

Standard or **simultaneous** multiple regression, as we have already discovered, is where all the predictor variables are entered together as a single set or block.

Example of standard multiple regression

Imagine a computing lecturer is teaching a course on the use of computer statistical packages. The lecturer is interested in discovering those variables that are related to success in the course. This would enable her to predict students' course success from knowledge of their scores on the relevant variables. Perhaps, also, it would indicate ways of modifying the course to assist those at risk of failure. On the basis of her experience and review of the literature the lecturer hypothesises that the criterion (DV) of course success (i.e., final percent grade in the computing course) will be related to four predictor variables: (a) prior experience with computers (measured in months), (b) logical reasoning ability, (c) gender, and (d) attitude to technology.

Assume logical thinking ability is measured on a scale with a possible range of 0 to 50, with high scores indicative of greater logical thinking ability; and attitude to technology is measured on a scale with a possible range of 0 to 10, with high scores indicative of more positive or favourable attitudes to technology.

Imagine the lecturer wishes to test this hypothesis using the 52 students in her course. With four IVs, 52 is less than the desirable sample size of 82 for testing the multiple correlation (i.e., $N \geq 50 + 8m$), and 108 for testing individual predictors ($N \geq 104 + m$), but it is close to the recommendation of Stevens (2002) for around15 times as many cases as variables. Were this to be a real study, a question would arise as to whether it is worth proceeding? In many respects, the answer would be "probably not", not unless the effect sizes are very large, assumptions are met, and the measures are very reliable. For the sake of the exercise, assume the lecturer decides to proceed on the understanding that she will have to interpret the results with caution, given the low power of the study.

Using SPSS to conduct a standard multiple regression

Step 1: Enter data in SPSS and save data file

First, enter the data (listed on the next page) into SPSS and save it in a file called ***MultReg.sav***.

ID	Participant number	***Nominal*** *variable*
Compgrd	Computing grade percent	***Scale*** *variable*
Logic	Logical reasoning	***Scale*** *variable*
Exper	Months of computing experience	***Scale*** *variable*
Attitude	Attitude to technology	***Scale*** *variable*
Gender	0=Males 1=Females	***Nominal*** *variable*

By now you should be skilled in entering data, so if you want to save yourself the trouble, the datafile is available at: www.pearson.com.au/9781442549821. Click on the link 'Student downloads' to download the datafile.

Step 2: Screen data and check assumptions

Univariate assumptions should be checked as usual. The following SPSS procedures and associated output can then be used to check the assumptions of multiple regression.

48 How to: Evaluate multiple regression assumptions

Analyze
Regression
Linear...
[Select variable(s) you want]
Computing grade percent [Compgrd] ▸ **Dependent:**

Months of computing experience [Exper]
Logical reasoning [Logic]
Gender
Attitude to technology [Attitude] ▸ **Independent(s):**

Statistics...
[From **Residuals** select]
☑ **Casewise diagnostics**
Continue
Plots...
***ZPRED** ▸ **X**
***ZRESID** ▸ **Y**
[From **Standardized Residual Plots** select]
☑ **Normal probability plot**
Continue
Save...
[From **Distances** select]
☑ **Mahalanobis**
Continue

OK

ID	Compgrd	Logic	Exper	Attitude	Gender
1	53	44	24	4.0	0
2	83	24	56	7.5	0
3	66	43	55	3.5	1
4	42	22	1	2.5	1
5	54	21	9	4.0	0
6	40	9	12	1.5	1
7	91	46	64	8.0	1
8	70	47	24	8.5	0
9	77	42	17	6.5	1
10	51	43	11	3.0	1
11	67	41	48	4.5	0
12	56	20	55	3.5	0
13	62	11	29	6.5	1
14	43	16	14	2.0	1
15	88	13	62	8.0	1
16	92	46	72	9.0	0
17	57	23	49	3.5	1
18	63	22	21	7.0	0
19	94	28	62	8.5	0
20	85	27	42	8.0	0
21	55	40	10	4.0	0
22	55	31	64	5.0	1
23	42	3	7	1.5	1
24	46	26	21	2.0	1
25	77	36	24	7.5	0
26	68	40	52	4.0	0
27	96	48	68	9.0	1
28	63	19	23	6.5	1
29	97	46	41	8.0	0
30	48	28	21	3.0	1
31	62	39	52	3.5	1
32	77	25	26	7.5	0
33	51	41	11	3.0	1
34	74	40	21	6.0	1
35	85	25	57	9.0	0
36	53	4	20	3.0	0
37	92	42	58	7.0	0
38	72	3	26	7.0	1
39	51	6	11	2.5	0
40	39	5	3	1.0	1
41	86	20	54	7.5	0
42	44	28	12	4.5	1
43	95	48	59	8.0	0
44	93	46	62	8.5	0
45	77	38	22	7.5	0
46	56	26	45	4.5	0
47	57	23	50	4.0	0
48	52	46	10	1.5	1
49	54	44	15	4.5	1
50	66	36	54	5	1
51	76	49	93	.5	1
52	98	5	4	1.0	1

The complete regression output will be produced by SPSS, but only the sections relevant to assumption checking are included here.

Casewise Diagnostics[a]

Case Number	Std. Residual	Computing grade percent	Predicted Value	Residual
52	5.231	98	40.85	57.145

a. Dependent Variable: Computing grade percent

1

Casewise Diagnostics, if present, identify outliers where there is an unusual relationship between the IVs and DV. Here it is Case Number 52. (Note this is the SPSS number at the left of the Data View window. It may or may not correspond with the ID that is entered.) Case Number 52 in this example does correspond with ID 52. She has a high computing grade score, despite being a female with little experience, poor logical reasoning, and a terrible attitude. You should easily see that this is very different to the other participants.

Residuals Statistics[a]

	Minimum	Maximum	Mean	Std. Deviation	N
Predicted Value	40.60	93.47	67.13	14.406	52
Std. Predicted Value	-1.842	1.828	.000	1.000	52
Standard Error of Predicted Value	2.276	6.774	3.316	.701	52
Adjusted Predicted Value	33.97	93.62	66.96	14.598	52
Residual	-18.943	57.145	.000	10.487	52
Std. Residual	-1.734	5.231	.000	.960	52
Stud. Residual	-1.814	5.537	.007	1.016	52
Deleted Residual	-20.725	64.031	.174	11.788	52
Stud. Deleted Residual	-1.861	9.290	.079	1.457	52
Mahal. Distance	1.233	18.625	3.923	2.551	52
Cook's Distance	.000	.739	.026	.106	52
Centered Leverage Value	.024	.365	.077	.050	52

a. Dependent Variable: Computing grade percent

2

Residual Statistics indicate that the highest Mahalanobis distance score is 18.625. This exceeds the critical value of 18.467 for four IVs. The next step is to look at the Data View window. SPSS has added a new variable **MAH_1** to the data file.

If you look through this variable you will find that it is ID 51 who has a Mahalanobis distance value > 18.467. Again, it is clear she has an unusual pattern among the IVs: a female with a poor attitude is OK, but with a lot of experience and very good logical reasoning ability?

To deal with these outliers:

Delete the multivariate outliers (IDs 51 and 52) and **run the assumption check again**. There should no longer be any outliers identified.

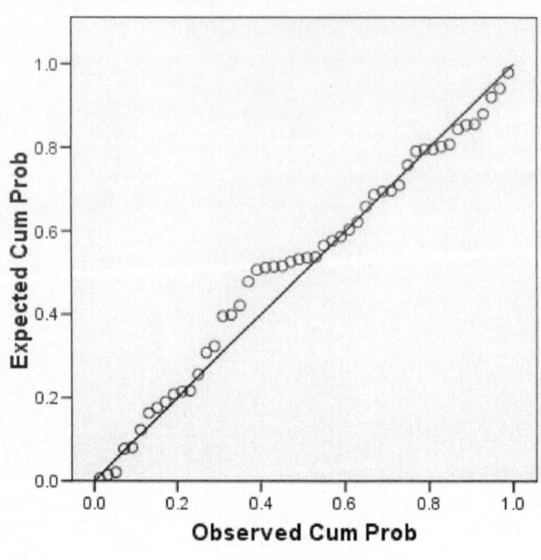

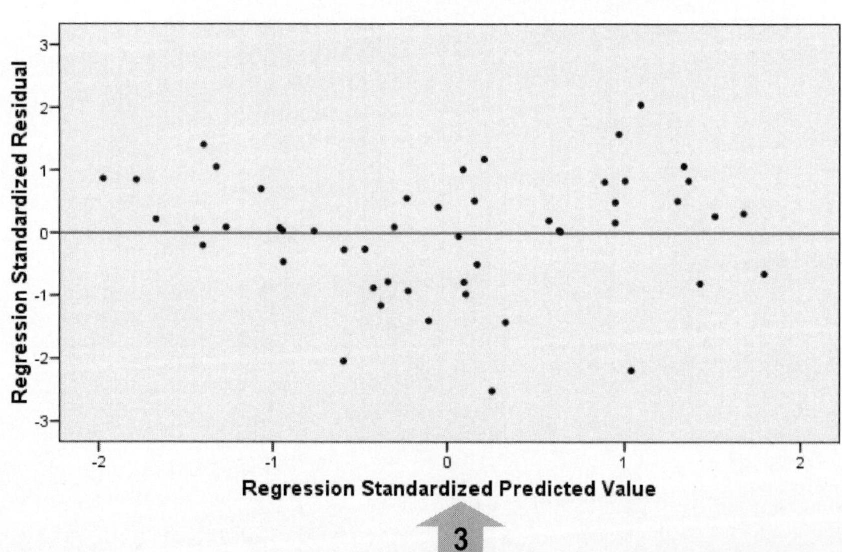

3

These are the plots after removing outliers 51 and 52.

The **normality plot** at the top suggests some departure from normality (it should follow the straight line).

The **residual scatterplot** at the bottom also suggests less than perfect assumptions. The points should be evenly spread either side of the centre line. However, with small sample sizes coherent plots are much less likely. It is difficult to say what is going on here—all the more reason why the results will need to be interpreted with caution. (You could investigate further, or try transforming variables.)

Note that I have added the 0 reference line to the plot. You should double-click on your own plot to open **Chart Editor** and find out how to do this yourself (*Hint:* Look under **Options**).

Step 3: Request the standard multiple regression analysis

Assuming the decision is to proceed with the data as they are, a standard multiple regression analysis, where all the independent variables are entered together in one block, is requested as follows, but remember to exclude those two outliers from the analysis:

49 How to: Conduct a standard multiple regression analysis

Analyze
Regression
Linear...
[Select variable(s) you want]
Computing grade percent [Compgrd] ▸ **Dependent:**

Months of computing experience [Exper]
Logical reasoning [Logic]
Gender
Attitude to technology [Attitude] ▸ **Independent(s):**

Statistics...
[From **Regression Coefficients** select in addition to default ☑ **Estimates** and ☑ **Model Fit**]
☑ **Confidence intervals**
[For **Level (%):** type] **95** [assuming it is not already entered]
☑ **Descriptives**
Continue
OK

Step 4: Examine and interpret the standard multiple regression output

Output: **Standard multiple regression**

Regression

Descriptive Statistics

	Mean	Std. Deviation	N
Computing grade percent	66.34	17.573	50
Months of computing experience	34.5200	21.16238	50
Logical reasoning	29.80	13.837	50
Gender	.50	.505	50
Attitude to technology	5.300	2.4349	50

1

Examine **Descriptive Statistics.**
Look especially at standard deviations to obtain an indication of any range restriction.

Correlations

		Computing grade percent	Months of computing experience	Logical reasoning	Gender	Attitude to technology
Pearson Correlation	Computing grade percent	1.000	.715	.432	-.408	.909
	Months of computing experience	.715	1.000	.317	-.269	.588
	Logical reasoning	.432	.317	1.000	-.117	.337
	Gender	-.408	-.269	-.117	1.000	-.407
	Attitude to technology	.909	.588	.337	-.407	1.000
Sig. (1-tailed)	Computing grade percent	.	.000	.001	.002	.000
	Months of computing experience	.000	.	.012	.029	.000
	Logical reasoning	.001	.012	.	.210	.008
	Gender	.002	.029	.210	.	.002
	Attitude to technology	.000	.000	.008	.002	.
N	Computing grade percent	50	50	50	50	50
	Months of computing experience	50	50	50	50	50
	Logical reasoning	50	50	50	50	50
	Gender	50	50	50	50	50
	Attitude to technology	50	50	50	50	50

2

Examine the **Correlation matrix**. Note that all the predictors have significant bivariate correlations with the criterion. (The Sig. level is **1-tailed** by default, and has to be doubled to give the 2-tailed significance level).

Note also the extent to which predictors are intercorrelated among themselves. Ensure there is no suggestion of multicollinearity (intercorrelations > .90).

Model Summary

Model	R	R Square	Adjusted R Square	Std. Error of the Estimate
1	.942[a]	.887	.877	6.159

a. Predictors: (Constant), Attitude to technology, Logical reasoning, Gender, Months of computing experience

3

Note the values of R and R Square (R^2) in the **Model Summary**.

Adjusted R Square

Due to shrinkage, R^2 in one sample will tend to be less in another sample. Before using a regression equation for predicting scores in other samples, it should be **cross-validated** in other samples (or by dividing the sample in half if it is large enough, and analysing each half separately) to ensure it has acceptable generalisability. Cross-validation is especially important for small samples.

Adjusted R Square relates to shrinkage. It is the estimated R^2 in the population. The sample value tends to overestimate the population value, more so in smaller samples.

The **Std. Error of the Estimate** is the standard error of the predicted criterion scores ($\hat{Y}$).

ANOVA[b]

Model		Sum of Squares	df	Mean Square	F	Sig.
1	Regression	13424.263	4	3356.066	88.475	.000[a]
	Residual	1706.957	45	37.932		
	Total	15131.220	49			

a. Predictors: (Constant), Attitude to technology, Logical reasoning, Gender, Months of computing experience

b. Dependent Variable: Computing grade percent

4

From the **ANOVA** table, establish if the multiple correlation coefficient (R) is significant or not. It is significant if Sig. < α (.05 unless otherwise stipulated). Clearly it is significant here.

Coefficients[a]

Model		Unstandardized Coefficients		Standardized Coefficients	t	Sig.	95.0% Confidence Interval for B	
		B	Std. Error	Beta			Lower Bound	Upper Bound
1	(Constant)	28.610	3.067		9.329	.000	22.434	34.787
	Months of computing experience	.211	.052	.254	4.044	.000	.106	.316
	Logical reasoning	.137	.068	.108	2.009	.051	.000	.275
	Gender	-1.380	1.909	-.040	-.723	.473	-5.225	2.465
	Attitude to technology	5.103	.481	.707	10.616	.000	4.135	6.072

a. Dependent Variable: Computing grade percent

5

From the **Coefficients** table, determine the regression equation: the values of the **Constant** (*a*) and the **regression coefficients** (B). Note their Standard Errors and standardised regression coefficients (Beta).

The significance of the *t* tests is important. These test the regression coefficients (B and Beta). If significant they indicate that the predictor makes a significant **unique** contribution to the regression (i.e., to predicting the criterion).

In this example, only **experience** and **attitude** are significant unique predictors. The other two variables add nothing significant to predicting the criterion.

However, the sample size is not large enough to provide reliable tests of the significance of the predictors. In the case of logical reasoning, for example, there is a strong chance of a Type II error (i.e., it **does** make a real unique contribution, but the study is not powerful enough to detect it). This is a good example of the problem that arises when samples are too small. The accuracy of conclusions can become highly doubtful.

Hierarchical multiple regression

Standard multiple regression is appropriate if all one is interested in doing is **predicting** the criterion, without necessarily wanting to understand too much about what is going on. It is also more straightforward the **less** intercorrelated the predictors.

However, theoretical explanation tends to be best served by **hierarchical** or **sequential** regression. Here, the researcher enters the predictor variables in a defined order, or sequence of steps, and assesses their predictive role at each step.

To illustrate, using the current example, suppose the lecturer had approached the whole topic differently from the very beginning. On the basis of theory and a practical approach to the topic, logical reasoning is hypothesised as a critical causal determinant of course success. Moreover, it may be possible to enhance performance by teaching an introductory course on logical reasoning. It may also be possible to teach introductory courses in computing to provide students with a basic level of prior experience. The question thus becomes: assuming logical reasoning is causally prior, how much of the variance in computing grades does it explain when it is entered as the first predictor variable; then how much variance does experience add to logical reasoning when entered second; then, after these two variables (which we might be able to enhance with training) have been entered and thus controlled for, how much additional variance is explained by attitude? Finally, after all three of these prior variables have been taken into account, does gender (about which we can do nothing) add anything significant to the prediction of computing grade? In other words, is there anything else about gender that is predictive of course success?

Note that one of the problems with hierarchical regression is that there are all sorts of possible orderings, based on all sorts of criteria (e.g., theoretical causation, practicalities, logic, controlling for nuisance variables). It is **not appropriate** to try them all and choose the "best" one. This defeats the whole purpose of testing a model according to a particular criterion determined at the outset. Worse still, it capitalises on chance. The "best" fitting model may only be the best fit because of chance characteristics of the **particular sample**. In general, the better a model fits a sample as a result of a "fishing expedition", the less likely it is to generalise to other samples, or to the population.

It can be appropriate, however, to test between limited numbers of specific, theoretically competing models

Using SPSS to conduct a hierarchical multiple regression

Step 1: Enter data in SPSS and save data file

The data, for the purposes of the exercise, are the same as for the standard example.

Step 2: Screen data and check assumptions

Again, for the purposes of the exercise, this will be the same as for the standard example. Remember to delete the multivariate outliers (IDs 51 and 52).

Step 3: Request the hierarchical multiple regression analysis

Note below how hierarchical regression differs from standard regression in that each predictor variable is entered separately in its own block. Depending on the structure of a hierarchical regression, predictors may be entered singularly in blocks (as here), or two or more predictors may be entered together in blocks.

50 How to: Conduct a hierarchical multiple regression analysis

Analyze
Regression
Linear...
[Select variable(s) you want]
Computing grade percent [Compgrd] ▶ **Dependent:**
Logical reasoning [Logic] **Block 1 of 1** ▶ **Independent(s):**
[Click on] **Next**
Months of computing experience [Exper] **Block 2 of 2** ▶ **Independent(s):**
[Click on] **Next**
Attitude to technology [Attitude] **Block 3 of 3** ▶ **Independent(s):**
[Click on] **Next**
Gender **Block 4 of 4** ▶ **Independent(s):**
[Click on] **Next**

Statistics...
[From **Regression Coefficients** select in addition to default ☑ **Estimates** and ☑ **Model Fit**]
☑ **Confidence intervals** [And for **Level (%):** type] **95** [assuming it is not already entered]
☑ **R squared change**
☑ **Descriptives**
Continue
OK

Step 4: Examine and interpret the hierarchical multiple regression output

Regression

1 2

Examine the **Descriptive Statistics** and the **correlation matrix**.

They are not reproduced here as they are the same in this particular example as for the standard regression.

Model Summary

					Change Statistics				
Model	R	R Square	Adjusted R Square	Std. Error of the Estimate	R Square Change	F Change	df1	df2	Sig. F Change
1	.432[a]	.186	.169	16.016	.186	10.988	1	48	.002
2	.747[b]	.558	.539	11.932	.371	39.475	1	47	.000
3	.941[c]	.886	.878	6.127	.328	132.270	1	46	.000
4	.942[d]	.887	.877	6.159	.001	.523	1	45	.473

a. Predictors: (Constant), Logical reasoning

b. Predictors: (Constant), Logical reasoning, Months of computing experience

c. Predictors: (Constant), Logical reasoning, Months of computing experience, Attitude to technology

d. Predictors: (Constant), Logical reasoning, Months of computing experience, Attitude to technology, Gender

3

Note the value of *R* and R^2 in **each step** of the **Model Summary**.

R Square Change

Note also the value of R Square Change, and whether or not it is significant. This specifies the **additional** proportion of variance that is **uniquely** explained by each variable **at its point of entry**, and whether or not it adds **significantly** to the variables already entered in previous steps.

The first three variables each explain significant amounts of additional variance, but gender at the final step adds nothing significant.

ANOVA[e]

Model		Sum of Squares	df	Mean Square	F	Sig.
1	Regression	2818.618	1	2818.618	10.988	.002[a]
	Residual	12312.602	48	256.513		
	Total	15131.220	49			
2	Regression	8439.218	2	4219.609	29.636	.000[b]
	Residual	6692.002	47	142.383		
	Total	15131.220	49			
3	Regression	13404.441	3	4468.147	119.028	.000[c]
	Residual	1726.779	46	37.539		
	Total	15131.220	49			
4	Regression	13424.263	4	3356.066	88.475	.000[d]
	Residual	1706.957	45	37.932		
	Total	15131.220	49			

a. Predictors: (Constant), Logical reasoning

b. Predictors: (Constant), Logical reasoning, Months of computing experience

c. Predictors: (Constant), Logical reasoning, Months of computing experience, Attitude to technology

d. Predictors: (Constant), Logical reasoning, Months of computing experience, Attitude to technology, Gender

e. Dependent Variable: Computing grade percent

From the **ANOVA** table, establish if the **multiple correlation coefficient** (R) is significant or not for the variables in the equation **at each step**.

Coefficients[a]

Model		Unstandardized Coefficients		Standardized Coefficients	t	Sig.	95.0% Confidence Interval for B	
		B	Std. Error	Beta			Lower Bound	Upper Bound
1	(Constant)	50.006	5.423		9.221	.000	39.102	60.910
	Logical reasoning	.548	.165	.432	3.315	.002	.216	.881
2	(Constant)	39.305	4.385		8.964	.000	30.484	48.126
	Logical reasoning	.289	.130	.228	2.224	.031	.028	.550
	Months of computing experience	.534	.085	.643	6.283	.000	.363	.705
3	(Constant)	27.320	2.481		11.012	.000	22.326	32.314
	Logical reasoning	.136	.068	.107	1.999	.052	-.001	.273
	Months of computing experience	.212	.052	.256	4.102	.000	.108	.317
	Attitude to technology	5.214	.453	.722	11.501	.000	4.301	6.126
4	(Constant)	28.610	3.067		9.329	.000	22.434	34.787
	Logical reasoning	.137	.068	.108	2.009	.051	.000	.275
	Months of computing experience	.211	.052	.254	4.044	.000	.106	.316
	Attitude to technology	5.103	.481	.707	10.616	.000	4.135	6.072
	Gender	-1.380	1.909	-.040	-.723	.473	-5.225	2.465

a. Dependent Variable: Computing grade percent

5

From the **Coefficients** table, determine the regression equation: the values of the Constant (*a*) and the regression coefficients (B), also Standard Errors and standardised regression coefficients (Beta) for variables in the equation **at each step**. Note that the final step, when all the variables are entered, will be the same as standard regression.

In this example, the big difference between the hierarchical model and the standard regression lies in what happens to **logical reasoning**. When entered **first** it is a **significant predictor**. It only ceases to make a significant **unique** contribution at Step 3, after attitude to technology is entered.

Different types of hierarchical model can sometimes dramatically change the results and interpretation, which is why the model to be tested must be **decided prior** to the study, and according to **sound criteria**.

Stepwise or statistical multiple regression

A final type of regression is stepwise or statistical regression. Essentially, it is used when statistical criteria (i.e., the sizes of correlations and semi-partial correlations) are used to determine the set of predictors that provides the **best prediction** of the criterion. It can be appropriate when prediction is all you want to achieve. Generally, stepwise is not the approach of choice when the goal is theoretical understanding and explanation.

Stepwise should be used with caution as it **capitalises on chance**, because the exact sizes of correlations reflect chance characteristics of the sample. For this reason, it requires even larger sample sizes (e.g., case to variable ratios of 40:1 or more) to ensure more reliable correlations. With adequate sample size and robust solutions it can be the case that standard, hierarchical, and stepwise regression lead to very similar conclusions.

Using SPSS to conduct a stepwise multiple regression

Step 1: Enter data in SPSS and save data file

The data, for the purposes of the exercise, are the same as for the standard regression.

Step 2: Screen data and check assumptions

Again, for the purposes of the exercise, this will be the same as for the standard example. Remember to delete the multivariate outliers (IDs 51 and 52).

Step 3: Request the stepwise multiple regression analysis

51 **How to:** **Conduct a stepwise multiple regression analysis**

Analyze
Regression
Linear...
[Select variable(s) you want]
Computing grade percent [compgrd] ▸ **Dependent:**

Months of computing experience [Exper]
Logical reasoning [Logic]
Gender
Attitude to technology [Attitude] ▸ **Independent(s):**

[For **Method** select] **Stepwise** ▾

Statistics...
[From **Regression Coefficients** select in addition to default ☑ **Estimates** and ☑ **Model Fit**]
☑ **Confidence intervals** [And for **Level (%):** type] **95** [assuming it is not already entered]
☑ **R squared change**
☑ **Descriptives**
Continue

OK

Step 4: Examine and interpret the stepwise multiple regression output

Regression

Examine the **Descriptive Statistics** and the **correlation matrix**.

They are not reproduced here as they are the same in this particular example as for the standard regression.

Model Summary

Model	R	R Square	Adjusted R Square	Std. Error of the Estimate	Change Statistics				
					R Square Change	F Change	df1	df2	Sig. F Change
1	.909[a]	.826	.823	7.398	.826	228.467	1	48	.000
2	.936[b]	.876	.871	6.319	.050	18.790	1	47	.000

a. Predictors: (Constant), Attitude to technology

b. Predictors: (Constant), Attitude to technology, Months of computing experience

In stepwise regression variables are entered in steps according to statistical criteria. Basically, the predictor with the highest correlation with the criterion is entered first, followed by the next most highly correlated predictor that has the lowest correlation with the first predictor, and so on.

Note the value of *R* and R^2 in **each step** of the **Model Summary**.

R Square Change

R Square Change will always be significant. Nonsignificant predictors will not enter the model.

ANOVA[c]

Model		Sum of Squares	df	Mean Square	F	Sig.
1	Regression	12504.147	1	12504.147	228.467	.000[a]
	Residual	2627.073	48	54.731		
	Total	15131.220	49			
2	Regression	13254.446	2	6627.223	165.965	.000[b]
	Residual	1876.774	47	39.931		
	Total	15131.220	49			

a. Predictors: (Constant), Attitude to technology

b. Predictors: (Constant), Attitude to technology, Months of computing experience

c. Dependent Variable: Computing grade percent

4

From the **ANOVA** table, the **multiple correlation coefficient** (R) will be significant for the variables in the equation at each step.

Coefficients[a]

Model		Unstandardized Coefficients		Standardized Coefficients	t	Sig.	95.0% Confidence Interval for B	
		B	Std. Error	Beta			Lower Bound	Upper Bound
1	(Constant)	31.568	2.527		12.491	.000	26.487	36.649
	Attitude to technology	6.561	.434	.909	15.115	.000	5.688	7.433
2	(Constant)	29.872	2.194		13.617	.000	25.459	34.286
	Attitude to technology	5.391	.459	.747	11.757	.000	4.469	6.314
	Months of computing experience	.229	.053	.275	4.335	.000	.123	.335

a. Dependent Variable: Computing grade percent

5

From the **Coefficients** table, determine the regression equation: the values of the Constant (*a*) and the regression coefficients (B), also Standard Errors and standardised regression coefficients (Beta) for variables in the equation **at each step**.

Again, all predictors in the model will be significant, as those predictors that do not add a significant proportion of variance do not enter the model at all.

Stepwise regression in this example leads to the conclusion that the best prediction of (and the best equation for) computing grade is made by the combination of **attitude** and **experience** in that order. Logical reasoning and gender are not even included, because they do not add anything significant to the prediction of computing grade.

While this may be useful for prediction, it may convey a misleading idea of the theoretical explanation for computing performance, and one that will not generalise to the population (especially in view of the theoretical importance of logical reasoning proposed in the hierarchical version).

Hierarchical multiple regression: Research report sample Results section

Results

With alpha set at .05, a hierarchical multiple regression was performed between computing grade as the criterion, and the four predictor variables entered in the order of logical reasoning, followed by months of computing experience, then attitude to technology, and finally gender. Results of evaluation of assumptions were satisfactory after one outlier with a standardised residual > 3, and another with an extreme Mahalanobis distance score ($p < .001$) were deleted from the analysis. The small sample size ($N = 50$) rendered it necessary to interpret the results with caution.

Table 1 shows descriptive statistics and variable intercorrelations. Table 2 displays the unstandardised regression coefficients (B), standardised regression coefficients (β), and R^2 change for the predictors at their step of entry, together with the final B and β. After step 4, with all predictors in the equation, $R = .94$, $R^2 = .89$, adjusted $R^2 = .88$, $F(4, 45) = 88.48$, $p < .001$.

After step 1, with logical reasoning in the equation, $R = .43$, $R^2 = .19$, $F(1,48) = 10.99$, $p = .002$. After step 2, with months of computing experience added to the equation, $R = .75$, $R^2 = .56$, $F(2,47) = 29.64$, $p < .001$; and after step 3, with attitude to technology added to the two prior variables in the equation, $R =.94$, $R^2 = .89$, $F(3,46) = 119.03$, $p < .001$. Addition of gender at step 4 did not reliably improve R^2.

Table 1

Descriptive Statistics and Intercorrelations Among Logical Reasoning, Months of Computing Experience, Attitude to Technology, Gender; and Computing Grade as the Criterion

Variables	Grade (Criterion)	Reasoning	Experience	Attitude	Gender
Logical reasoning ability	.43**				
Computing experience	.72***	.32*			
Attitude to technology	.91***	.34*	.59***		
Gender	-.41**	-.12	-.27	-.41**	
M	66.34	29.80	34.52	5.30	0.50
SD	17.57	13.84	21.16	2.44	0.51

*** $p < .001$. ** $p < .01$. * $p < .05$.

Table 2

Hierarchical Multiple Regression of Logical Reasoning, Months of Computing Experience, Attitude to Technology, and Gender as Predictors of Computing Grade[1]

Step	Variables	B at Step	β at Step	R^2 change at Step	Final B	Final β
1	Logical reasoning ability	0.55	0.43	.19*	0.14	0.11
2	Computing experience	0.53	0.64	.37**	0.21	0.25**
3	Attitude to technology	5.21	0.72	.33**	5.10	0.71**
4	Gender	-1.38	-0.04	.00	-1.38	-0.04
	Constant				28.61	

** p < .001. * p < .01.

[1] See pages 144-145 of *APA Manual* for other examples of regression tables, including an example of how to report confidence intervals.

General comment about using multiple regression

Tabachnick and Fidell (2007) provide much more detail on multiple regression, and give further examples of how to write up Results. If you ever need to use multiple regression in your research it is recommended that you consult this text. Stevens (2002) and also Field (2005) are worth consulting as well, not to mention many other specialised texts that are available.

Path analysis (and structural equation modelling)

Correlation does not equal causation, but an extension of correlational design involves constructing causal models on the basis of theory and testing how well they fit observed data. When such structural models provide a good fit to the data the model receives empirical support. Such modelling is known as **structural equation modelling** or SEM (see Tabachnick & Fidell, 2007, Ch. 14).

A simplified version, modelling observed variables, is known as **path analysis**. A variable can affect another variable **directly**, or **indirectly** through other **mediator** variables. A path analysis involves developing a theoretical model of direct and indirect paths among a set of variables, and testing which paths are significant.

Example of path analysis

Figure 16.4 is an example of a path model of the relationship between coping humour and negative life events with depression, mediated by positive and negative affect (Hewson, 1999).

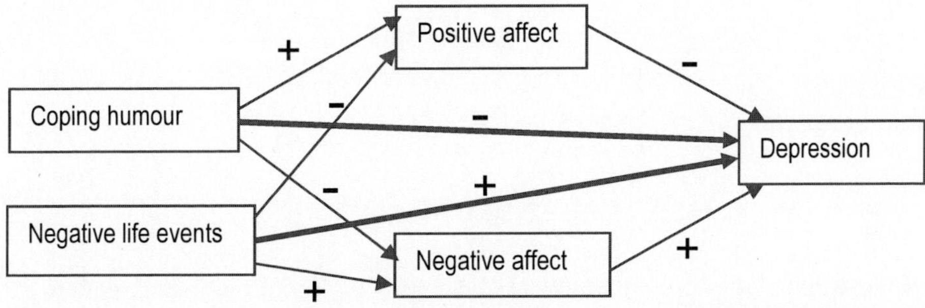

Figure 20.4. Model of the hypothesised association between coping humour and negative life events (predictors), and depression (outcome), via the mediating variables of positive and negative affect.
Note: Plus signs indicate hypothesised positive associations, and minus signs negative associations.

In Figure 20.4 the bold lines represent **direct** paths from coping humour to depression and negative life events to depression. Other lines refer to **indirect** paths.

Conducting a path analysis

In order to conduct a path analysis there must first be a significant correlation between the predictors and outcome variables, in this case between coping humour and depression and between negative life events and depression.

Then, a series of standard multiple regression analyses are performed in which variables are regressed on **prior variables in the model**. In this particular example, there would be three regressions:

- Regressing **positive affect** on the prior variables of **coping humour** and **negative life events**; the regression coefficients are the path coefficients for the paths from coping humour to positive affect, and negative life events to positive affect.
- Regressing **negative affect** on the prior variables of **coping humour** and **negative life events**; the regression coefficients are the path coefficients for the paths from coping humour to negative affect, and negative life events to negative affect.
- Regressing **depression** on the prior variables of **positive affect**, **negative affect**, **coping humour**, and **negative life events**; the regression coefficients are the path coefficients for the paths from positive affect to depression, negative affect to depression, coping humour to depression, and negative life events to depression.

When there is only one predictor or when the predictors are uncorrelated it is possible to decompose the correlations between the predictors and outcome variables into their **direct** and **indirect** paths, and to establish which of these paths are significant.

In this study by Hewson (1999), the results were as follows.

The correlation between coping humour and depression (-.26) and negative life events and depression (.34) were both significant ($p < .001$). The regression analyses provided the following path coefficients:

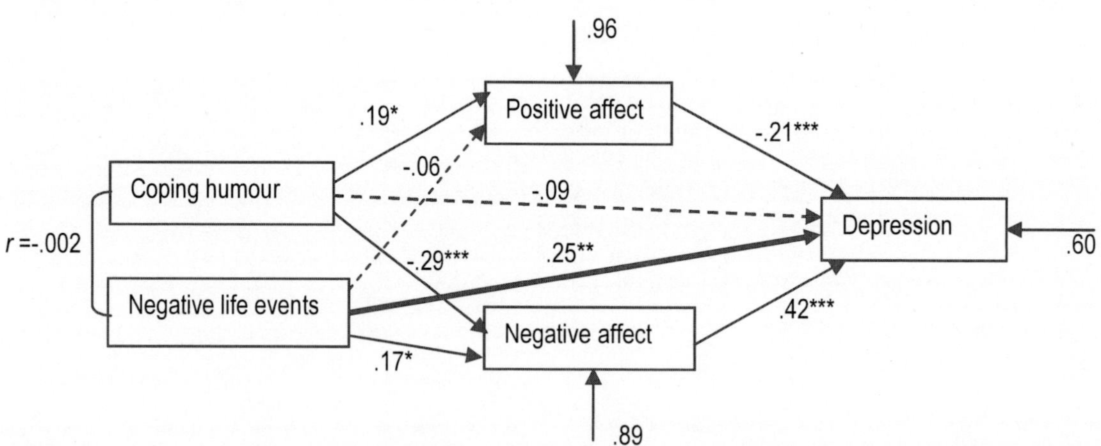

Figure 20.5. Path coefficients for the associations between coping humour and negative life events and depression, mediated by positive and negative affect. Standardised betas are shown for all paths.
* $p < .05$, ** $p < .01$, *** $p < .001$. $N = 186$.
Note: Broken lines indicate non-significant paths; curved lines represent correlations.

Figure 20.5 shows that all the hypothesised paths from Figure 20.4 were significant, with the exception of the path from negative life events to positive affect, and the direct path from coping humour to depression.

Indirect path coefficients are found by multiplying the coefficients for the relevant paths. For example, the indirect path from coping humour to depression via positive affect is .19 (coping humour-positive affect) multiplied by -.21 (positive affect-depression), that is, -.0399.

Table 20.1 shows the correlations between coping humour and depression (-.26) and negative life events and depression (.34) decomposed into their direct and indirect paths. Correlations equal total effects, but only when the predictors are uncorrelated (i.e., zero), as is very nearly the case here.

The conclusion in this study was that there was a small but significant negative association between coping humour and depression, but it was an **indirect** one that operated through the intermediaries of negative and positive affect. That is, high coping humour tended to accompany low negative affect and high positive affect, which in turn tended to accompany lower depression. There was no **direct** association between coping humour and depression.

The association between negative life events and depression was both a **direct** one (unrelated to affect) and an **indirect** one **mediated** by negative affect, but not by positive affect.

In Figure 20.5 depression has an external arrow leading to it with a value of .60. This is error variance, the variance in depression unexplained by the variables in the model. That is, 60% of the variance in depression has nothing to do with the variables in the model, and only 40% is explained by the variables in the model. Similarly 89% of the variance in negative affect and 96% of the variance in positive affect is unexplained by the prior variables of coping humour and negative life events.

Table 20.1
Effects of Coping Humour and Negative Life Events on Depression, as Mediated by Positive and Negative Affect (N = 186)

	Path Coefficients		Effect Coefficient
Coping Humour- Depression			
Indirect via Positive Affect	.19	-.21	-.04*
Indirect via Negative Affect	-.29	.42	-.12*
Indirect Effect			-.16*
Direct Effect			-.09
Total Effect			-.25**
Negative Life Events- Depression			
Indirect via Positive Affect	-.06	-.21	.01
Indirect via Negative Affect	.17	.42	.07*
Indirect Effect			.08*
Direct Effect			.25*
Total Effect			.33**

* p < .05. **p < .001.

Analysis programs specially designed for structural equation modelling (e.g., LISREL, and AMOS which is available in SPSS) automatically perform a path analysis, as well as more complex analyses, and automatically determine the significance of both direct and indirect paths. However, path analysis can also be performed by a series of multiple regressions, as described, and the significance of indirect paths can be assessed by supplying the necessary information (i.e., the regression coefficients and their standard errors) to the following internet site: http://www.danielsoper.com/statcalc/calc31.aspx

Sample size

Path analysis ideally requires large samples. Recommendations vary, but one rule-of-thumb is that there should be at least 10 participants for each path parameter to be estimated, but this assumes large effect sizes. Ideally, try to at least double this recommendation for smaller effect sizes.

Mediation and moderation

For a useful discussion of moderation and mediation see Baron and Kenny (1986).

Mediator variables can be tested using path analysis. **Mediation** occurs when the effect of one variable on another actually occurs through intervening variables (mediators).

Moderation occurs when the effect of one variable on another is different at the different levels of another variable. **Moderation** is **interaction** and is tested in multiple regression by including an interaction term.

In this example, one possibility was that negative life events moderated the relationship between coping humour and depression, that is, that the relationship between coping humour and depression is different for people experiencing high negative life events compared to those experiencing low negative life events. This was tested initially in a multiple regression with depression as the criterion; and coping humour, negative life events, and their interaction (coping humour x negative life events) as predictors. The interaction was not significant so both variables were included in the path analysis. Had the interaction been significant it would have been necessary to perform the path analysis of coping humour, positive and negative affect, and depression; separately for samples of high and low negative life events (possibly using a **median split**[1] of negative life events).

[1] A median split involves splitting a sample on a continuous variable at the **median** to produce equal sized "high" and "low" groups. Scores exactly on the median can be randomly assigned to the groups to achieve equal *n*. Median splits are often used to form IV groups for ANOVA designs. They are problematic, however, if there are no really high or low scores, so that all you end up with is a middling-high group versus a middling-low group. In such an event the IV might be too weak to produce significant effects.

21 Reliability Analysis

Measuring instruments, measuring scales, or psychological tests need to have acceptable reliability and validity. There are two main types of reliability: test-retest and internal consistency.

Test-retest reliability

This refers to a scale's **temporal** reliability, or the consistency with which it measures whatever it measures **over time**, that is, from one occasion to the next.

Test-retest reliability is relatively easy to estimate by simply measuring a group of participants on the same scale on two occasions, then correlating the two sets of scores. A **high positive** correlation indicates that participants have a similar rank ordering on the two occasions (i.e., high scorers on the first administration of the scale tend to be higher scorers on the second, and low scorers on the first administration tend to be low scorers on the second).

Acceptable test-retest reliability is usually considered to be in the range of .80 to .90 or above, although correlations above .70 are usually acceptable for research purposes.

Issues to be considered in test-retest reliability are the nature of the construct being measured, especially the time period over which it is expected to remain stable; and the appropriateness of the time interval between the two administrations of the scale. If the interval is too long there may have been real changes in participants' scores over that time period. If the interval is too short participants may remember their earlier responses, or practice or other **carry-over** effects may influence their scores. Intervals of two weeks to two months are often appropriate, although this depends on the construct being measured. It would not be appropriate, for instance, for something like mood, which we know can change over short time periods.

Parallel or **alternate** forms of a scale are useful in counteracting some of the problems with test-retest reliability. These are versions of the same instrument that have been shown to be equivalent, but because they contain different items they overcome the problem of participants simply remembering their prior response.

Internal consistency reliability

This refers to the extent to which items in a measuring instrument or scale are all measuring the same thing. It requires only one administration of the instrument.

Split-half reliability

Measures of internal consistency reliability derive from splitting a scale in half and correlating scores on the two halves. If the test is internally consistent the resulting correlation coefficient should be in the acceptable range of .80 and above (or above .70 for research purposes).

How to split the scale can be an issue. One method is to split according to order, that is, in a 12-item scale the first 6 items are split from the last 6 items, but this can be problematic if participants improve with practice on items, or if the items increase or decrease in difficulty. Another method is to make the split according to even versus odd numbered items. Alternatively a random split may be made.

The Spearman-Brown formula is frequently used as a measure of split-half reliability. It is based on the correlation between the two split halves, but contains an adjustment that takes into account the number of items in the scale. Other things being equal, the longer the scale, the higher the internal consistency reliability. The **Kuder-Richardson** formula is an alternative used when items are scored dichotomously (e.g., correct or incorrect).

Coefficient alpha or Cronbach's alpha

This is probably the most commonly used measure of a scale's internal consistency reliability. Cronbach's alpha corresponds approximately to the "mean of all the split-half coefficients resulting from different splittings of a test" (Anastasi, 1982, p. 116), or as de Vaus (2002) explains, it is an index of all the item-item correlations. A useful explanation of Cronbach's alpha and some cautionary comments is provided by Field (2005, pp. 666-670).

A feature of Cronbach's α is that it automatically becomes larger the more items there are in an instrument. However, simply having lots of items is not how one should go about achieving high internal consistency. The general guideline is first to ensure that items are **distinct indicators** of the latent construct to be measured (not just the same item written in several different ways), then to aim for the **shortest, most reliable** scale that still samples an appropriate range of indicators.

When describing a measuring scale in the Method section of a research report it is usual to report Cronbach's alpha for the scale, as well as test-retest correlations if they are available. This information can usually be obtained from the scale manual, test catalogues, or from published studies that have used the scale. If you then go on to use the scale yourself in another study with another sample, Cron-bach's α with that sample should also be reported, as it will almost certainly vary from sample to sample. The generalisability of the original coefficient depends on the extent to which the original sample was representative of the population as a whole.

Example of assessment of internal consistency reliability

Internal consistency reliability estimates are used in the **construction** and **evaluation** of measurement scales.

As reasonably large samples are desirable in the scale construction process, this example uses an existing dataset that contains the hypothetical responses of 130 participants to 10 items from an instrument being constructed to measure *meaning in life*. The example is taken from an actual research area I was exploring some years ago (i.e., the elusive area of life meaning), but the data themselves are from a created dataset. The variable names and labels, and the items in the instrument are listed in Table 21.1.

The data are available in a file called ***21Reliability.sav*** at: www.pearson.com.au/9781442549821. Click on the link 'Student downloads' to download the datafile.

Imagine that the participants responded to each item on the following rating scale:

0	1	2	3	4
Not at all	Yes, slightly	Yes, moderately	Yes, very much so	Yes, totally

In most instruments it is necessary to **reverse code** some items to achieve consistency in the way they are scored. High scores in this instrument are intended to indicate a strong sense of meaning while low scores are intended to indicate a sense of meaninglessness. Immediately we can see that items Q1 and Q6 are worded oppositely to the others, so that anyone who ticks 4 (Yes, totally) for these is clearly not very happy with their lot. We therefore need to reverse code these items so that a high score has a positive meaning, consistent with all the other items, and so that it will then make sense to add across items to obtain a total score on the instrument.

Reverse coding can be achieved (the hard way) in SPSS by using one of the **Transform**, then **Recode** options to change the values for these two items as follows: 0=4, 1=3, 3=1, 4=0. Or it can be done by using **Transform, Compute Variable...** to create new variables (Q1R and Q6R), according to the formula: New variable = Maximum score – Old variable (i.e., 4 – Old variable, so for a score of 3, 4-3 = 1

etc.). Be careful, though; this formula only works for rating scales that start at zero. If a rating scale starts at 1, the formula is: New variable = (Maximum score **+1**) – Old variable.
Reverse coding has already been done in the dataset for this example (as indicated by the items with an R following the question number).

Table 21.1

Items in the Meaning in Life Scale

Name	Label	Item
Q01R	Regretful	Are you regretful about the choices you have made, or about missed opportunities?
Q02	ClearIdea	Do you have a sense of, or clear idea, of what you want to live for?
Q03	Commit	Are you invested in or committed to the activities in your life?
Q04	Purpose	Do you have a sense of purpose, mission, or calling in life?
Q05	Difference	Do you feel that your life makes a difference, even if only in small ways?
Q06R	Change	Are you waiting or hoping that something will change or that things will get better in your life?
Q07	Person	Are you becoming, or have you become, the kind of person you wish to be?
Q08	Satisfied	Are you generally satisfied with the way things are?
Q09	Values	Do your life circumstances enable you to give expression to your values or cherished beliefs?
Q10	Content	Are you content to live your life the way you are doing now?

Using SPSS to determine a scale's internal consistency reliability

Step 1: Enter data in SPSS and save data file

By now you should be thoroughly familiar with defining and entering data in SPSS and saving the data file as a first step in any analysis. Have a look at the data being used here, and how they have been defined in the data file.

Step 2: Screen data and check assumptions

Univariate assumptions[1] should be checked as usual, together with assumptions appropriate to correlational designs, including the identification of multivariate outliers using Mahalanobis distance, as explained in the chapter on multiple regression.

[1] If you check for normality you will find it is violated for all the variables according to the Shapiro-Wilk test; although there is less of a problem when skewness and kurtosis are evaluated as discussed in the chapter on data screening. With large samples significant nonnormality is less important than the actual values of skewness and kurtosis and the extent they depart from zero (see Tabachnick & Fidell, 2007, p. 80). I would suggest there is no real cause for concern in this example.

To identify multivariate outliers you need to run a multiple regression using the 10 scale items as predictors and ID as the criterion. The criterion is irrelevant to the identification of multivariate outliers among the predictors. Make sure to request Mahalanobis distance. With 10 variables and $\alpha = .001$, the critical χ^2 for Mahalanobis distance is 29.5883. If you perform this check you should find that ID number 6 exceeds this value.

Step 3: Request the analysis

Conduct the reliability analysis as shown below, but remember to exclude the multivariate outlier (ID 6).

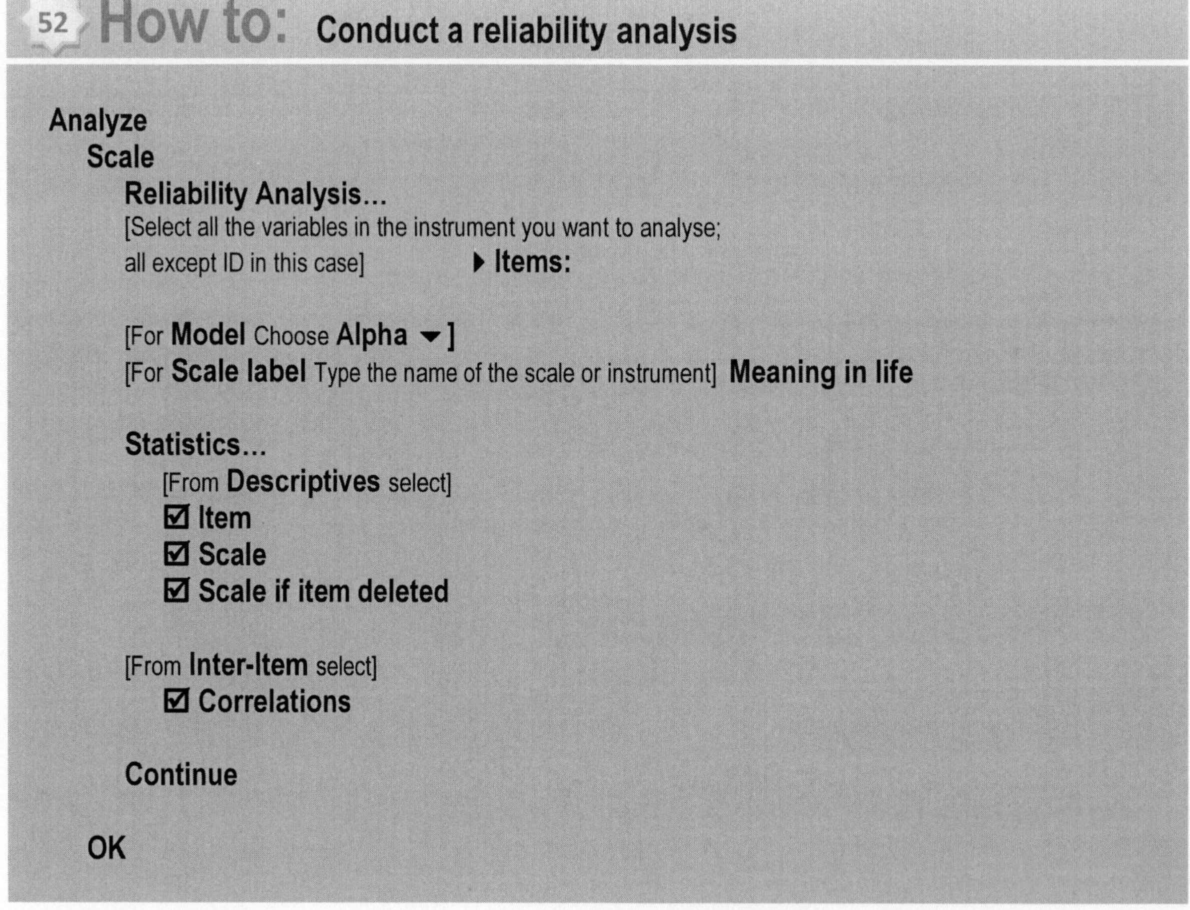

52 **How to: Conduct a reliability analysis**

Analyze
Scale
Reliability Analysis...
[Select all the variables in the instrument you want to analyse; all except ID in this case] ▸ **Items:**

[For **Model** Choose **Alpha** ▾]
[For **Scale label** Type the name of the scale or instrument] **Meaning in life**

Statistics...
[From **Descriptives** select]
☑ **Item**
☑ **Scale**
☑ **Scale if item deleted**

[From **Inter-Item** select]
☑ **Correlations**

Continue

OK

Step 4: Examine and interpret the output

Reliability

Scale: Meaning in life

Case Processing Summary

		N	%
Cases	Valid	124	96.1
	Excluded[a]	5	3.9
	Total	129	100.0

a. Listwise deletion based on all variables in the procedure.

1

With exclusion of ID 6, and listwise deletion of missing data the sample is reduced to 124.

Obviously 5 cases had missing data on one or more items, and when this happens they have to be deleted from the analysis as a total score cannot be computed for them.

Reliability Statistics

Cronbach's Alpha	Cronbach's Alpha Based on Standardized Items	N of Items
.793	.791	10

2

Cronbach's Alpha is provided in the **Reliability Statistics** table.

The 10-item scale has acceptable internal consistency reliability with a Cronbach's alpha of .79. The standardized alpha (based on standard, not raw, scores for the items) is also .79.

Item Statistics

	Mean	Std. Deviation	N
Regretful	1.89	1.211	124
ClearIdea	2.06	1.261	124
Commit	1.98	1.259	124
Purpose	2.27	1.224	124
Difference	1.49	1.071	124
Change	1.94	1.185	124
Person	2.00	1.141	124
Satisfied	2.06	1.095	124
Values	2.03	1.242	124
Content	2.23	1.195	124

3

Examine **Items Statistics** for any items that seem very different.

Although not reproduced here you should also examine the **Inter-Item Correlation Matrix** that appears next in the SPSS output.

4

Item-Total Statistics

	Scale Mean if Item Deleted	Scale Variance if Item Deleted	Corrected Item-Total Correlation	Squared Multiple Correlation	Cronbach's Alpha if Item Deleted
Regretful	18.06	43.524	.277	.346	.797
ClearIdea	17.89	38.995	.561	.440	.762
Commit	17.98	40.723	.442	.420	.778
Purpose	17.69	38.250	.638	.564	.753
Difference	18.46	41.860	.462	.342	.775
Change	18.01	42.967	.325	.339	.791
Person	17.95	39.689	.586	.533	.760
Satisfied	17.89	44.589	.248	.230	.798
Values	17.92	39.424	.541	.571	.765
Content	17.73	39.436	.569	.450	.762

Scale Statistics

Mean	Variance	Std. Deviation	N of Items
19.95	49.412	7.029	10

5

Two columns are particularly important in the **Item-Total Statistics** table at the top of the page.

Corrected Item-Total Correlation

The Corrected Item-Total Correlation is the Pearson correlation between each individual item and the sum of all the other items. If an item is highly consistent with the other items this correlation will be moderate to high. As a general rule-of-thumb .30 is used as the minimum for deeming an item to be internally consistent with the other items in the scale.

When reliability analysis is used in scale construction, as it is in this example, we would normally eliminate items with item-total correlations below .30. When more than one variable is below .30, it is advisable to delete them **one at a time** (starting with the one with the lowest item-total correlation), then rerun the analysis to see if the other or others still have low correlations.

Alpha if Item Deleted

Alpha if Item Deleted indicates how the overall alpha will change if the individual item is deleted from the scale. If using reliability in scale construction, one would consider deleting those items whose deletion would result in a considerably higher alpha.

As we are using reliability in a scale construction process in this example, we would want to delete the two items: Satisfied and Regretful, both of which have item-total correlations below .30.

We will delete both together here, as I have already checked that the outcome is the same even when they are deleted one at a time.

If we repeat the analysis deleting items Q08(Satisfied) and Q01R (Regretful), we obtain the following results:

Reliability Statistics

Cronbach's Alpha	Cronbach's Alpha Based on Standardized Items	N of Items
.813	.811	8

Item-Total Statistics

	Scale Mean if Item Deleted	Scale Variance if Item Deleted	Corrected Item-Total Correlation	Squared Multiple Correlation	Cronbach's Alpha if Item Deleted
ClearIdea	14.00	30.937	.529	.379	.791
Commit	14.09	31.134	.513	.377	.794
Purpose	13.78	29.237	.681	.541	.768
Difference	14.56	33.534	.427	.306	.805
Change	14.09	35.451	.222	.273	.832
Person	14.04	30.546	.638	.497	.776
Values	14.01	30.040	.617	.556	.778
Content	13.81	30.408	.617	.456	.778

6

With the number of items reduced to 8, Cronbach's α has now improved to .813, but now the item, Change, has an item-total correlation < .30, and its deletion will result in a marginal improvement in Cronbach's α.

If we repeat the analysis a second time, deleting item Q06R (Change), we obtain the following results:

Reliability Statistics

Cronbach's Alpha	Cronbach's Alpha Based on Standardized Items	N of Items
.832	.831	7

Item-Total Statistics

	Scale Mean if Item Deleted	Scale Variance if Item Deleted	Corrected Item-Total Correlation	Squared Multiple Correlation	Cronbach's Alpha if Item Deleted
ClearIdea	12.04	27.340	.494	.324	.824
Commit	12.13	26.831	.537	.376	.817
Purpose	11.82	25.007	.714	.538	.787
Difference	12.60	29.845	.383	.243	.838
Person	12.08	26.406	.655	.497	.798
Values	12.05	25.410	.680	.509	.793
Content	11.85	26.573	.606	.421	.806

7

With the number of items reduced to 7, Cronbach's α has now improved to be quite acceptable at .832, and all of the items have item-total correlations exceeding .30.

We could still improve Cronbach's α slightly by deleting the item, Difference; but the improvement is slight and not really warranted.

Inter-rater reliability

Another type of reliability is inter-rater reliability. This applies when human observers are required to determine scores on a construct or variable. It can arise when observers are rating the incidence of particular behaviours in observational research, or when researchers are content-coding open-ended data in a questionnaire, or even qualitative data such as an essay (e.g., coding parts of the text as representing different themes). To ensure the reliability of such rating or coding a measure of agreement between at least two raters is required.

A common measure of inter-rater agreement is **Cohen's kappa (κ)**, which can range from 0 to 1, and is interpreted similar to correlation. Again, a Cohen's kappa of .70 or above indicates acceptable inter-rater reliability, but in any clinical context we would ideally want it to be .80 or above.

Example of inter-rater reliability

Let us use an example of a variable that is often relevant in organisational psychology. Generally it is not a good idea to be too passive, or too aggressive in your interpersonal interactions (for fairly obvious reasons). Ideally one needs to be assertive without being aggressive.

Suppose then, we have two raters observing a participant and scoring the interactions that person has over a given period according to definitions of the three interaction styles. Do the raters exhibit inter-rater agreement?

Imaging the data below are the ratings by the two raters of 15 interactions they have observed.
The scoring key is:

1 = Assertive
2 = Passive
3 = Aggressive

Step 1: Enter data in SPSS and save data file

The data from the two raters for the 15 interactions are shown below. Enter these data into SPSS and save them in a file that you might call ***Kappa.sav***.

Interaction	Rater A	Rater B
1	2	1
2	2	1
3	2	2
4	2	2
5	2	2
6	2	2
7	3	3
8	1	1
9	2	1
10	3	3
11	2	1
12	2	2
13	2	1
14	2	2
15	2	1

Step 2: Request the analysis

53 How to: Conduct an inter-rater reliability analysis (Cohen's kappa)

Analyse
Descriptive Statistics ▸
Crosstabs...
[Select the variables(s) you want]
Rater A ▸ Row(s):
Rater B ▸ Column(s):
Cells...
[Select from] **Counts ☑ Observed**
[Select from] **Percentages ☑ Total**
Continue
Statistics...
[Select] **☑ Kappa**
Continue
OK

Step 3: Examine and interpret the output

Output: **Inter-rate reliability (Kappa)**

RaterA * RaterB Crosstabulation

			RaterB			Total
			Assertive	Passive	Aggressive	
RaterA	Assertive	Count	1	0	0	1
		% of Total	6.7%	.0%	.0%	6.7%
	Passive	Count	6	6	0	12
		% of Total	40.0%	40.0%	.0%	80.0%
	Aggressive	Count	0	0	2	2
		% of Total	.0%	.0%	13.3%	13.3%
Total		Count	7	6	2	15
		% of Total	46.7%	40.0%	13.3%	100.0%

Symmetric Measures

		Value	Asymp. Std. Error[a]	Approx. T[b]	Approx. Sig.
Measure of Agreement	Kappa	.366	.165	2.836	.005
N of Valid Cases		15			

a. Not assuming the null hypothesis.

b. Using the asymptotic standard error assuming the null hypothesis.

We are interested in the **Kappa** value, which at only **.366** indicates a **lack** of inter-rater agreement (the significance of Kappa is not relevant in this context).

The **Crosstabulation** table above shows that there was agreement between the raters on one assertive interaction, and also on the two aggressive interactions, but they were not in agreement for passive interactions: Rater A rated 12 interactions as passive, but Rater B rated 6 of them as assertive. (It may be necessary here to refine the definition of passive and assertive so they are more distinct; then to repeat the study, ideally with different raters).

22 Exploratory Factor Analysis

Factor analysis (not to be confused with factorial ANOVA) is a generic term that covers a number of different but related analysis techniques, most importantly **principal components analysis (PCA)** and **factor analysis** proper **(FA)**. Both are concerned with identifying the underlying **structure** in a set of variables, so that a large number of variables can be reduced to a smaller set of underlying factors that summarise the essential information contained in the variables. In the process, the researcher's understanding of the research domain should be enhanced. The researcher aims to identify the **dimensions** (i.e., factors or components) that underlie the domain. In essence, these techniques group together those variables that are highly intercorrelated with one another, but uncorrelated with other variables. Each variable grouping is then examined in a subjective process to identify the underlying **latent variable**—the underlying thing they have in common.

The essential difference between PCA and FA is that PCA produces **components** that account for **all** the variance in the variables, while FA produces **factors** that only account for that variance that is **shared** by the variables. The term *factors* is used frequently as a generic term covering both components and factors, and the term *factor analysis* is similarly used for both PCA and FA.

Factor analysis is most often used in **exploratory** research where one wants to **identify** and summarise the structure of a set of variables by grouping together those variables that are highly intercorrelated. **Confirmatory factor analysis**, on the other hand, is a more specialised technique used to **test prior theories** about the structure of a particular domain, that is, to test (i.e., confirm) whether a particular construct does indeed produce the factors it should according to theory.

Factor analysis is often used in test construction where the goal is to develop a **reliable** test, that is, one with high internal consistency reliability—a test in which the items are all measuring the same thing. If a test has high internal consistency reliability the items should reduce to one factor, as opposed to several factors. At the very least there should be one major factor, perhaps with one or two minor factors. Items belonging to factors others than the main one are likely to be removed from the final test, because they are measuring something different. Alternatively, factor analysis provides information on the extent to which a construct is **unidimensional** (with only one aspect to it, or only one underlying factor), as opposed to **multidimensional** (with several aspects, or more than one underlying factor).

Factor analysis can also be useful in establishing the **construct validity** of a test, by ascertaining whether or not it has the structure it ought to have according to the theory behind the construct.

Assumptions

Factor analysis is based on correlation, and so has the same assumptions as correlation. As always, these need to be addressed **before** the analysis is conducted. The principal components and factor analysis chapter in Tabachnick and Fidell (2007) should be consulted for a detailed coverage of the issues involved in the use of these techniques.

Sample size: There must be more participants than variables in factor analysis. A bare minimum is five participants per variable and ideally at least 100 participants, with preferably something in the region of 200 to 1000.

Be careful with missing data. Participants with data missing on **any** of the variables are eliminated from the analysis if the **listwise** exclusion option is in effect, but this can sometimes produce a dramatic reduction in *N*. For this reason, always ensure that the SPSS option chosen only eliminates missing data from the relevant correlations (i.e., **pairwise exclusion**). Another option is to substitute the variable mean for random occurrences of missing data. (Nonrandom or **systematic** missing data must be investigated.)

Normality: The individual variables should be examined to assess normality, as the solution is greatly enhanced when this is present.

Linearity: A linear relationships among variables is very important to factor analysis. The scatterplots should be inspected for all pairs of variables to ensure that this assumption is met. This can be a time consuming exercise when there is a large number of variables, so an option can be to just inspect a sample of scatterplots, investigating further if any suggestion of nonlinearity is encountered.

Outliers: Univariate outliers need be modified (or deleted) as explained in the chapter dealing with data screening. Multivariate outliers need to be identified and deleted. They are identified using Mahalanobis distance, as explained in the chapter on multiple regression. **Outlying variables** should also be deleted from the analysis. These are variables that are unrelated to any others in the set, and there is information on how to identify them later in this chapter.

Very important cautions about the use of factor analysis

One has to be very careful with factor analysis, because it is easy to produce misleading information. Factor analysis is often misused! It is important, for example, that there is a wide spread in the scores returned by participants on the variables, in order to produce enough variance for intercorrelations to emerge. At the same time, choosing groups of participants who are likely to have extreme scores on the variables and factors can distort the results. Where there are **subgroups** of participants, separate analyses should be conducted for the subgroups to ascertain if they have the same structure.

Also, choice of variables can determine the result to a considerable extent. Including a number of "variables" that are really just semantic variations of the same thing virtually guarantees that they will emerge as a “factor”; however, they can hardly be said to represent an underlying dimension.

As de Vaus (2002, p. 187) points out, one also needs to avoid selecting variables that are **causally** related, as opposed to being actual components of an underlying dimension or factor.

Finally, beware of factors that are defined by only one or two variables. They may not represent stable factors at such, but rather variables that are outliers in relation to the other variables in the set. Alternatively, they may be indicative of a legitimate factor for which more variables need to be identified, measured, and included.

A high level of conceptual and analytic skill on the part of the researcher is critical to factor analysis—despite appearances to the contrary, it is **not** a technique where one can mindlessly rely on the statistics to produce magical answers. It is very much a case of “garbage in-garbage out”, and certainly not “garbage in-roses out”!

If using factor analysis it is essential to read specialised texts on the subject, so one is fully conversant with the issues, problems, and pitfalls, including the different extraction methods, and the different kinds of rotation. This chapter serves merely as an introduction.

An example of factor analysis or, more correctly, principal components analysis

Some years ago I conducted exploratory research into constructing a scale of attitudes toward space exploration (i.e., outer space), and this example is drawn from that research, although the dataset used here is a created one.

Table 22.1 shows the eight items from a space exploration scale. Assume that participants responded to each item on an agree-disagree rating scale that ranged from -3 to 3, with positive numbers indicating a positive view with respect to space exploration, and negative numbers indicating a negative view. The data from 113 participants are in a file called ***22FactorAnalysis.sav*** which is available on the internet site: www.pearson.com.au/9781442549821. Click on the link 'Student downloads' to download the datafile.

Reverse coding of the two oppositely worded items (Relevance and Interest) has already been carried out on the data in this data file.

Table 22.1

The Space Exploration Questionnaire

Variable Name	Item
Destiny	It is human destiny to move out into the universe.
Relevance	Space exploration has absolutely no relevance to my life.
Explore	The exploration of space is vital to the future of humanity.
Colonies	We should be working to establish human colonies on other planets, such as Mars.
Frontier	Outer space truly is a final frontier and should be explored.
Interest	Television programs or magazine articles on space exploration are of no interest to me.
Moon	The landing on the moon was one of the most significant events in human history.
Excited	I feel a sense of excitement when new discoveries are made in space.

Using SPSS to conduct a principal components analysis

Step 1: Enter data in SPSS and save data file

By now you should be thoroughly familiar with defining and entering data in SPSS and saving the data file as a first step in any analysis. Have a look at these data and how they have been defined in the data file.

Step 2: Screen data and check assumptions

Univariate assumptions should be checked as usual, together with assumptions appropriate to correlational designs, including the identification of multivariate outliers using Mahalanobis distance, as explained in the chapter on multiple regression. If you check for normality you will find it is violated for all the variables according to the Shapiro-Wilk test. With large samples significant nonnormality is less important than the actual values of skewness and kurtosis and the extent they depart from zero (see Tabachnick & Fidell, 2007, p. 80). There are, however, some quite distinct departures on some variables, so if this was a real study, we might need to think about transforming variables to address the problem. To keep things simple in this example we will proceed as is, but mindful of the fact that the reliability of any solution will be compromised to some extent.

To identify multivariate outliers you need to run a multiple regression using the 8 scale items as predictors and ID as the criterion. The criterion is irrelevant to the identification of multivariate outliers among the predictors. Make sure to request Mahalanobis distance. With 8 variables and $\alpha = .001$, the critical χ^2 for Mahalanobis distance is 26.1245. If you perform this check you should find that ID number 2 exceeds this value.

Step 3: Request the analysis

Conduct the principal components analysis as shown below, but remember to exclude the multivariate outlier (ID 2).

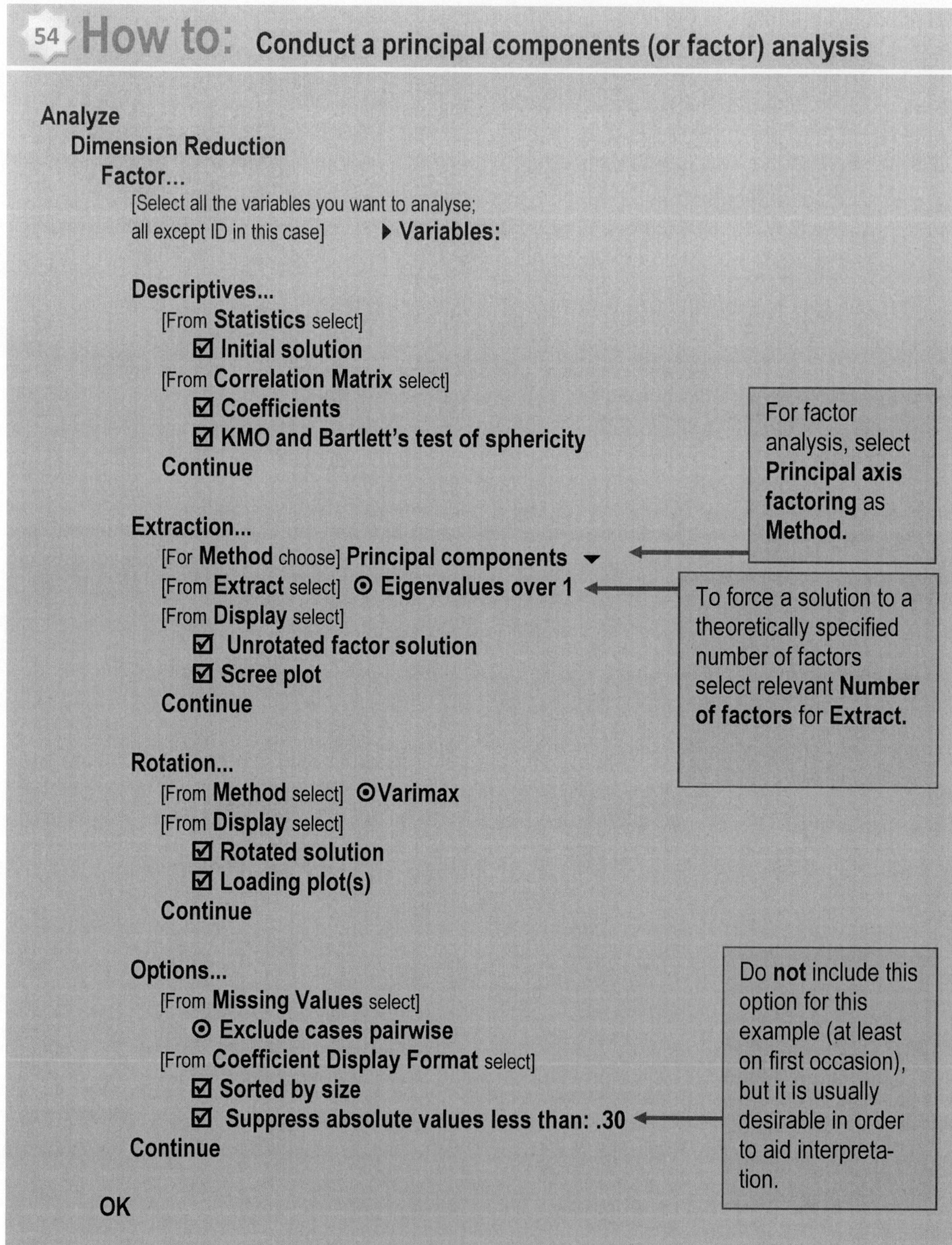

54 How to: **Conduct a principal components (or factor) analysis**

Analyze
Dimension Reduction
Factor...
[Select all the variables you want to analyse; all except ID in this case] ▸ **Variables:**

Descriptives...
[From **Statistics** select]
☑ **Initial solution**
[From **Correlation Matrix** select]
☑ **Coefficients**
☑ **KMO and Bartlett's test of sphericity**
Continue

Extraction...
[For **Method** choose] **Principal components** ▾
[From **Extract** select] ⊙ **Eigenvalues over 1**
[From **Display** select]
☑ **Unrotated factor solution**
☑ **Scree plot**
Continue

Rotation...
[From **Method** select] ⊙ **Varimax**
[From **Display** select]
☑ **Rotated solution**
☑ **Loading plot(s)**
Continue

Options...
[From **Missing Values** select]
⊙ **Exclude cases pairwise**
[From **Coefficient Display Format** select]
☑ **Sorted by size**
☑ **Suppress absolute values less than: .30**
Continue

OK

For factor analysis, select **Principal axis factoring** as **Method.**

To force a solution to a theoretically specified number of factors select relevant **Number of factors** for **Extract.**

Do **not** include this option for this example (at least on first occasion), but it is usually desirable in order to aid interpretation.

Step 4.1: Examine and interpret the output for factor extraction

The factor (or component) extraction phase

There are two distinct phases in factor analysis. The first is about determining how many reliable factors to extract, and the second is about rotating the factors to try to achieve **simple structure**, which facilitates interpretation. Let us look at factor extraction first.

Factor Analysis

Correlation Matrix

		Destiny	Relevance	Explore	Colonies	Frontier	Interest	Moon	Excited
Correlation	Destiny	1.000	.215	.427	.311	.301	.101	.242	.029
	Relevance	.215	1.000	.237	.256	.226	.507	.351	.192
	Explore	.427	.237	1.000	.418	.420	.189	.278	.047
	Colonies	.311	.256	.418	1.000	.411	.152	.273	-.035
	Frontier	.301	.226	.420	.411	1.000	.197	.149	.168
	Interest	.101	.507	.189	.152	.197	1.000	.204	.292
	Moon	.242	.351	.278	.273	.149	.204	1.000	.202
	Excited	.029	.192	.047	-.035	.168	.292	.202	1.000

Examine the **Correlation Matrix**. If it is going to be possible to perform a factor analysis some of the variables should be moderately or highly intercorrelated (with correlations of at least .30). Few of the correlations in this matrix are very high, although there are a number above .30.

KMO and Bartlett's Test

Kaiser-Meyer-Olkin Measure of Sampling Adequacy.		.735
Bartlett's Test of Sphericity	Approx. Chi-Square	161.165
	df	28
	Sig.	.000

2

The **KMO Measure of Sampling Adequacy** assesses the factorability of the correlation matrix. Values of .60 or above are considered satisfactory.

Bartlett's Test should be **significant** (the null hypothesis is that correlations are zero); however, it is a very liberal test.

Communalities

	Initial	Extraction
Destiny	1.000	.477
Relevance	1.000	.585
Explore	1.000	.605
Colonies	1.000	.558
Frontier	1.000	.458
Interest	1.000	.622
Moon	1.000	.351
Excited	1.000	.469

Extraction Method: Principal Component Analysis.

3

Examine the extraction **communalities** for each variable. (*Note:* The variables may be in alphabetical or variable entry order, depending on how you have set up SPSS.)

Initial communalities in PCA are 1, as **all** the variance is to be explained.

Extraction communalities

These are what matter—they indicate how much variance is explained by the "true" components that emerge.

A high communality indicates that the variable is well-represented by the components; a low communality indicates an outlying variable that should be eliminated from the analysis.

Important Definition

The **communality** for a variable is the proportion of variance in **that variable** accounted for by the **components** or **factors**.

Communality has the symbol h^2.

Total Variance Explained

Component	Initial Eigenvalues			Extraction Sums of Squared Loadings			Rotation Sums of Squared Loadings		
	Total	% of Variance	Cumulative %	Total	% of Variance	Cumulative %	Total	% of Variance	Cumulative %
1	2.754	34.421	34.421	2.754	34.421	34.421	2.288	28.603	28.603
2	1.370	17.130	51.551	1.370	17.130	51.551	1.836	22.948	51.551
3	.873	10.918	62.469						
4	.827	10.333	72.801						
5	.705	8.816	81.617						
6	.543	6.784	88.401						
7	.497	6.219	94.620						
8	.430	5.380	100.000						

Extraction Method: Principal Component Analysis.

Deciding the Number of Factors: The Initial Factor Extraction

There are a number of techniques for deciding how many "true" factors exist, that is, how many factors to extract. These can produce different results, which is one of the reasons why some people criticise factor analysis ("you can get whatever you like, by juggling the various techniques!").

Principal components extraction (PCA) is the most common extraction technique, and strictly speaking it produces **components** from a principal components analysis rather than **factors** from a factor analysis. Principal components extraction always produces as many components as there are variables, so how does it help in reducing the data to the underlying dimensions? The question that arises is: how many components should be retained as being important or genuine, as opposed to trivial and probably comprised of error variance?

The **eigenvalue** of a factor or component is used to answer this question. In the initial component extraction the eigenvalue represents the proportion of variance in the set of variables accounted for by a component. The eigenvalue divided by the number of variables in the set gives this % of Variance.

For example, for Component 1 above:

2.754 / 8 = .34425 = 34.425% (the slight difference is due to rounding errors).

Any component with an eigenvalue less than 1 explains less of the variance than does an individual variable, so it can hardly constitute an underlying dimension. So, to make the decision of "how many components to retain" the most commonly used criterion is to retain those components with eigenvalues greater than 1. This is also known as **Kaisers's criterion**.

Another method is to examine the scree plot (on next page).

Important Definition

An **eigenvalue** represents the proportion of variance in **the set of variables** accounted for by **a component** or **factor**.

Output: **Factor or principal components analysis**

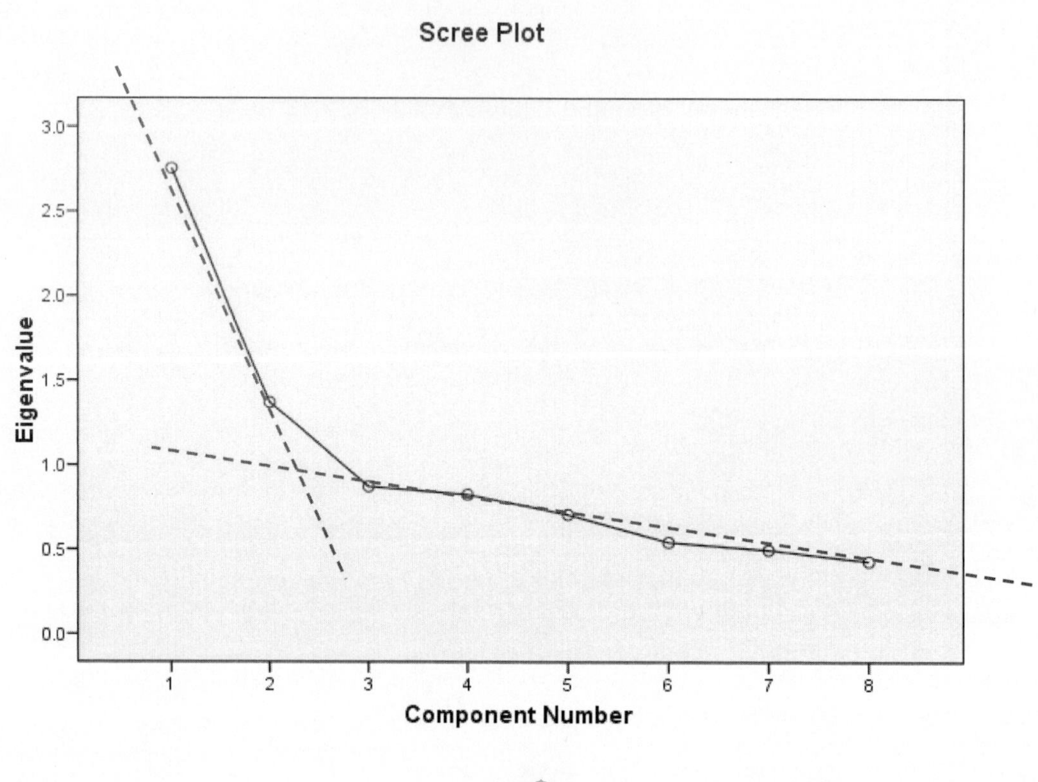

5

According to the **Scree Plot**, factors or components defining the "slope" are retained, and factors or components defining the "scree" (debris at the bottom of the slope) are discarded.

Note in the above scree plot that SPSS does **not** include the two dashed lines that represent the "slope" and "scree". I have added these to illustrate how by drawing such lines one can see by their point of intersection how many **reliable** components are in the slope. In this case, consistent with the eigenvalue > 1 method, it appears that two components should be retained.

With a large number of variables, the scree plot is often the more valid criterion to use.

In this example, both methods indicate that **two** components should be retained; together accounting for 51.551% of the variance in the set of eight variables.

Component Matrix[a]

	Component	
	1	2
Explore	.690	-.358
Relevance	.643	.413
Colonies	.641	-.384
Frontier	.637	-.228
Destiny	.585	-.368
Moon	.570	.162
Excited	.303	.614
Interest	.538	.577

Extraction Method: Principal Component Analysis.

a. 2 components extracted.

6

The **Component Matrix** is a matrix of **loadings** or **correlations** between the **variables** and **components**. It is used to help the researcher interpret the components, but **not** in this initial extraction phase.

The table below (which is **not** part of SPSS output) shows the relationships among loadings or correlations, communalities, and eigenvalues, for the initial extraction phase.

Columns A and B are from the component matrix above. The squared loadings (good old familiar r^2) are in columns C and D. When these are added they give the communality for the variables (column E), which agree with the communalities shown earlier at output section 3. Columns C, D, and E sum down to the values shown in the Sum of squared loadings row (third from bottom), and the sum of columns C and D equals the sum of column E. The last two rows show how these sums, as a proportion of the total of 8 variables, give the percentage of variance explained, that corresponds with the output section 4. (Slight differences with the output are due to rounding errors.)

Variable	A Loading (Correlation) with Component 1	B Loading (Correlation) with Component 2	C Squared Component 1 Loading	D Squared Component 2 Loading	E Sum of squared loadings **Communality**
Explore	.690	-.358	.476	.128	**.604**
Relevance	.643	.413	.413	.171	**.584**
Colonies	.641	-.384	.411	.148	**.559**
Frontier	.637	-.228	.406	.052	**.458**
Destiny	.585	-.368	.342	.135	**.477**
Moon	.570	.162	.325	.026	**.351**
Excited	.303	.614	.092	.377	**.469**
Interest	.538	.577	.289	.333	**.622**
Sum of squared loadings (SSL) or **Eigenvalues**			**2.754**	**1.370**	**4.124**
% of variance in total of 8 variables			2.755/8 34.425%	1.370/8 17.125%	4.124/8 51.550%

The factor (or component) rotation phase

Factor analysis is a geometric technique that maps variables to dimensions (or components) in a mathematical space. It starts with an arbitrary positioning of the axes for the components, and the loadings in the Component Matrix are the coordinates on those arbitrary initial axes. To make the Component Matrix more interpretable it is usual to **rotate** the axes (and therefore the components), to redistribute the variance among them to try to achieve **simple structure**, where variables load highly on one and only one component, and where each component has high loadings for some variables, and low loadings for the rest.

Each component can then be interpreted in terms of what is common to the high loading variables. Loadings of .30 and above are usually considered important enough to take into account. **Complex**, as opposed to **pure**, variables have relatively high loadings on more than one component. They make interpretation more difficult, and result in a less than ideal solution.

This is the second phase of factor analysis, and is only necessary when there is more than one factor or component extracted.

Rotation methods

There are a number of different rotation methods that can produce somewhat different factors or components, the two most common being:

Orthogonal rotation:

This is where the component axes are kept at **right angles** to one another. In other words the components are **independent**; they are **unrelated**. There are a number of different kinds of orthogonal rotation method; the most common one used in SPSS is known as **Varimax**.

If varimax provides a poor solution (i.e., simple structure is not achieved and many variables still have high loadings on more than one factor), then other rotation methods should be tried. Lack of simple structure may indicate **correlated** factors, for which an oblique rotation method is appropriate. (Only varimax, orthogonal rotation will be demonstrated in this chapter.)

Oblique rotation:

When the components are clearly related to one another, **oblique rotation** is used. Here, the axes are not at right angles to one another. The interpretation of oblique rotation is more complex, one reason being that the Component Matrix no longer consists of correlations, but measures of the **unique** relationship between variables and components (similar to regression coefficients) **Direct oblimin** is often the rotational method of choice in SPSS.

After rotation

The communalities

After rotation, the communality (h^2) for each variable remains the same. Remember, this is the variance in the variable accounted for by the factors or components.

Sum of the squared loadings (SSL)

After rotation, the SSL across variables for a component is the proportion of variance in the set of variables accounted for by the **rotated** components. It will be different to the SSL before rotation. SPSS provides the SSL for rotated components in the second part of the **Total Variances Explained table** (see output section 4), under the heading of **Rotation Sums of Squared Loadings.** These sums of squared loadings equal the proportion of variance in the variables accounted for by each component in the **rotated component solution**.

Step 4.2: Examine and interpret the output for factor rotation

Rotated Component Matrix[a]

	Component	
	1	2
Explore	.770	.109
Colonies	.744	.059
Destiny	.690	.039
Frontier	.651	.183
Interest	.104	.782
Relevance	.285	.710
Excited	-.110	.676
Moon	.371	.462

Extraction Method: Principal Component Analysis.
Rotation Method: Varimax with Kaiser Normalization.

a. Rotation converged in 3 iterations.

Factor or component rotation and interpretation:

The **Rotated Component Matrix** is the matrix of loadings or correlations between the variables and **rotated** components. The variance in the variables is now redistributed between the components to produce the best simple structure, and the loadings have been sorted in order of magnitude for each component. This redistribution is shown in the table below (compare this table with that for the initial **Component Matrix** two pages back).

The Rotated Component Matrix can now be used to interpret the components, because each variable in a good solution will load highly (i.e., above .30) on only one component. Components can then be interpreted in terms of the variables that load highly on them.

Variable	A Loading (Correlation) with Component 1	B Loading (Correlation) with Component 2	C Squared Component 1 Loading	D Squared Component 2 Loading	E Sum of squared loadings **Communality**
Explore	.770	.109	.593	.012	**.605**
Colonies	.744	.059	.554	.003	**.557**
Destiny	.690	.039	.476	.002	**.478**
Frontier	.651	.183	.424	.033	**.457**
Interest	.104	.782	.011	.612	**.622**
Relevance	.285	.710	.081	.504	**.585**
Excited	-.110	.676	.012	.457	**.469**
Moon	.371	.462	.138	.213	**.351**
Sum of squared loadings (SSL) **after rotation**			**2.288**	**1.836**	**4.124**
% of variance in total of 8 variables			2.288/8 28.600%	1.836/8 22.950%	4.124/8 51.550%

Rotated Component Matrix[a]

	Component	
	1	2
Explore	.770	
Colonies	.744	
Destiny	.690	
Frontier	.651	
Interest		.782
Relevance		.710
Excited		.676
Moon	.371	.462

Extraction Method: Principal Component Analysis.
Rotation Method: Varimax with Kaiser Normalization.

a. Rotation converged in 3 iterations.

The final step in factor or component analysis is the **subjective** process of interpreting the components. The researcher looks at the variables that load highly on each component, and makes a subjective decision about what those variables have in common. It is vital that this means **identifying an underlying construct**, rather than just attaching a convenient summary name to a grouping of variables. The component is then identified and labelled as this construct. A good solution has simple structure that is convincingly interpretable.

It is in this final interpretation phase that it is useful to suppress the printing of loadings less than .30 to make the structure more obvious. When this option is requested the **Rotated Component Matrix** appears as above. **Do not forget that suppressed loadings still exist; they are simply not printed.**

It can now be clearly seen that quite a good simple structure has emerged, except for the complex variable **Moon**. One might conclude from this analysis (provided the variables were comprehensive enough to begin with) that there are two unrelated dimensions underlying attitudes toward space exploration. Because they are **unrelated** or **orthogonal**, knowledge of a person's position on one dimension tells nothing about their position on the other. A person might have a favourable attitude on both dimensions or a favourable attitude on one but not the other.

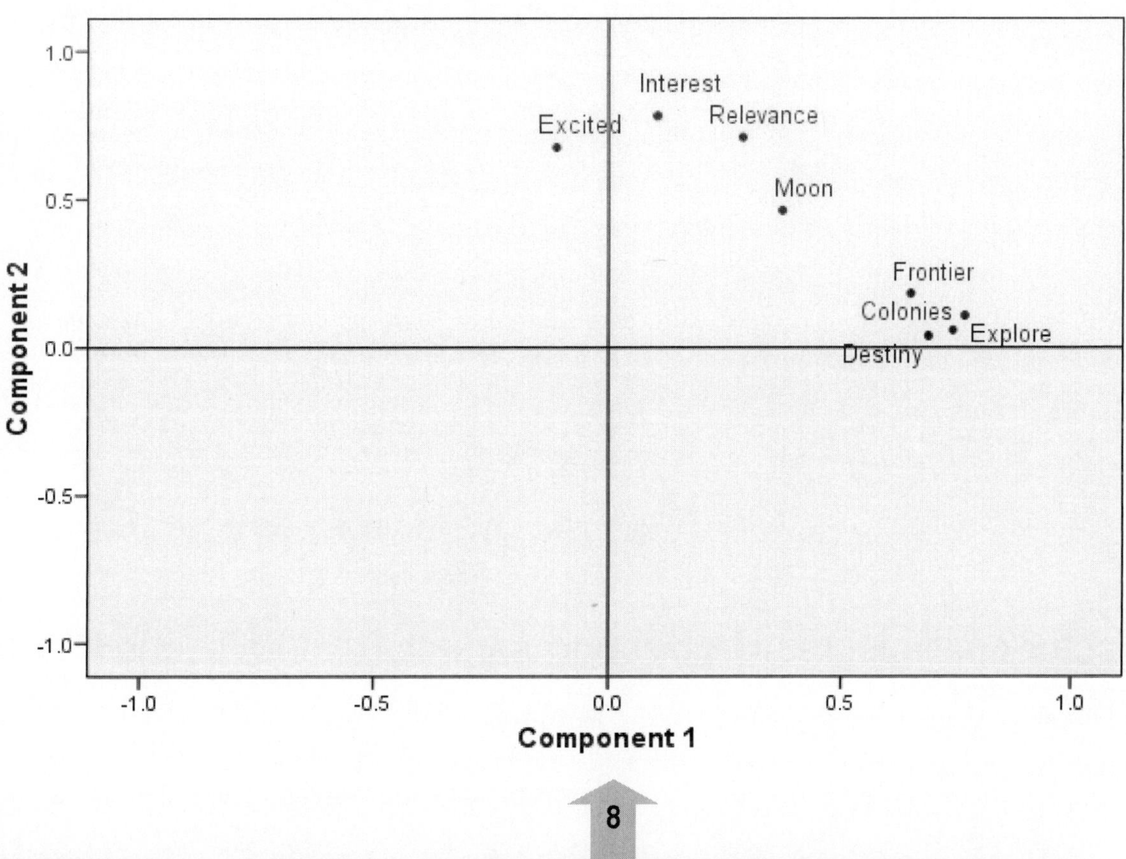

8

The component plot can sometimes be useful as an additional aid in interpreting the components.

All that remains is for the researcher to provide a subjective interpretation of what the variables loading on each component have in common, and to label the components accordingly. To do this it is necessary to look carefully at the actual items back in Table 22.1.

My interpretation is as follows:

Component 1 concerns attitudes about the importance of space exploration, and so might be labelled **perceived importance**.

Component 2 seems to be more about the level of interest a person has in space and space exploration, and so might be labelled **personal interest**.

Can you see a different interpretation? Would you label the components differently?

The complex variable, Moon, clearly falls between the two, having to do both with attitudes to exploration and interest in space.

Factor scores

Are factor scores required?

It is possible to request factor scores for each participant. Participants' standardised scores on **all** the variables are weighted according to the component or factor loadings, then added to give a predicted "score" for the factor. Factor scores are standard scores with a mean of zero, and a standard deviation of 1 (after PCA).

As de Vaus (2002, pp. 191-192) has indicated, factor scores can then be used as scale scores if factor analysis is being used in scale construction, although this is not often done. (Often, only those variables that load highly on the factor are used, and their raw scores are added to give the scale score.)

Factor scores can also be subjected to further analysis, for example, in ANOVA to ascertain group differences.

Factor scores for each participant are added to the dataset if the following option is included in the SPSS procedure:

Scores...
☑ Save as variables

Factor analysis: research report sample Results section

Results

A principal components analysis with varimax rotation was performed on the eight variables measuring attitudes to space exploration. There were no missing data, but one multivariate outlier with an extreme Mahalanobis distance score ($p < .001$) was deleted from the analysis, reducing the sample size to 112. Four of the variables were negatively skewed and two positively skewed, so that a weakened analysis was to be expected. Two components with eigenvalues greater than one were extracted, accounting for 51.55% of the variance. The component loadings, communalities (h^2), and percentages of variance explained after varimax rotation are shown in Table 1. Component loadings less than .30 have been suppressed to aid interpretation.

As can be seen from Table 1, variables loading on component 1 seemed to be concerned with attitudes toward the importance of space exploration, while those loading on component 2 were concerned with personal interest in space and space exploration. The two factors were thus labelled as perceived importance and personal interest respectively.

Table 1

Varimax Rotated Component Loadings for Attitude to Space Exploration Items

	Components[a]		
Item description	1	2	h^2
Exploration of space vital to future	**.77**	.11	.61
Need to establish colonies on planets	**.74**	.06	.56
Human destiny to move out into universe	**.69**	.04	.48
Outer space is final frontier to explore	**.65**	.18	.46
Interest in programs or articles on space	.10	**.78**	.62
Personal relevance of space exploration	.29	**.71**	.59
Feel excitement at new space discoveries	-.11	**.68**	.47
Moon landing significant historical event	**.37**	**.46**	.35
Percentage of variance	28.60	22.95	51.55
Label	Perceived importance	Personal interest	

Note: Component loadings > .30 are in boldface.

23 Chi-Square

We have talked about the chi-square distribution in previous chapters, but now it is time to look at the chi-square test, specifically Pearson's chi-square.

When do you use chi-square (symbol: χ^2)?

Chi-square is used when there are nominal data, that is, data in the form of frequencies or head counts. Research looking at mock jury verdicts (i.e., number of people voting guilty versus not guilty) is a good example of research that generates frequency data. None of the statistics so far considered are appropriate for nominal data.

An important **nonparametric** test that deals with **nominal** or **categorical** data is **chi-square** (χ^2). It is used for data (the DV) in the form of **frequencies** or **head counts** for two or more different categories (that represent the IV).

There are two types of chi-square: chi-square for **one-way designs** (also known as the goodness of fit test), and chi-square for **two-way designs** (also known as contingency table analysis).

One-way chi-square

Example of a one-way chi-square

To demonstrate the use of chi-square, imagine a hypothetical study in which a researcher interested in human factors in road design wants to know if there is a difference in the number of car accidents occurring at three different types of intersection (traffic lights, roundabouts, and stop signs). Suppose the researcher selects a sample of 20 traffic lights, 20 roundabouts, and 20 stop signs (comparable on things like road size, visibility, traffic volume, and other potential confounds). The researcher then counts the number of accidents occurring at each type of intersection over a 12-month period. One can think of the count or frequency of accidents as the DV, and the type of intersection as the IV. (Ignore the fact that it would be rather unrealistic to actually attempt this study, because of all the problems it would present.)

The **null hypothesis** is that the **observed frequencies** and the **expected frequencies due to chance** will be the same. Expected frequencies due to chance are usually equal across categories, as they are in our example, although this can sometimes be different when we are expecting frequencies to be equal to **known population frequencies**. The expectation of equal frequencies in this example is equivalent to a null hypothesis of no difference in the number of accidents at each type of intersection.

Chi-square involves comparing the expected frequencies under the null hypothesis with the actual observed frequencies.

One-way chi-square tests the **goodness-of-fit** between the observed and expected frequencies. If the difference between them is sufficiently large, the null hypothesis is rejected, and the alternative hypothesis is accepted, in this case, that there is a **real** (nonchance) difference in the number of accidents at each type of intersection.

Assumptions

The accuracy of chi-square depends on the following assumptions:

1. Observations fall in one and only one category.
2. Observations are **independent** of one another.
3. The observations are measured as frequencies.
4. The **expected frequency** in each category should be ≥ 5, so you need to have a sufficiently large sample for this to be possible.

Manual calculation of a one-way chi-square

For our hypothetical study, suppose the data collected at the end of 12-months are as follows:

Traffic lights 105 accidents
Roundabouts 64 accidents
Stop signs 161 accidents

Assume, further, that the assumption of independence is satisfactory, because none of the accidents involve the same car or driver.

Step 1: Specify hypotheses and decision criteria

H_0: There is no difference in the population in the number of accidents at each type of intersection; the proportions (P) for each are equal

$P_{\text{Lights}} = P_{\text{Round}} = P_{\text{Stop}} = .3333$ (or 1/3)

H_1: There is a difference in the population in the number of accidents at each type of intersection (i.e., a real, nonchance, difference): the three proportions are not equal.

$\alpha = .05$ (always nondirectional with three or more categories)

Retain H_0 if observed $\chi^2 <$ critical χ^2
Reject H_0 if observed $\chi^2 \geq$ critical χ^2

Step 2: Set down and calculate the observed chi-square using a computational table

Chi-square formula: $$\chi^2 = \Sigma \frac{(O-E)^2}{E}$$

Category	Observed Frequency (O)	Expected Proportion (P)	Expected Frequency (E) $P \times \Sigma O$	$O - E$	$(O - E)^2$	$\frac{(O-E)^2}{E}$
Traffic lights	105	1/3	110	-5	25	$\frac{25}{110}$ = 0.2273
Roundabouts	64	1/3	110	-46	2116	$\frac{2116}{110}$ = 19.2364
Stop signs	161	1/3	110	51	2601	$\frac{2601}{110}$ = 23.6455
Totals	$\Sigma O = 330$	1.00	330	0		χ^2 = 43.1092

Step 3: Calculate degrees of freedom

$df = k - 1$ (where k = number of categories)
$= 3 - 1$
$= \mathbf{2}$

Step 4: Look up a table of critical values of χ^2

Find the row for 2 *df*. Look along this row until you find the column for α of .05. This is the critical χ^2 of 5.99. Alternatively, refer to this internet site to directly obtain the critical value:
http://www.danielsoper.com/statcalc/calc12.aspx

Step 5: Draw statistical conclusion

The observed χ^2 of 43.1092 > the critical χ^2 of 5.99; therefore **reject** H_0: There is no difference in the number of accidents at each type of intersection, and **accept** H_1: There **is** a real difference in the number of accidents at each type of intersection.

Step 6: Specify the research conclusion

There was a statistically significant difference in the number of accidents occurring at the three different types of intersection during the 12-month period.

Note that when there are three or more categories we cannot specify exactly which is different from which from the chi-square result. All we can say is that they are not equal. By reference to the frequencies, one may go on to say something like: the least number of accidents occurred at roundabouts, while the greatest number occurred at stop signs. You could, of course, perform post hoc comparisons with separate chi-squares looking at pairs of categories, using a Bonferroni correction to alpha for the number of tests performed.

Using SPSS to conduct one-way chi-square

Step 1: Enter data in SPSS and save data file

In the SPSS data window define your variables as follows, and then enter the data:

Type	Type of intersection 1=Traffic lights 2=Roundabouts 3=Stop signs	*Set **Measure** column to **Nominal***
Count	Number of accidents	*Set **Measure** column to **Nominal***

Your data will be very simple; just two columns as follows:

1	105
2	64
3	161

Remember to save this data in a file that might be called ***OnewayChi.sav***.

Step 2: Screen data and check assumptions

There is no formal data screening, just a visual check of data accuracy and a check that expected frequencies in the final output are ≥ 5.

Step 3: Request the analysis

When the data are in the form of frequency counts, it is necessary in older versions of SPSS to indicate that the numbers are **frequencies**, not scores. This is done by requesting the **Weight Cases** option, as follows.

55 How to: Use Weight Cases when data are frequencies, not scores

Data
  Weight Cases...
    [Select] ⊙ **Weight cases by**
    [Select variable as follows]
    Number of accidents [count] ▸ **Frequency Variable**
    OK

(***Note:*** A Weight on message appears in the bar at the bottom right of the data window.)

Turning Weight Cases OFF:

Should you want to do later analyses that do not have data in the form of frequencies, it is necessary to turn Weight Cases off, as follows (the **Weight on** message will then disappear):

Data
  Weight Cases...
  [Select] ⊙ **Do not weight cases**
  OK

56 How to: Conduct a one-way chi-square

Analyze
  Nonparametric Tests ▸
    Legacy Dialogs ▸
      Chi-Square...
        [Select variable(s) you want]
        Type of intersection [type] ▸ **Test Variable List**
        [Select from **Expected Values**]
        ⊙ **All categories equal**
      OK

See Appendix 1 for the new way to perform nonparametric tests in PASW 18. For one-way Chi-square you select:
  Analyze
    Nonparametric Tests ▸
      One Sample...

Then proceed as shown in Appendix 1.

To obtain more information, roll your mouse over the graph that appears in the final output window.

Step 4: Examine and interpret the output

Output: **One-way chi-square**

NPar Tests

Chi-Square Test

Frequencies

Type of intersection

	Observed N	Expected N	Residual
Traffic lights	105	110.0	-5.0
Roundabouts	64	110.0	-46.0
Stop signs	161	110.0	51.0
Total	330		

Test Statistics

	Number of accidents
Chi-Square[a]	43.109
df	2
Asymp. Sig.	.000

a. 0 cells (.0%) have expected frequencies less than 5. The minimum expected cell frequency is 110.0.

1

The **Asymp. Sig**. is the *p* value. It is < α of .05; therefore, the test result **is significant.**

One-way chi-square: Research report sample Results section

Results

With alpha set at .05, a one-way chi-square revealed a statistically significant difference in the number of accidents occurring over the 12-month period at the three different types of intersection, χ^2 (2, N = 330) = 43.11, $p < .001$. The greatest number of accidents (161, 48.79%) occurred at stop signs, while the least number occurred at roundabouts (64, 19.39%). One hundred and five accidents (31.82%) occurred at traffic lights.

Note: The general format for reporting χ^2 is:

χ^2 (df, N = number of cases) = chi-square value, p = probability value.

Two-way chi-square

When do you use two-way chi-square?

Two-way chi-square is also known as a **test of independence**. It is not unlike correlation in that there are frequencies for **two variables** and the researcher wants to know if there is a **relationship** between them, or if they are **independent** of one another.

Assumptions

The assumptions are virtually the same as for one-way chi-square.

1. Observations fall in one and only one category.
2. Observations are independent of one another.
3. The observations are measured as frequencies.
4. The **expected frequency** in each category should be ≥ 5.

   Note: If the assumption regarding the required expected frequency is violated it may be possible to collapse small categories together.

When you have a 2 x 2 chi-square some argue that Type I error levels become inflated and **Yate's correction for continuity** should be employed, however, not all sources agree that this is necessary, or even desirable (see Field, 2005, pp. 685-686). Howell (2002, pp. 151-152) makes a case for **not** using either Yates correction or another test for 2x2 chi-square, namely, Fisher's Exact Test.

Example of a two-way chi-square

To demonstrate a two-way chi-square let us extend the example we are already using. Suppose our researcher wants to know if there is a relationship between type of intersection and accident severity. In other words, is the proportion of accidents occurring at the different types of intersection the same for minor accidents as it is for major accidents?

If intersection and severity are **independent**, then the proportions of minor and major accidents will **not** depend on type of intersection, but will be comparable (i.e., the same) across the three types of intersection, and vice versa.

If they are **related**, then the proportions of minor and major accidents **will** depend on type of intersection, that is, they will be **different** across the three types of intersection, and vice versa.
The extent of minor versus major accidents will **depend upon—will be related to**—the type of intersection involved, and vice versa..

Manual calculation of a two-way chi-square

For our hypothetical study, suppose the following additional data are collected at the end of 12 months:

	Total Accidents	**Minor**	**Major**
Traffic lights	**105**	55	50
Roundabouts	**64**	50	14
Stop signs	**161**	90	71
Total	**330**	195	135

Step 1: Specify hypotheses and decision criteria

H_0: There is no relationship in the population between type of intersection and accident severity.

H_1: There is a relationship in the population between type of intersection and accident severity.

$\alpha = .05$ (always nondirectional with three or more categories)

Retain H_0 if observed $\chi^2 <$ critical χ^2
Reject H_0 if observed $\chi^2 \geq$ critical χ^2

Step 2: Calculate expected frequencies (*E*)

Row and column frequencies (*f*) are obtained from the raw data, and used to calculate *E* according to the following formula:

$$E = \frac{f_{row} \; f_{column}}{N}$$

For example, the expected frequency of minor accidents at traffic lights is:

$$E = \frac{(105)(195)}{330}$$

$$= \mathbf{62.0455}$$

Type of Intersection:	Accident Severity Minor	 Major	Row *f*	Row *P*
Traffic lights	62.0455	42.9545	105	.318181818
Roundabouts	37.8182	26.1818	64	.193939393
Stop signs	95.1364	65.8636	161	.487878787
Column *f*	195	135	330	1.0
Column *P*	.59090909	.409090909	1.0	

It is not necessary to calculate row and column proportions (*P*), but it has been done here to show a conceptually clearer, but more cumbersome, way to calculate *E*.

The proportion of minor accidents overall is 195 / 330 = .59090909. If intersection and severity are independent then this proportion will be the same for each type of intersection:

Traffic lights	.59090909 of 105 accidents	=	62.0455
Roundabouts	.59090909 of 64 accidents	=	37.8182
Stop signs	.59090909 of 161 accidents	=	95.1364

Similarly, the proportion of all accidents at traffic lights is 105 / 330 = .318181818. If intersection and severity are independent then this proportion will be the same for each accident severity:

Minor	.318181818 of 195 accidents	=	62.0455
Major	.318181818 of 135 accidents	=	42.9545

This approach can be used to calculate *E* for each cell.

Step 3: Set down and calculate the observed chi-square using a computational table

Chi-square formula: $$\chi^2 = \Sigma \frac{(O-E)^2}{E}$$

Cell	Observed Frequency (O)	Expected Frequency (E)	$O - E$	$(O - E)^2$	$\frac{(O-E)^2}{E}$
Traffic lights Minor	55	62.0455	-7.0455	49.6391	$\frac{49.6391}{62.0455}$ = 0.8000
Traffic lights Major	50	42.9545	7.0455	49.6391	$\frac{49.6391}{42.9545}$ = 1.1556
Roundabouts Minor	50	37.8182	12.1818	148.3963	$\frac{148.3963}{37.8182}$ = 3.9239
Roundabouts Major	14	26.1818	-12.1818	148.3963	$\frac{148.3963}{26.1818}$ = 5.6679
Stop signs Minor	90	95.1364	-5.1364	26.3826	$\frac{26.3826}{95.1364}$ = 0.2773
Stop signs Major	71	65.8636	5.1364	26.3826	$\frac{26.3826}{65.8636}$ = 0.4006
Totals	330	330	0		χ^2 = **12.2253**

Step 4: Calculate degrees of freedom

df = $(r - 1)(c - 1)$ (where r = number of rows in Step 2 Table)
(where c = number of columns in Step 2 Table)

= $(3 - 1)(2 - 1)$
= $(2)(1)$
= **2**

Step 5: Look up a table of critical values of χ^2

Find the row for 2 *df*. Look along this row until you find the column for α of .05. This is the critical χ^2 of 5.99. Alternatively, refer to this internet site to directly obtain the critical value:
http://www.danielsoper.com/statcalc/calc12.aspx

Step 6: Draw statistical conclusion

The observed χ^2 of 12.2253 > the critical of χ^2 of 5.99 (and $p < .001$); therefore **reject** H_0: There is no relationship in the population between type of intersection and accident severity, and **accept** H_1: There **is** a relationship in the population between type of intersection and accident severity.

Step 7: Specify your research conclusion

There was a statistically significant relationship between accident severity and type of intersection. That is, the proportion of minor to major accidents was different for the three different types of intersection.

Again, when you have three or more categories for any IV you cannot specify exactly where the differences are from the chi-square result alone. This issue will be addressed a little later on.

Using SPSS to conduct two-way chi-square

Step 1: Enter data in SPSS and save data file

Type	Type of intersection 1=Traffic lights 2=Roundabouts 3=Stop signs	*Set **Measure** column to **Nominal***
Severity	Severity of accident 1=Minor 2=Major	*Set **Measure** column to **Nominal***
Count	Number of accidents	*Set **Measure** column to **Nominal***

Then, enter your data. They will again be quite simple, just three columns as follows:

1	1	55
1	2	50
2	1	50
2	2	14
3	1	90
3	2	71

Remember to save the data in a file that might be called ***TwowayChi.sav***.

Step 2: Screen data and check assumptions

There is no formal data screening, just a visual check of data accuracy and a check that expected frequencies in the final output are ≥ 5. In large contingency tables it can be acceptable to have up to 20% of cells violate this requirement, although the power of the test is reduced accordingly (Field, 2005, p. 686). Cells should never have zero frequency. If the assumption is seriously violated it means more data are needed, or in some cases it may be possible to combine small cells.

Step 3: Request the analysis

As the data are in the form of frequency counts be sure to again request the appropriate **Weight Cases** option. A **Weight on** message should be present in the bottom bar of the data window.

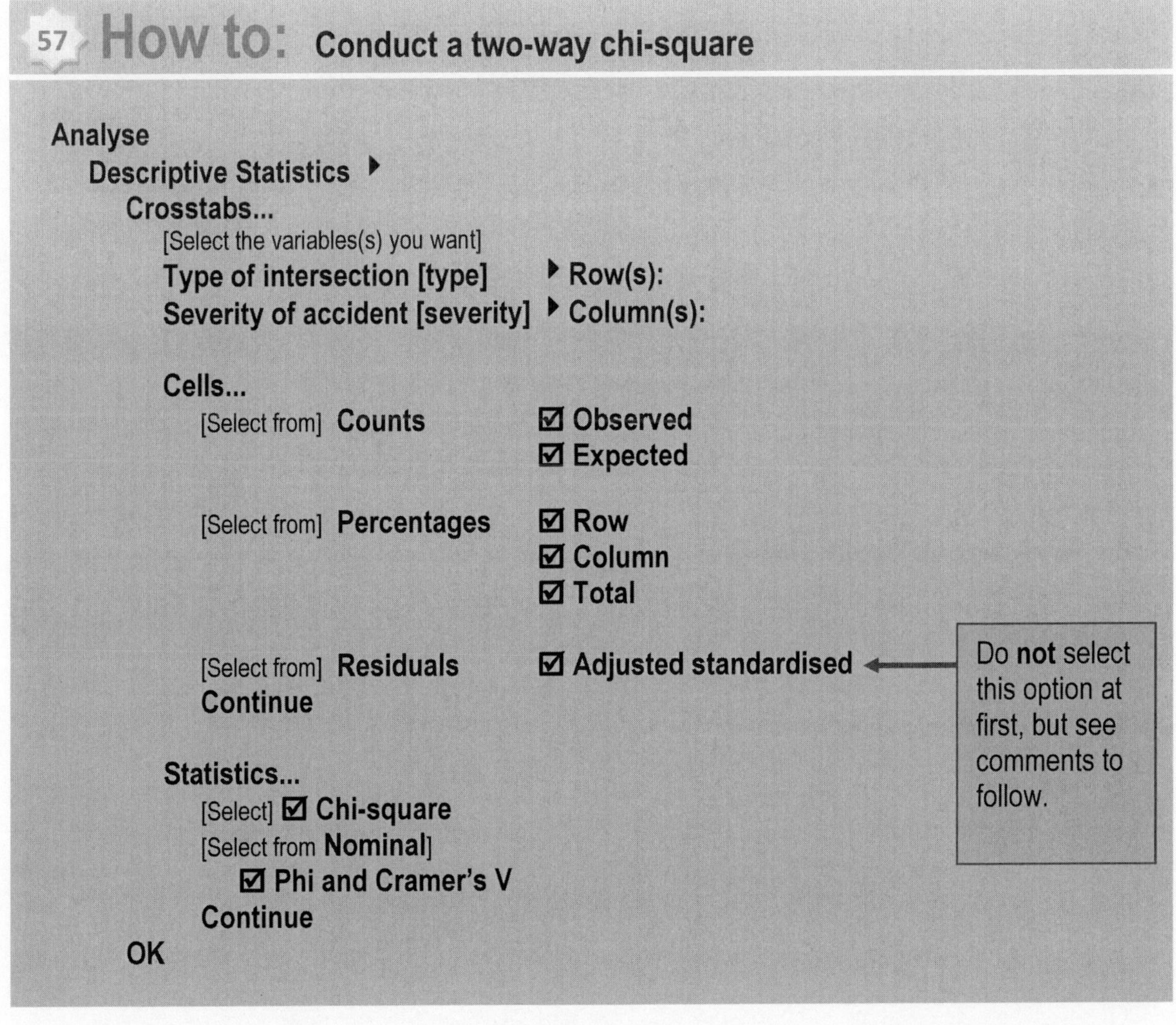

Step 4: Examine and interpret the output

Output: **Two-way chi-square**

Crosstabs

Type of intersection * Severity of accident Crosstabulation

			Severity of accident		
			Minor	Major	Total
Type of intersection	Traffic lights	Count	55	50	105
		Expected Count	62.0	43.0	105.0
		% within Type of intersection	52.4%	47.6%	100.0%
		% within Severity of accident	28.2%	37.0%	31.8%
		% of Total	16.7%	15.2%	31.8%
	Roundabouts	Count	50	14	64
		Expected Count	37.8	26.2	64.0
		% within Type of intersection	78.1%	21.9%	100.0%
		% within Severity of accident	25.6%	10.4%	19.4%
		% of Total	15.2%	4.2%	19.4%
	Stop signs	Count	90	71	161
		Expected Count	95.1	65.9	161.0
		% within Type of intersection	55.9%	44.1%	100.0%
		% within Severity of accident	46.2%	52.6%	48.8%
		% of Total	27.3%	21.5%	48.8%
Total		Count	195	135	330
		Expected Count	195.0	135.0	330.0
		% within Type of intersection	59.1%	40.9%	100.0%
		% within Severity of accident	100.0%	100.0%	100.0%
		% of Total	59.1%	40.9%	100.0%

1

It is very important to examine and understand the **Crosstabs** table, so as to appreciate what might be happening. In particular make sure you understand the **% within** figures.

For example, taking the top row for Traffic lights

% within Type of intersection is read **across the row**. It tells us that of the 100% of accidents (105) at **traffic lights**, 52.4% (55) were minor ones, and 47.6% (50) were major ones.

% within Severity of accident is read **down the column**. It tells us that of the 100% of **Minor** accidents (195), 28.2% (55) occurred at traffic lights; and of the 100% of **Major** accidents (135), 37.0% (50) occurred at traffic lights.

% within total tells us that of the grand total of 330 accidents, 16.7% (55) were minor ones at traffic lights, and 15.2% (50) were major ones at traffic lights, resulting in a combined 31.8% of accidents (105) at traffic lights.

And so on for the other intersections.

Output: **Two-way chi-square**

Chi-Square Tests

	Value	df	Asymp. Sig. (2-sided)
Pearson Chi-Square	12.225[a]	2	.002
Likelihood Ratio	12.997	2	.002
Linear-by-Linear Association	.059	1	.809
N of Valid Cases	330		

a. 0 cells (.0%) have expected count less than 5. The minimum expected count is 26.18.

2

The **Pearson Chi-Square** is the statistic you want.

The **Asymp. Sig. (2-sided)** is the *p* value.

It is < α of .05; therefore the test result **is** significant.

Symmetric Measures

		Value	Approx. Sig.
Nominal by Nominal	Phi	.192	.002
	Cramer's V	.192	.002
N of Valid Cases		330	

a. Not assuming the null hypothesis.

b. Using the asymptotic standard error assuming the null hypothesis.

3

These are measures of effect size. **Phi** (phi coefficient) is used with **2x2 tables**.

With tables **larger than 2x2** (as here, with a 3x2) **Cramer's V** is used.

Both are interpreted similar to correlation

Conducting follow-up tests when a chi-square has more than two categories

We know from the results that there is a relationship between type of intersection and accident severity, but just as in ANOVA, to identify exactly what is happening we need to conduct three 2x2 follow-up comparisons: for traffic lights with roundabouts, traffic lights with stop signs, and roundabouts with stop signs; using a Bonferroni correction to alpha of .017 (.05/3) to maintain familywise error at .05 across the three tests conducted. We can do this by running three further analyses, using **Data**, **Select Cases . . .** prior to each one, to remove stop signs for the first analysis, roundabouts for the second analysis, and traffic lights for the third analysis.

If you do these analyses you will find no relationship between intersection and accident severity for traffic lights and stop signs, χ^2 (1, N = 266) = 0.32, p = .57; but a significant relationship for traffic lights and roundabouts, χ^2 (1, N = 169) = 11.20, p = .001; and also for roundabouts and stop signs, χ^2 (1, N = 225) = 9.62, p = .002. From these results and the crosstabulation table it is clear that the relationship between intersection and accident severity is driven by the increased proportion of minor to major accidents at roundabouts.

Using standardised residuals

Although rarely reported in text books standardised residuals, or preferably **adjusted** standardised residuals, are another way to pinpoint where differences are occurring in crosstabulation tables where one or other variable has more than two categories. If you repeat the main analysis for this example asking for **adjusted standardised residuals** (as noted in the SPSS commands for conducting a two-way chi-square), you will obtain the table on the next page. It is the same as previously, but has the extra **Adjusted Residual** row.

Adjusted standardised residuals for each cell are the observed minus expected frequency divided by an estimate of the standard error. These are standard scores (i.e., *z*-scores), so we can obtain their probability in order to determine which cells have observed frequencies significantly different from the expected frequency. It is usual to evaluate standardised residuals at alpha = .05, in which case we automatically know from the properties of the normal curve that any standardised residual is significant if it has an absolute value ≥ 1.96. To find out the exact probability for a standardised residual, refer either to *z*-score tables or to the internet site:
http://www.danielsoper.com/statcalc/calc21.aspx

The only cells that exceed 1.96 for the adjusted residual in the table below are those for roundabout, indicating a **greater** than expected frequency of minor accidents (positive residual) and a corresponding **less** than expected frequency of major accidents (negative residual). The residual of 3.4 has a *p* value of < .001. If we were concerned about familywise error we could again use a Bonferroni corrected alpha of .017 (.05/3 categories), in which case this residual would still be significant.

Type of intersection * Severity of accident Crosstabulation

			Severity of accident		
			Minor	Major	Total
Type of intersection	Traffic lights	Count	55	50	105
		Expected Count	62.0	43.0	105.0
		% within Type of intersection	52.4%	47.6%	100.0%
		% within Severity of accident	28.2%	37.0%	31.8%
		% of Total	16.7%	15.2%	31.8%
		Adjusted Residual	-1.7	1.7	
	Roundabouts	Count	50	14	64
		Expected Count	37.8	26.2	64.0
		% within Type of intersection	78.1%	21.9%	100.0%
		% within Severity of accident	25.6%	10.4%	19.4%
		% of Total	15.2%	4.2%	19.4%
		Adjusted Residual	3.4	-3.4	
	Stop signs	Count	90	71	161
		Expected Count	95.1	65.9	161.0
		% within Type of intersection	55.9%	44.1%	100.0%
		% within Severity of accident	46.2%	52.6%	48.8%
		% of Total	27.3%	21.5%	48.8%
		Adjusted Residual	-1.2	1.2	
Total		Count	195	135	330
		Expected Count	195.0	135.0	330.0
		% within Type of intersection	59.1%	40.9%	100.0%
		% within Severity of accident	100.0%	100.0%	100.0%
		% of Total	59.1%	40.9%	100.0%

Odds ratio

Another useful indicator of effect size is the **odds ratio**, which can be calculated as shown below.

If we were to focus on minor accidents, the odds of having a **minor accident** versus a major accident in this example are as follows:

$$\text{Odds}_{\text{Traffic lights}} = \text{Number minor accidents/number major accidents} = 55/50 = 1.10$$

$$\text{Odds}_{\text{Roundabouts}} = \text{Number minor accidents/number major accidents} = 50/14 = 3.5714$$

$$\text{Odds}_{\text{Stop signs}} = \text{Number minor accidents/number major accidents} = 90/71 = 1.2676$$

The odds ratio for a minor accident at roundabouts compared to traffic lights is:

$$\text{Odds ratio} = \text{Odds}_{\text{Roundabouts}} / \text{Odds}_{\text{Traffic lights}} = 3.5714/1.1 = 3.25$$

The odds ratio for a minor accident at roundabouts compared to stop signs is:

$$\text{Odds ratio} = \text{Odds}_{\text{Roundabouts}} / \text{Odds}_{\text{Stop signs}} = 3.5714/1.2676 = 2.82$$

In other words, you are 3.25 times more likely to have a minor accident (as opposed to a major one) at roundabouts than at traffic lights, and 2.82 times more likely to have minor accident at roundabouts than at stop signs (according to this fictitious study!)

Two-way chi-square: research report sample Results section

Results

With α set at .05, a two-way chi-square revealed a significant relationship between type of intersection and accident severity. The proportions of minor to major accidents were different for the three different types of intersection over the 12-month period, χ^2 (2, N = 330) = 12.23, p = .002, Cramer's V = .19. The frequencies are shown in Table 1.

Table 1

Frequencies, Percentages, and Adjusted Standardised Residuals (ASR) for Minor and Major Accidents at Three Different Types of Intersection

	Minor Accidents				Major Accidents				
Intersection	f	%	f_e	ASR[a]	f	%	f_e	ASR[a]	Total
Traffic lights	55.0	52.4	62.0	-1.7	50.0	47.6	43.0	1.7	105
Roundabouts	50.0	78.1	37.8	3.4*	14.0	21.9	26.2	-3.4*	64
Stop signs	90.0	55.9	95.1	-1.2	71.0	44.1	65.9	1.2	161

Note. [a] The adjusted standardised residual is the observed frequency – expected frequency/estimated standard error. * $p < .001$.

While there was no significant difference in the proportions of minor to major accidents at traffic lights or at stop signs, minor accidents more likely than major accidents at roundabouts. As can be seen from Table 1, the proportions at roundabouts were 78% to 22% respectively. Minor accidents were significantly more likely to occur at roundabouts than they were at either stop signs (2.82 times more likely at roundabouts) or traffic lights (3.25 times more likely at roundabouts), while the opposite pattern was found for major accidents. There was no significant relationship between type of intersection and accident severity for traffic lights and stop signs.

Note:
f is the symbol for frequency, and f_e is the symbol for expected frequency.

I have chosen to rely on adjusted standardised residuals and odds ratios in reporting these results, but an alternative would have been to report the three 2x2 follow-up comparisons. Decisions in this regard can be made according to preferred approaches in the research area in which you are working.

Performing chi-square with scores, not frequencies

Chi-square analyses can be performed when nominal data are recorded as scores rather than frequency counts—SPSS adds up the score frequencies for you. In this example, imagine we have randomly allocated 36 people to one of two groups. They are asked to act as mock jurors. The first group watches a videotape of a trial where the defendant acts with a nervous demeanour. The second group watches the same videotape of a trial but in this version the defendant acts with a calm demeanour. The participants are asked to assign a verdict of guilty or not guilty. The research hypothesis is that the proportion of "Guilty" versus "Not guilty" verdicts will be different for the two groups.

The raw data are as follows. There are three variables: ID, Group and Verdict, and the data are in the form of alphabetic characters not numbers. SPSS can handle this provided that when defining the variables you define them as **String** variables in the **Type...** option:

ID	Participant number	*Set **Measure** column to **Nominal***
Group	N=Nervous C=Calm	*Set **Measure** column to **Nominal***
Verdict	G=Guilty N=Not guilty	*Set **Measure** column to **Nominal***

ID	Group	Verdict	ID	Group	Verdict	ID	Group	Verdict
1	C	G	13	N	G	25	N	N
2	N	N	14	C	G	26	N	N
3	C	N	15	C	G	27	C	N
4	N	G	16	N	G	28	C	G
5	C	N	17	N	G	29	N	G
6	N	G	18	C	N	30	N	G
7	C	G	19	N	G	31	C	N
8	C	N	20	C	N	32	N	G
9	N	N	21	N	N	33	C	N
10	N	G	22	C	N	34	N	N
11	C	G	23	N	N	35	C	N
12	C	N	24	C	N	36	N	G

Using SPSS to conduct two-way chi-square with scores, not counts

Step 1: Enter data in SPSS and save data file

Enter the data as above and save them in a file called ***ChiScores.sav***.

Step 2: Screen data and check assumptions

There is no formal data screening, just a visual check of data accuracy and a check that expected frequencies in the final output are ≥ 5.

Step 3: Request the analysis

As the data are **not** in the form of frequency counts be sure to have turned **off** any **Weight Cases** option. A **Weight on** message should **not** be present in the bottom bar of the data window.

58 How to: Conduct a two-way chi-square on scores, not frequencies

Analyse
  Descriptive Statistics ▸
    Crosstabs...
      [Select the variables(s) you want]
      Group ▸ **Row(s):**
      Verdict ▸ **Column(s):**
      Cells...
        [Select from] **Counts** ☑ **Observed**
        ☑ **Expected**
        [Select from] **Percentages** ☑ **Row**
        ☑ **Column**
        ☑ **Total**
        [Select from] **Residuals** ☑ **Adjusted standardised**
        Continue
      Statistics...
        [Select] ☑ **Chi-square**
        [Select from **Nominal**]☑ **Phi and Cramer's V**
      Continue
    OK

Step 4: Examine and interpret the output

Examine the output that follows on the next page.

Group * Verdict Crosstabulation

			Verdict		Total
			Guilty	Not guilty	
Group	Calm	Count	6	12	18
		Expected Count	8.5	9.5	18.0
		% within Group	33.3%	66.7%	100.0%
		% within Verdict	35.3%	63.2%	50.0%
		% of Total	16.7%	33.3%	50.0%
		Adjusted Residual	-1.7	1.7	
	Nervous	Count	11	7	18
		Expected Count	8.5	9.5	18.0
		% within Group	61.1%	38.9%	100.0%
		% within Verdict	64.7%	36.8%	50.0%
		% of Total	30.6%	19.4%	50.0%
		Adjusted Residual	1.7	-1.7	
Total		Count	17	19	36
		Expected Count	17.0	19.0	36.0
		% within Group	47.2%	52.8%	100.0%
		% within Verdict	100.0%	100.0%	100.0%
		% of Total	47.2%	52.8%	100.0%

1

As always with Chi-Square it is very important to examine and **understand** the crosstabs table—to obtain an indication of what might be happening.

Make sure you are careful to work out which percentage is which, otherwise you can seriously misinterpret the table.

Note that none of the adjusted standardised residuals exceed 1.96; thus none of the observed frequencies are significantly different to the expected.

Chi-Square Tests

	Value	df	Asymp. Sig. (2-sided)	Exact Sig. (2-sided)	Exact Sig. (1-sided)
Pearson Chi-Square	2.786[b]	1	.095		
Continuity Correction[a]	1.783	1	.182		
Likelihood Ratio	2.824	1	.093		
Fisher's Exact Test				.181	.091
N of Valid Cases	36				

a. Computed only for a 2x2 table

b. 0 cells (.0%) have expected count less than 5. The minimum expected count is 8.50.

2

Even though, as the researcher might have expected, there are relatively more guilty verdicts for the nervous as opposed to calm demeanour, this effect does not achieve significance. **The Asymp. Sig. (2-sided)** of .095 is > α of .05; therefore, the null hypothesis must be retained, and we conclude there is no statistically significant relationship between demeanour and verdict. (Note how this is consistent with the adjusted standardised residuals.)

Of course we may have a Type II error—if there is a real relationship we will need a more powerful test (i.e., a larger sample size) to detect it.

Following the advice of Howell (2002) we can ignore the Continuity Correction and Fisher's Exact Test (although some texts will recommend otherwise).

Symmetric Measures

		Value	Approx. Sig.
Nominal by Nominal	Phi	-.278	.095
	Cramer's V	.278	.095
N of Valid Cases		36	

a. Not assuming the null hypothesis.

b. Using the asymptotic standard error assuming the null hypothesis.

3

As this is a 2x2 table, the **phi** coefficient is appropriate for effect size.

24 Multiway Frequency Analysis (Loglinear Analysis)

Multiway frequency analysis (also known as **loglinear analysis**) is best conceptualised as an extension of two-way chi-square to a multiway analysis when **frequencies** are obtained on **more than two** variables. It is analogous to a factorial ANOVA for frequencies. Main effects and interactions can be examined as for ANOVA, and using model selection the simplest effect that fits the observed data can be identified. At other times a particular theoretical model is proposed (e.g., a three-way interaction may be hypothesised from theory) and the **goodness-of-fit** of this model is assessed (i.e., how well does the model fit the observed data?).

There are variants of loglinear analysis (e.g., where one variable is considered the DV, and hierarchical and nonhierarchical approaches) for which the reader is referred to Tabachnick and Fidell (2007).

The following hypothetical example demonstrates one approach to the analysis of multiway frequency designs.

A hypothetical example of multiway frequency analysis

A question that has been considered in forensic psychology is whether high status is a shield or liability for defendants. That is, are high status defendants more or less likely to be found guilty compared to low status defendants? A forensic psychologist conducts a study in which mock jurors read a transcript of a trial in which the defendant is accused of stealing. The evidence is ambiguous. Sixty-two mock jurors are members of the judiciary (judges and magistrates), and 57 are jury-eligible members of the public. Each participant reads one of four versions of the transcript, in which the defendant is depicted as either a male or female, either of high status (a doctor) or low status (a shopping centre cleaner).

The study constitutes a 2 x 2 x 2 x 2 multiway frequency design, that is, group (judges versus public) x defendant status (high versus low) x defendant gender (male versus female) x verdict (guilty versus not guilty).

Using SPSS to conduct a multiway frequency analysis

Step 1: Enter data in SPSS and save data file

The data in this example are as follows:

ID	Participant number	*Set* ***Measure*** *column to* ***Nominal***
Group	1=judges 2=public	*Set* ***Measure*** *column to* ***Nominal***
Dstatus	Defendant status 1=high status 2=low status	*Set* ***Measure*** *column to* ***Nominal***
Dgender	Defendant gender 0=male 1=female	*Set* ***Measure*** *column to* ***Nominal***
Verdict	0=not guilty 1=guilty	*Set* ***Measure*** *column to* ***Nominal***

The data from 119 participants are in a file called ***24Multiway.sav*** which is available on the internet site: www.pearson.com.au/9781442549821. Click on the link 'Student downloads' to download the datafile.

Step 2: Screen data and check assumptions

Multiway frequency analysis is a nonparametric procedure with few assumptions. Its assumptions are those of two-way chi-square:observations fall in one and only one category; observations are inde-pendent of one another; observations are measured as frequencies; and the **expected frequency** in each category should be ≥ 5.

In order to have a chance of satisfying the expected frequency requirement always aim to have a sample size that exceeds five cases per cell. In this example there are 16 cells, so a sample exceeding 80 is necessary, and the actual sample size of 119 should be sufficient. It can be acceptable to have up to 20% of cells violate this requirement, although the power of the test is reduced accordingly (Field, 2005, p. 686). Cells should never have zero frequency. Where adequate sample sizes cannot be achieved it may be necessary to simplify the design by collapsing across cells, or alternatively, the researcher accepts the reduced power of the analysis and interprets accordingly (see Tabachnick & Fidell, 2007, p. 862)

Sometimes there are **outlier cells** (with large discrepancies between observed and expected frequencies) that do not fit well with any effect or any model. It may be necessary to delete them or collapse across cells.

Step 3: Conduct the multiway frequency analysis

59 How to: Produce a crosstabs table for multiway frequency analysis

This table provides the descriptive statistics of cell frequencies and percentages, together with expected frequencies, to check the requirement of expected frequencies > 5.

Analyze
Descriptive Statistics
Crosstabs...
[Select variables, as follows:]
Defendant status [Dstatus] ▸ **Row[s]:**
Verdict [Verdict] ▸ **Column(s):**

Defendant gender [Dgender] ▸ **Layer 1 of 1**
[Click **Next** button]
Group [Group] ▸ **Layer 2 of 2**

Cells...
[From **Counts** select:]
☑ **Observed**
☑ **Expected**
[From **Percentages** select:]
☑ **Row**
☑ **Column**
Continue

OK

60 How to: Conduct a multiway frequency analysis

Analyze
  Loglinear
    Model Selection...

[Select variables, as follows:]
Verdict [Verdict] ▸ **Factor(s):**
[Click **Define Range** button and type the code ranges for **Verdict]**
Minimum: 0
Maximum: 1
Continue

Group [Group] ▸ **Factor(s):**
[Click **Define Range** button and type the code ranges for **Group]**
Minimum: 1
Maximum: 2
Continue

Defendant status [Dstatus] ▸ **Factor(s):**
[Click **Define Range** button and type the code ranges for **Dstatus]**
Minimum: 1
Maximum: 2
Continue

Defendant gender [Dgender] ▸ **Factor(s):**
[Click **Define Range** button and type the code ranges for **Dgender]**
Minimum: 0
Maximum: 1
Continue

Options...
[From **Display for Saturated Model** select:]
☑ **Parameter estimates**
☑ **Association table**
Continue

OK

Note:
There is a **Models...** button with options for model selection. Normally, if testing for all main effects and interactions, the default **Saturated Model** is used. If a researcher wants to test a particular hypothesised model then the appropriate **Custom** model needs to be specified.

Step 4: Examine and interpret crosstabs and multiway output

Defendant status * Verdict * Defendant gender * Group Crosstabulation

Group	Defendant gender				Verdict: not guilty	Verdict: guilty	Total
judges	male	Defendant status	High status	Count	10	3	13
				Expected Count	6.1	6.9	13.0
				% within Defendant status	76.9%	23.1%	100.0%
				% within Verdict	71.4%	18.8%	43.3%
			Low status	Count	4	13	17
				Expected Count	7.9	9.1	17.0
				% within Defendant status	23.5%	76.5%	100.0%
				% within Verdict	28.6%	81.3%	56.7%
		Total		Count	14	16	30
				Expected Count	14.0	16.0	30.0
				% within Defendant status	46.7%	53.3%	100.0%
				% within Verdict	100.0%	100.0%	100.0%
	female	Defendant status	High status	Count	9	8	17
				Expected Count	8.5	8.5	17.0
				% within Defendant status	52.9%	47.1%	100.0%
				% within Verdict	56.3%	50.0%	53.1%
			Low status	Count	7	8	15
				Expected Count	7.5	7.5	15.0
				% within Defendant status	46.7%	53.3%	100.0%
				% within Verdict	43.8%	50.0%	46.9%
		Total		Count	16	16	32
				Expected Count	16.0	16.0	32.0
				% within Defendant status	50.0%	50.0%	100.0%
				% within Verdict	100.0%	100.0%	100.0%
public	male	Defendant status	High status	Count	10	4	14
				Expected Count	5.9	8.1	14.0
				% within Defendant status	71.4%	28.6%	100.0%
				% within Verdict	76.9%	22.2%	45.2%
			Low status	Count	3	14	17
				Expected Count	7.1	9.9	17.0
				% within Defendant status	17.6%	82.4%	100.0%
				% within Verdict	23.1%	77.8%	54.8%
		Total		Count	13	18	31
				Expected Count	13.0	18.0	31.0
				% within Defendant status	41.9%	58.1%	100.0%
				% within Verdict	100.0%	100.0%	100.0%
	female	Defendant status	High status	Count	6	7	13
				Expected Count	5.5	7.5	13.0
				% within Defendant status	46.2%	53.8%	100.0%
				% within Verdict	54.5%	46.7%	50.0%
			Low status	Count	5	8	13
				Expected Count	5.5	7.5	13.0
				% within Defendant status	38.5%	61.5%	100.0%
				% within Verdict	45.5%	53.3%	50.0%
		Total		Count	11	15	26
				Expected Count	11.0	15.0	26.0
				% within Defendant status	42.3%	57.7%	100.0%
				% within Verdict	100.0%	100.0%	100.0%

Hierarchical Loglinear Analysis

Design 1

Convergence Information

Generating Class	Verdict*Group*Dstatus*Dgender
Number of Iterations	1
Max. Difference between Observed and Fitted Marginals	.000
Convergence Criterion	.250

Cell Counts and Residuals

Verdict	Group	Dstatus	Dgender	Observed Count[a]	Observed %	Expected Count	Expected %	Residuals	Std. Residuals
not guilty	judges	High status	male	10.500	8.8%	10.500	8.8%	.000	.000
			female	9.500	8.0%	9.500	8.0%	.000	.000
		Low status	male	4.500	3.8%	4.500	3.8%	.000	.000
			female	7.500	6.3%	7.500	6.3%	.000	.000
	public	High status	male	10.500	8.8%	10.500	8.8%	.000	.000
			female	6.500	5.5%	6.500	5.5%	.000	.000
		Low status	male	3.500	2.9%	3.500	2.9%	.000	.000
			female	5.500	4.6%	5.500	4.6%	.000	.000
guilty	judges	High status	male	3.500	2.9%	3.500	2.9%	.000	.000
			female	8.500	7.1%	8.500	7.1%	.000	.000
		Low status	male	13.500	11.3%	13.500	11.3%	.000	.000
			female	8.500	7.1%	8.500	7.1%	.000	.000
	public	High status	male	4.500	3.8%	4.500	3.8%	.000	.000
			female	7.500	6.3%	7.500	6.3%	.000	.000
		Low status	male	14.500	12.2%	14.500	12.2%	.000	.000
			female	8.500	7.1%	8.500	7.1%	.000	.000

a. For saturated models, .500 has been added to all observed cells.

Goodness-of-Fit Tests

	Chi-Square	df	Sig.
Likelihood Ratio	.000	0	.
Pearson	.000	0	.

1

The first thing to do is to become familiar with the **Crosstabulation table** on the **previous page**. Is anything apparent from just an inspection of the frequencies?

Is the requirement of expected count > 5 met for all cells? If not, especially if more than 20% of cells fail, does the table suggest any options for collapsing cells across any of the variables?

The output on this page is of no great relevance, as it tests the **saturated** model, which **must** fit the observed data perfectly because everything is included.

K-Way and Higher-Order Effects

	K	df	Likelihood Ratio Chi-Square	Likelihood Ratio Sig.	Pearson Chi-Square	Pearson Sig.	Number of Iterations
K-way and Higher Order Effects[a]	1	15	21.939	.109	21.773	.114	0
	2	11	20.425	.040	19.418	.054	2
	3	5	7.344	.196	7.256	.202	3
	4	1	.000	.989	.000	.989	3
K-way Effects[b]	1	4	1.514	.824	2.355	.671	0
	2	6	13.080	.042	12.163	.058	0
	3	4	7.344	.119	7.255	.123	0
	4	1	.000	.989	.000	.989	0

a. Tests that k-way and higher order effects are zero.

b. Tests that k-way effects are zero.

The first of the two tables in this section of the output is the test of K-way and Higher Order Effects.

Note that two methods are used for these tests, the likelihood ratio (L.R., or G^2) and Pearson chi-squares. Tabachnick and Fidell (2007, p. 888) suggest that consistency favours the likelihood ratio.

The row tests are as follows:
K = 1 row tests if any main effects **or** interactions are significant (no, as this is not significant);
K = 2 tests if any two-way **or** higher order interactions are significant (this is significant, so at least one of them is);
K = 3 tests if any three-way **or** four-way interactions are significant (no, as this is not significant);
K = 4 tests if there is a significant four-way interaction (no, this is not significant).

There is some inconsistency here, which can occur in multiway frequency analysis. When it does occur, further investigation becomes all the more important. The K = 1 row suggests that nothing is significant, but the K=2, 3, and 4 rows suggest that there is at least one significant **two-way** interaction.

The second table is the test of K-way Effects.
K = 1 row tests if any main effects are significant (no, as this is not significant);
K = 2 tests if any two-way interactions are significant (this is significant, so at least one of them is);
K = 3 tests if any three-way are significant (no, as this is not significant);
K = 4 tests if there is a significant four-way interaction (no, this is not significant).

This table supports the conclusion that there is at least one **two-way** interaction.

Testing the highest order interaction

The most important aspect of these tables is their test of the **highest order interaction** (the four-way in this case).

In this example, the four-way interaction of group x defendant status x defendant gender x verdict is **not** significant.

Partial Associations

Effect	df	Partial Chi-Square	Sig.	Number of Iterations
Verdict*Group*Dstatus	1	.007	.931	3
Verdict*Group*Dgender	1	.000	.983	3
Verdict*Dstatus*Dgender	1	7.200	.007	3
Group*Dstatus*Dgender	1	.187	.665	2
Verdict*Group	1	.474	.491	3
Verdict*Dstatus	1	11.467	.001	2
Group*Dstatus	1	.023	.879	3
Verdict*Dgender	1	.002	.964	2
Group*Dgender	1	.421	.516	3
Dstatus*Dgender	1	.613	.434	2
Verdict	1	1.018	.313	2
Group	1	.210	.647	2
Dstatus	1	.210	.647	2
Dgender	1	.076	.783	2

3

Given that the highest order interaction (the four-way) is not significant, the **Partial Associations** table becomes relevant to identify which specific lower order effects are significant. It indicates, in this example, that the only significant **two-way** interaction is the verdict x defendant status interaction.

However, there is more inconsistency. In contrast to the previous tables, there is also a **significant three-way interaction of verdict x defendant status x defendant gender**.

Parameter Estimates

Effect	Parameter	Estimate	Std. Error	Z	Sig.	95% Confidence Interval	
						Lower Bound	Upper Bound
Verdict*Group*Dstatus*Dgender	1	-.001	.097	-.010	.992	-.191	.189
Verdict*Group*Dstatus	1	-.008	.097	-.082	.935	-.198	.182
Verdict*Group*Dgender	1	.001	.097	.006	.995	-.190	.191
Verdict*Dstatus*Dgender	1	.246	.097	2.536	.011	.056	.436
Group*Dstatus*Dgender	1	-.039	.097	-.403	.687	-.229	.151
Verdict*Group	1	.071	.097	.733	.463	-.119	.261
Verdict*Dstatus	1	.312	.097	3.217	.001	.122	.502
Group*Dstatus	1	-.015	.097	-.152	.879	-.205	.175
Verdict*Dgender	1	.001	.097	.012	.991	-.189	.191
Group*Dgender	1	-.055	.097	-.571	.568	-.246	.135
Dstatus*Dgender	1	-.053	.097	-.549	.583	-.243	.137
Verdict	1	-.073	.097	-.751	.452	-.263	.117
Group	1	.046	.097	.479	.632	-.144	.237
Dstatus	1	-.018	.097	-.191	.849	-.209	.172
Dgender	1	-.049	.097	-.505	.613	-.239	.141

4

An alternative way to test individual effects is through **Parameter Estimates**, tested as z-scores. These yield the same conclusions as the Partial Associations table (and include another test of the four-way interaction).

One of the difficulties with multiway frequency analysis is that results can sometimes be ambiguous in this way. As with ANOVA, higher order effects render lower order effects of little interest. With inconsistent results it is all the more necessary to move onto the next step of modelling, which selects the simplest model that best fits the observed data. This output begins on the next page and is interpreted quite differently.

Model selection is a step-by-step backward elimination process. All the steps are shown in the **Step Summary** table on the next page.

In model selection the **null hypothesis** is that the data **fit** the model, and unlike most other statistics we are interested in the null hypothesis. We want to identify the model that provides the best fit to the data, so a **nonsignificant result** is desired, although one may want to use a more liberal alpha level (e.g., .10 or even .25).

How does SPSS identify this model using backward elimination?

Step 0: It begins by deleting the four-way interaction (**Deleted Effect**) and asks whether the data still fit the model after it has been deleted. The answer in this case is "yes", because with the four-way interaction removed there is still a good fit, Sig. = .989. Thus, there is no necessary four-way interaction here, consistent with the previous results.

Step 1: It now looks at the three-way interactions, and at what happens to the model if each one is deleted (Deleted Effect 1, 2, 3, 4). It then actually deletes the one, whose deletion has the least effect on the model. This is Verdict x Group x Dgender. With this removed the model is little changed, Sig. = .983. (Note that if Verdict x DStatus x Dgender is removed the data no longer fit the model, Sig. = .007).

5

Step 2: Generating Class at Step 2 no longer includes Verdict x Group x Dgender. Deleted Effect 1, 2, 3 identifies Verdict x Group x Dstatus as the one to be removed next, leaving Sig. = .931.

Step 3: Generating Class at Step 3 no longer includes Verdict x Group x Dstatus, but now we start including the two-way interactions. Deleted Effect 1, 2, 3 identifies Group x Dstatus x Dgender as the one to be removed next, leaving Sig. = .651.

Steps 4 – 7: This hierarchical process continues, moving successively down through the effects, deleting one at a time, until at step 7 we still have the Verdict x DStatus x Dgender effect, to which is added the one main effect not contained within the interaction, namely, Group. However, the Deleted Effect 1, 2 identifies Group as the next one to go.

Step 8: At this stage only Verdict x DStatus x Dgender is left, and if we delete it the model no longer fits the data, Sig. = .008.

Step 9: At this step, therefore, the model selection process arrives at the best fitting model. In this example we have a significant **three-way interaction of Verdict x DStatus x Dgender.** The ambiguity in the previous sections of output is finally resolved.

Step Summary

Step[b]			Effects	Chi-Square[a]	df	Sig.		Number of Iterations
0	Generating Class[c]		Verdict*Group*Dstatus*Dgender	.000	0	.		
	Deleted Effect	1	Verdict*Group*Dstatus*Dgender	.000	1	.989		3
1	Generating Class[c]		Verdict*Group*Dstatus, Verdict*Group*Dgender, Verdict*Dstatus*Dgender, Group*Dstatus*Dgender	.000	1	.989		
	Deleted Effect	1	Verdict*Group*Dstatus	.007	1	.931		3
		2	Verdict*Group*Dgender	.000	1	.983	Delete	3
		3	Verdict*Dstatus*Dgender	7.200	1	.007		3
		4	Group*Dstatus*Dgender	.187	1	.665		2
2	Generating Class[c]		Verdict*Group*Dstatus, Verdict*Dstatus*Dgender, Group*Dstatus*Dgender	.001	2	1.000		
	Deleted Effect	1	Verdict*Group*Dstatus	.007	1	.931	Delete	3
		2	Verdict*Dstatus*Dgender	7.210	1	.007		2
		3	Group*Dstatus*Dgender	.204	1	.652		3
3	Generating Class[c]		Verdict*Dstatus*Dgender, Group*Dstatus*Dgender, Verdict*Group	.008	3	1.000		
	Deleted Effect	1	Verdict*Dstatus*Dgender	7.263	1	.007		3
		2	Group*Dstatus*Dgender	.205	1	.651	Delete	2
		3	Verdict*Group	.606	1	.436		2
4	Generating Class[c]		Verdict*Dstatus*Dgender, Verdict*Group, Group*Dstatus, Group*Dgender	.213	4	.995		
	Deleted Effect	1	Verdict*Dstatus*Dgender	7.131	1	.008		3
		2	Verdict*Group	.474	1	.491		2
		3	Group*Dstatus	.023	1	.879	Delete	2
		4	Group*Dgender	.421	1	.516		2
5	Generating Class[c]		Verdict*Dstatus*Dgender, Verdict*Group, Group*Dgender	.237	5	.999		
	Deleted Effect	1	Verdict*Dstatus*Dgender	7.131	1	.008		3
		2	Verdict*Group	.455	1	.500		2
		3	Group*Dgender	.409	1	.522	Delete	2
6	Generating Class[c]		Verdict*Dstatus*Dgender, Verdict*Group	.646	6	.996		
	Deleted Effect	1	Verdict*Dstatus*Dgender	7.131	1	.008		3
		2	Verdict*Group	.473	1	.491	Delete	2
7	Generating Class[c]		Verdict*Dstatus*Dgender, Group	1.119	7	.993		
	Deleted Effect	1	Verdict*Dstatus*Dgender	7.131	1	.008		3
		2	Group	.210	1	.647	Delete	2
8	Generating Class[c]		Verdict*Dstatus*Dgender	1.329	8	.995		
	Deleted Effect	1	Verdict*Dstatus*Dgender	7.131	1	.008		3
9	Generating Class[c]		Verdict*Dstatus*Dgender	1.329	8	.995		

a. For 'Deleted Effect', this is the change in the Chi-Square after the effect is deleted from the model.

b. At each step, the effect with the largest significance level for the Likelihood Ratio Change is deleted, provided the significance level is larger than .050.

c. Statistics are displayed for the best model at each step after step 0.

Cell Counts and Residuals

				Observed		Expected			Std.
Verdict	Group	Dstatus	Dgender	Count	%	Count	%	Residuals	Residuals
not guilty	judges	High status	male	10.000	8.4%	10.000	8.4%	.000	.000
			female	9.000	7.6%	7.500	6.3%	1.500	.548
		Low status	male	4.000	3.4%	3.500	2.9%	.500	.267
			female	7.000	5.9%	6.000	5.0%	1.000	.408
	public	High status	male	10.000	8.4%	10.000	8.4%	.000	.000
			female	6.000	5.0%	7.500	6.3%	-1.500	-.548
		Low status	male	3.000	2.5%	3.500	2.9%	-.500	-.267
			female	5.000	4.2%	6.000	5.0%	-1.000	-.408
guilty	judges	High status	male	3.000	2.5%	3.500	2.9%	-.500	-.267
			female	8.000	6.7%	7.500	6.3%	.500	.183
		Low status	male	13.000	10.9%	13.500	11.3%	-.500	-.136
			female	8.000	6.7%	8.000	6.7%	.000	.000
	public	High status	male	4.000	3.4%	3.500	2.9%	.500	.267
			female	7.000	5.9%	7.500	6.3%	-.500	-.183
		Low status	male	14.000	11.8%	13.500	11.3%	.500	.136
			female	8.000	6.7%	8.000	6.7%	.000	.000

Goodness-of-Fit Tests

	Chi-Square	df	Sig.
Likelihood Ratio	1.329	8	.995
Pearson	1.323	8	.995

6

This output shows the results for the best fitting model. In the table of **Cell Counts and Residuals** the standardised residuals (Std Residuals) indicate the goodness-of-fit between observed frequencies and those expected from the model for each cell. They indicate any cells for which the fit is not good (>1.96 absolute would be a significant difference at $p = .05$ between observed and expected frequencies). In this example, the fit is very good indeed and all the standardised residuals are very small.

The **Goodness-of-Fit Tests** table tells us that the model (i.e., the three-way interaction) has a likelihood ratio $\chi^2(8) = 1.33$, $p = .995$, also indicating an extremely good fit between the observed and expected frequencies generated by the model.

Step 5: Conduct analytical comparisons

As with ANOVA, having identified the significant effects (or best fitting model) analytical comparisons of some sort may be needed to pinpoint the precise nature of differences. There are a number of ways to do this, including using contrasts that can be generated by SPSS. Other SPSS output, namely parameter estimates, can also be useful to assess the significance of departures from expected frequencies in each cell and in cells collapsed across variables. Researchers intending to use multiway frequency analysis must take the time to study more specialised texts, such as Tabachnick and Fidell (2007), in order to become thoroughly familiar with its many aspects.

One approach that has the advantage of ready interpretation is to conduct relevant 2 x 2 chi-square analyses, analogous to pairwise comparisons in ANOVA (collapsing across cells where appropriate). A Bonferroni correction to alpha should be made to control familywise error rates, and researchers should take care to select the most parsimonious set of comparisons. This can be difficult with complex designs with more than two levels for variables.

In this example, analytical comparisons are relatively easy. One could run the crosstabs again, leaving out Group, which played no significant role, and requesting chi-square analyses. The SPSS commands would be as follows:

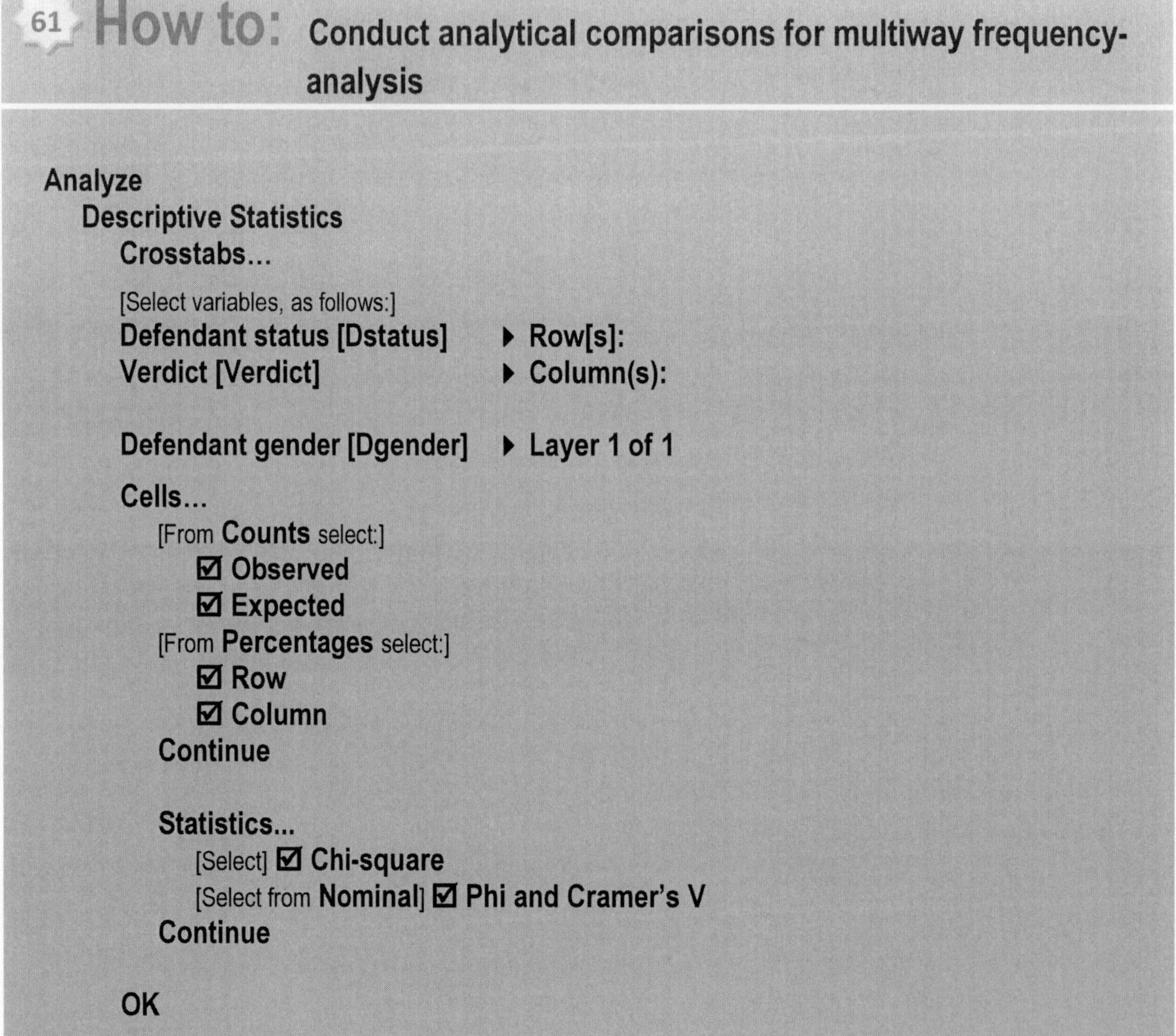

61 How to: Conduct analytical comparisons for multiway frequency-analysis

Analyze
Descriptive Statistics
Crosstabs...
[Select variables, as follows:]
Defendant status [Dstatus] ▸ Row[s]:
Verdict [Verdict] ▸ Column(s):

Defendant gender [Dgender] ▸ Layer 1 of 1

Cells...
[From Counts select:]
☑ Observed
☑ Expected
[From Percentages select:]
☑ Row
☑ Column
Continue

Statistics...
[Select] ☑ Chi-square
[Select from Nominal] ☑ Phi and Cramer's V
Continue

OK

Crosstabs

Defendant status * Verdict * Defendant gender Crosstabulation

Defendant gender				Verdict		Total
				not guilty	guilty	
male	Defendant status	High status	Count	20	7	27
			Expected Count	12.0	15.0	27.0
			% within Defendant status	74.1%	25.9%	100.0%
			% within Verdict	74.1%	20.6%	44.3%
		Low status	Count	7	27	34
			Expected Count	15.0	19.0	34.0
			% within Defendant status	20.6%	79.4%	100.0%
			% within Verdict	25.9%	79.4%	55.7%
	Total		Count	27	34	61
			Expected Count	27.0	34.0	61.0
			% within Defendant status	44.3%	55.7%	100.0%
			% within Verdict	100.0%	100.0%	100.0%
female	Defendant status	High status	Count	15	15	30
			Expected Count	14.0	16.0	30.0
			% within Defendant status	50.0%	50.0%	100.0%
			% within Verdict	55.6%	48.4%	51.7%
		Low status	Count	12	16	28
			Expected Count	13.0	15.0	28.0
			% within Defendant status	42.9%	57.1%	100.0%
			% within Verdict	44.4%	51.6%	48.3%
	Total		Count	27	31	58
			Expected Count	27.0	31.0	58.0
			% within Defendant status	46.6%	53.4%	100.0%
			% within Verdict	100.0%	100.0%	100.0%

Chi-Square Tests

Defendant gender		Value	df	Asymp. Sig. (2-sided)	Exact Sig. (2-sided)	Exact Sig. (1-sided)
male	Pearson Chi-Square	17.450[b]	1	.000		
	Continuity Correction[a]	15.350	1	.000		
	Likelihood Ratio	18.281	1	.000		
	Fisher's Exact Test				.000	.000
	Linear-by-Linear Association	17.164	1	.000		
	N of Valid Cases	61				
female	Pearson Chi-Square	.297[c]	1	.586		
	Continuity Correction[a]	.079	1	.778		
	Likelihood Ratio	.297	1	.586		
	Fisher's Exact Test				.610	.389
	Linear-by-Linear Association	.292	1	.589		
	N of Valid Cases	58				

a. Computed only for a 2x2 table

b. 0 cells (.0%) have expected count less than 5. The minimum expected count is 11.95.

c. 0 cells (.0%) have expected count less than 5. The minimum expected count is 13.03.

Symmetric Measures

Defendant gender			Value	Approx. Sig.
male	Nominal by Nominal	Phi	.535	.000
		Cramer's V	.535	.000
	N of Valid Cases		61	
female	Nominal by Nominal	Phi	.072	.586
		Cramer's V	.072	.586
	N of Valid Cases		58	

a. Not assuming the null hypothesis.

b. Using the asymptotic standard error assuming the null hypothesis.

The three-way interaction of verdict x defendant status x defendant gender can be interpreted adequately from this chi-square output.

A three-way interaction means that the two-way interaction is different for the different levels of the third variable.

Chi-Square Tests table (previous page)

Here, we see there is no significant association between verdict and defendant status when the defendant is depicted as a female (Asymp.Sig. = .586). From the **Crosstabulation** table we can see that for both low and high status females the proportions of guilty and not guilty verdicts are fairly similar.

However, there is a significant association when the defendant is depicted as a male. The cell frequencies show that the low status male was more likely to be found guilty (79.4% of the low status verdicts were guilty), whereas the high status male was more likely to have been found not guilty (74.1% of the high status verdicts were not guilty).

Remember, too, that given there was no significant four-way interaction, these findings hold for both groups of participants (judges and public).

Multiway frequency analysis: Research report sample Results section

Results

A four-way frequency analysis was performed to assess the relationship between mock juror verdict (guilty or not guilty), defendant status (high versus low), defendant gender, and participant group (judges versus jury eligible members of the public). With a sample size of 119, expected frequencies exceeded five in all cells of the design, and there were no outlying cells.

None of the effects involving participant group was significant (partial χ^2, $p > .49$), indicating that both judges and the public rendered similar mock verdicts across all other conditions of the design. Tests of partial associations for all possible effects resulted in a significant verdict by defendant status interaction, partial $\chi^2(1, N=119) = 11.47, p < .001$; and a significant verdict by defendant status by defendant gender interaction, partial $\chi^2(1, N = 119) = 7.20, p = .007$. Using backward elimination, the best fitting model was found to be the three-way interaction of verdict by defendant status by defendant gender, which had a likelihood ratio of $\chi^2(8, N = 119) = 1.33, p = .995$. This indicated an extremely good fit between the observed frequencies and the frequencies expected from the model.

To investigate the three-way interaction, 2 x 2 (verdict by defendant status) chi-square analyses were performed separately for male and female defendants, using a Bonferroni adjusted α of .025 for each. There was no significant association between verdict and defendant status when the offender was depicted as female, $\chi2(1, N =58) = 0.30, p = .59, \phi = .07$. As can be seen from the frequencies reported in Table 1, the verdicts for the female defendant were consistent with the ambiguous evidence presented to participants, in that around half were guilty and half not guilty. This occurred for both levels of defendant status: the high status doctor and the low status cleaner.

However, when the defendant was depicted as male there was a significant association between verdict and status, $\chi^2(1, N = 61) = 17.45, p < .001, \phi = .54$. From Table 1 it can be seen that the majority of mock verdicts were not guilty (74.1%) for the high status doctor, whereas this was reversed for the low status cleaner for whom the majority of verdicts were guilty (79.4%). Based on the odds ratio the low status male defendant was 11 times more likely to be found guilty than the high status male defendant.

Table 1

Frequency of Mock Verdicts as a Function of Gender and Status of Defendant

	Not Guilty			Guilty			
	f_e	f	%	f_e	f	%	Total
Male Defendant:							
High status	12	20	74.1	15	7	25.9	27
Low status	15	7	20.6	19	27	79.4	34
Female Defendant:							
High status	14	15	50.0	16	15	50.0	30
Low status	13	12	42.9	15	16	57.1	28

Note:

Several aspects of the Results section may require clarification.
Symbols used are as follows: ϕ = Phi, f = frequency, and f_e = expected frequency.

If you have forgotten how to calculate an odds ratio, refer back to the previous chapter.

APPENDICES

Appendix 1: The new style for nonparametric tests

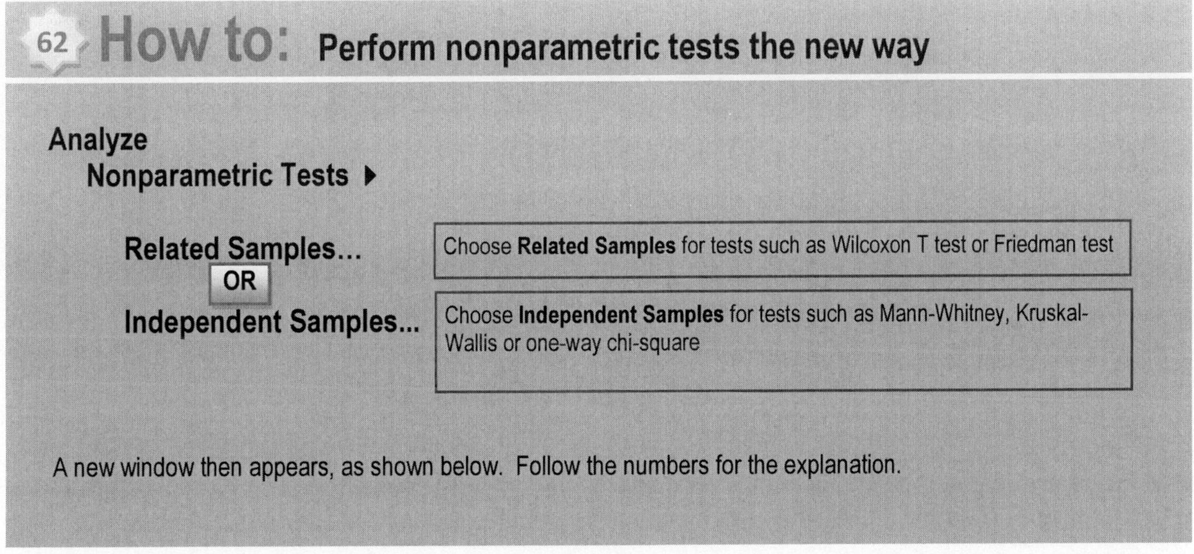

1. The window opens in the **Objective** tab. You click on the other two tabs (Fields and Settings) for further options.
2. The tests available are listed under **Description,** in this case for dependent samples.

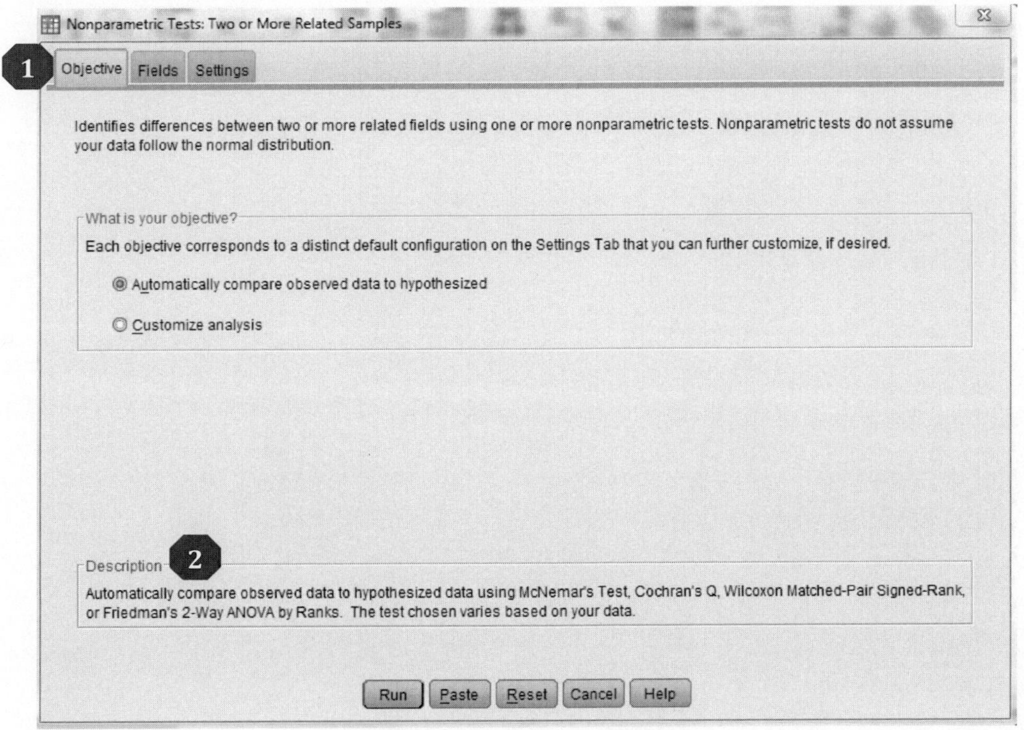

3. Clicking on the **Fields** tab opens a new window. **3A** shows the window for a dependent samples Wilcoxon T test (from Chapter 8), where you select the required variables from **Fields:**, then click on the arrow to move them to the **Test Fields:** box.

3B shows the window for an independent samples Kruskal-Wallis test (from Chapter 10).

Note that in both cases you might be asked to change the variable's **Measure** option, that is, to nominal, ordinal, or scale (continuous).

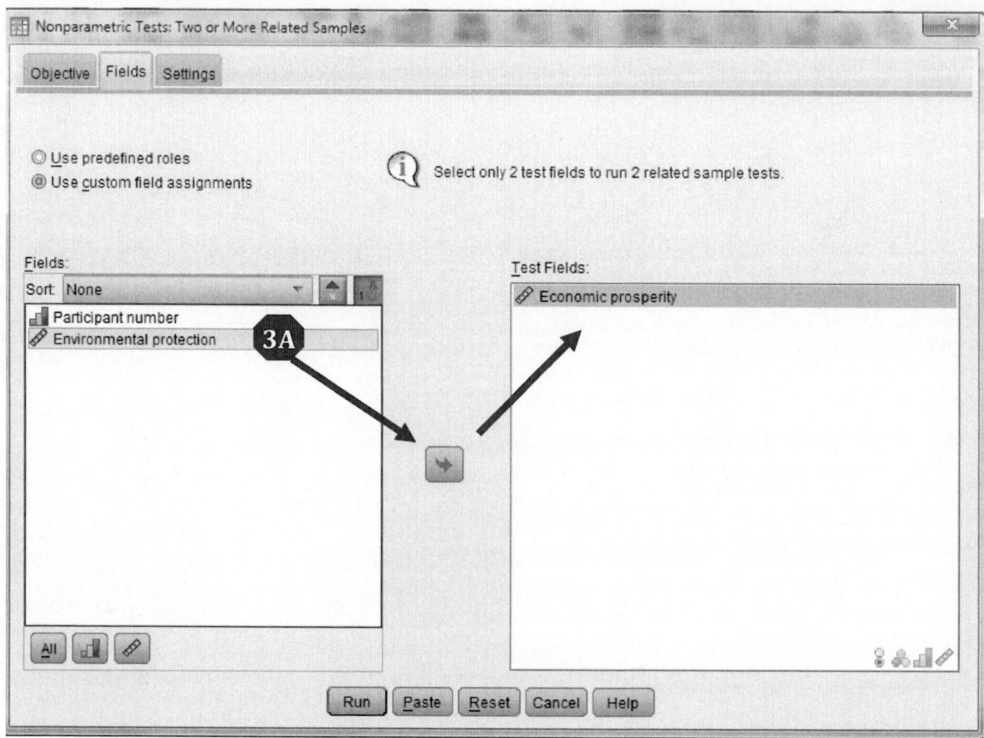

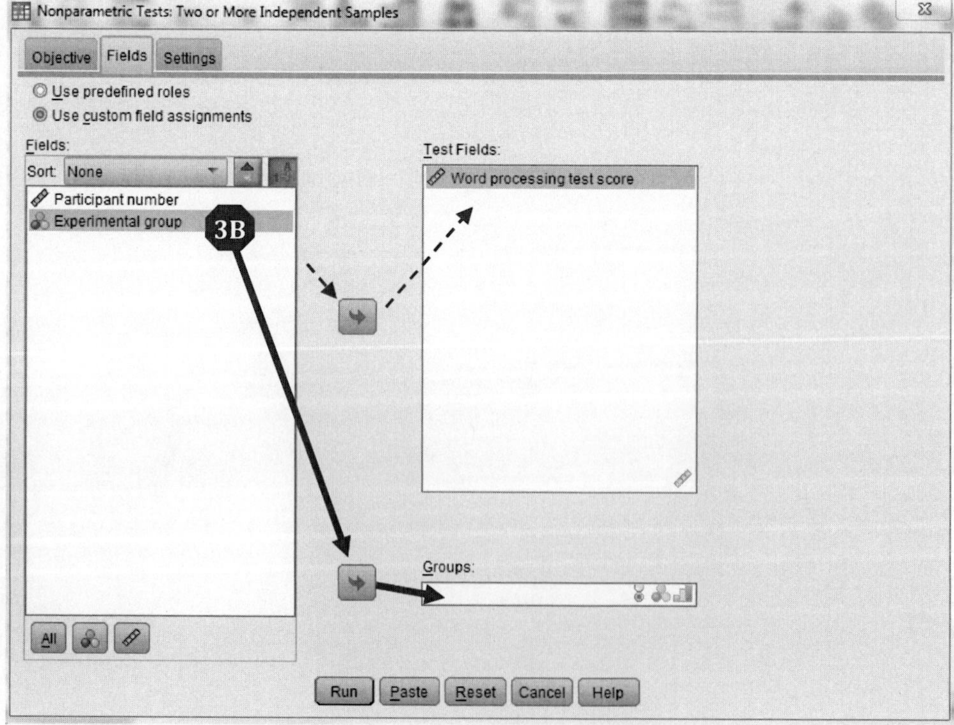

4. When you click on the **Settings** tab, **4A** shows the window for a Wilcoxon T test (from Chapter 8). You could accept the option: ⊙ **Automatically choose the test based on the data.**

The **Select an item**: column down the left side allows you to do things like change the significance level (under **Test Options**).

4B shows the window for an independent samples Kruskal-Wallis test (from Chapter 10). It also shows how to choose ⊙ **Customize tests**, then select Kruskal-Wallis with pairwise comparisons. You would choose Mann-Whitney if you had only 2 independent samples (as per Chapter 9).

5. You are then ready to click on **Run** to run the analysis and produce the output.

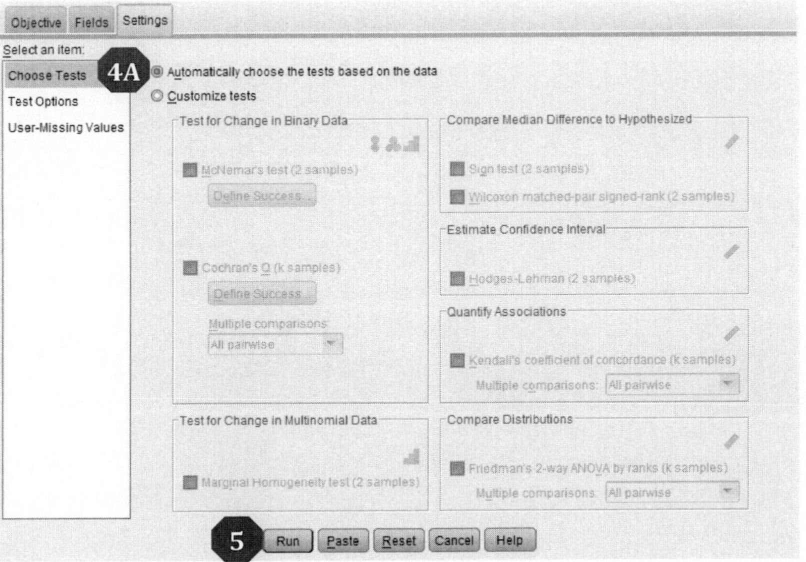

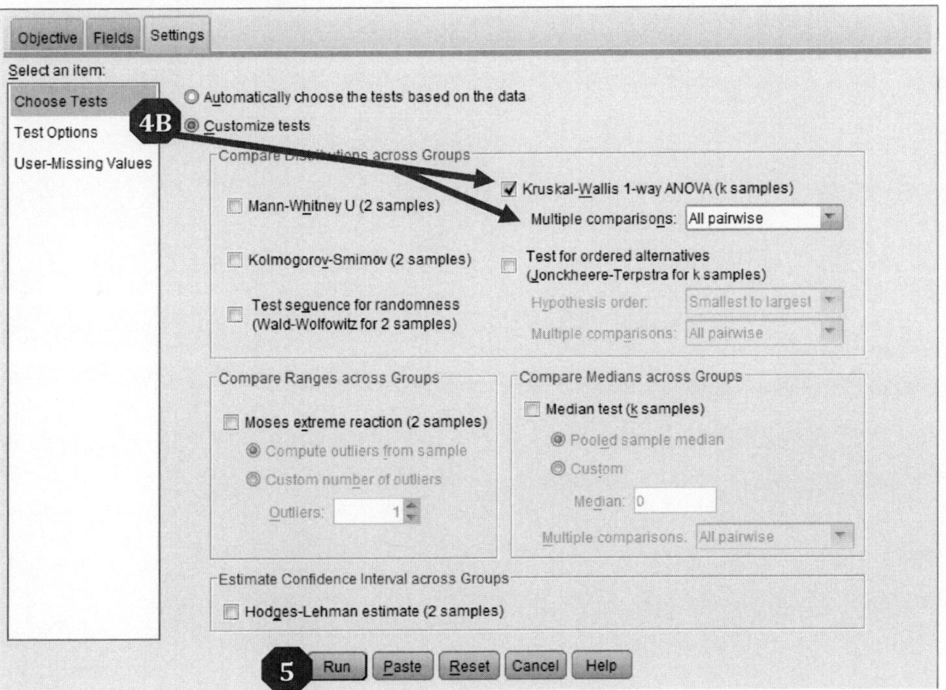

Output for the Wilcoxon T test

Hypothesis Test Summary

	Null Hypothesis	Test	Sig.	Decision
1	The median of differences between Economic prosperity and Environmental protection equals 0.	Related-Samples Wilcoxon Signed Ranks Test	.035	Reject the null hypothesis.

Asymptotic significances are displayed. The significance level is .05.

This is not very helpful you might think, but what you need to do is **double-click** on the **Hypothesis Test Summary** box, to open the new output window below. Compare this output to the old style of output shown in Chapter 8.

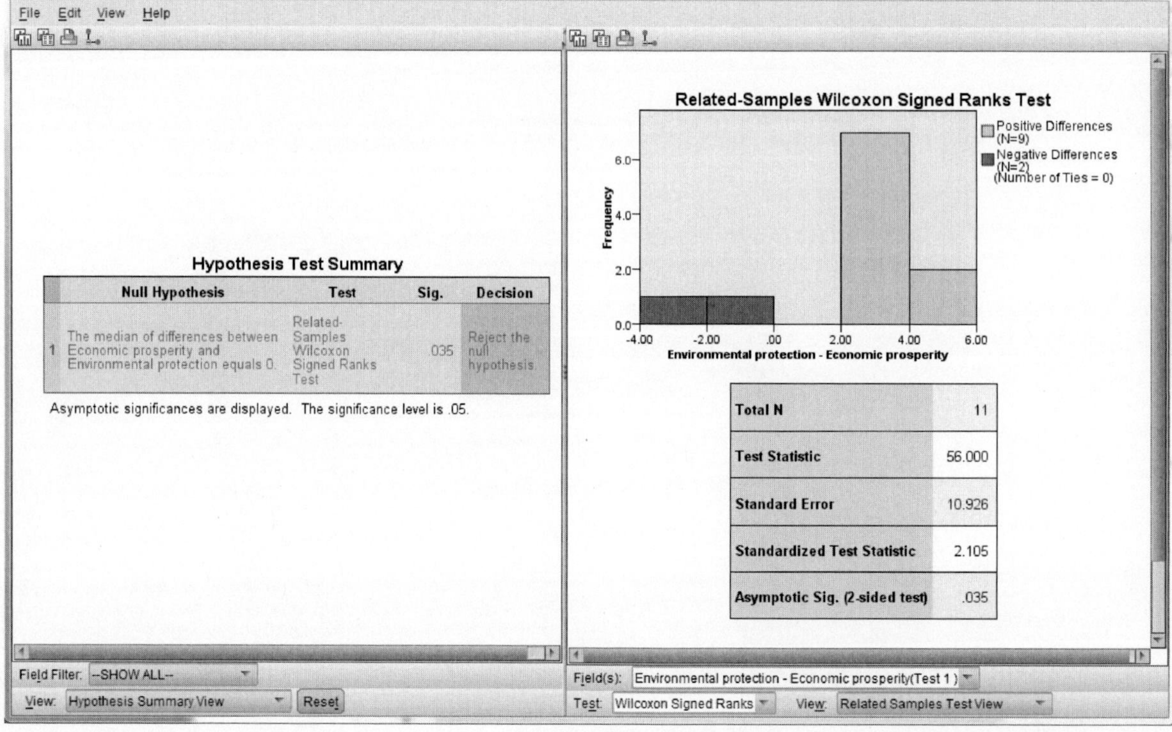

Output for Kruskal-Wallis test

Double click on the **Hypothesis Test Summary** box, as explained on the previous page, to obtain the detailed output window.

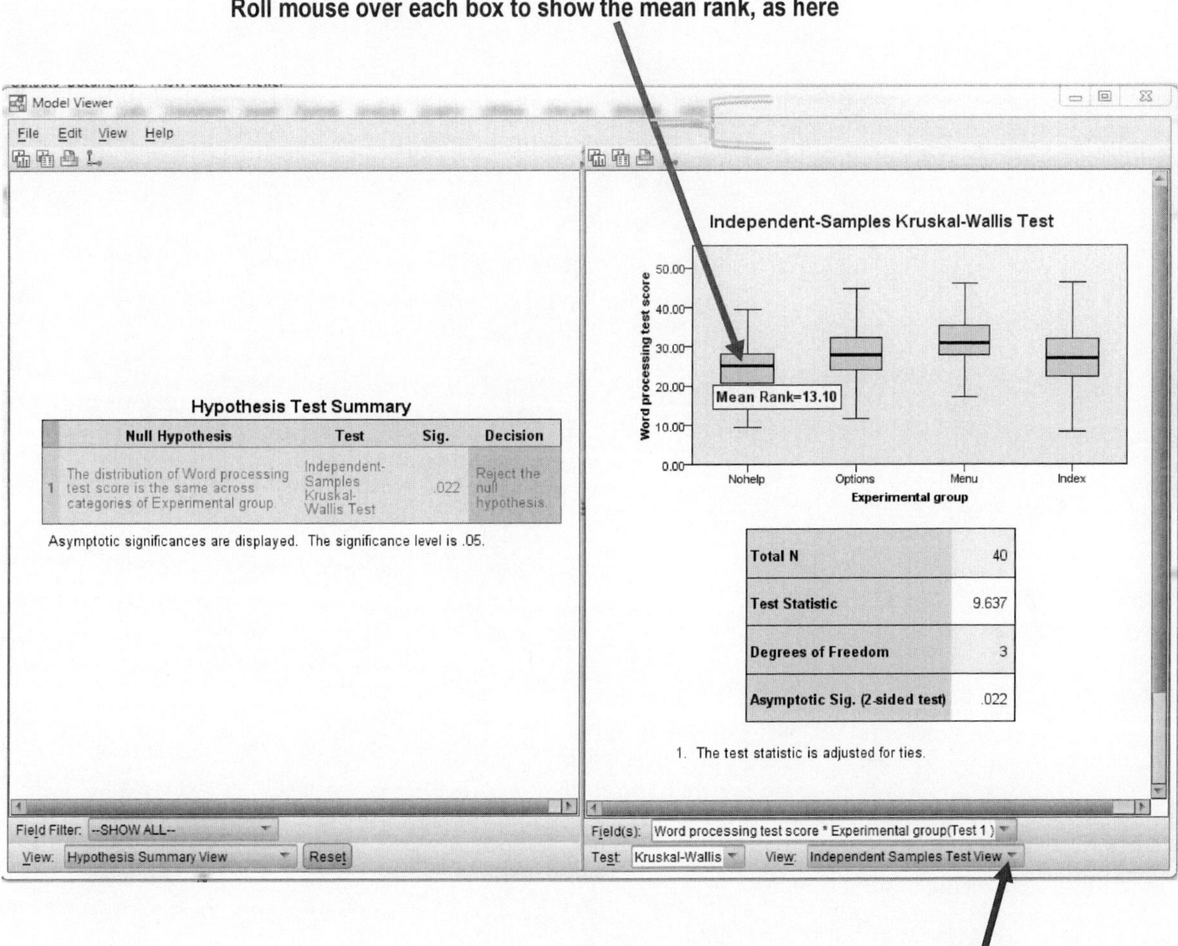

Click here to select the *Pairwise Comparisons* view

The Pairwise Comparisons view gives you output for post hoc comparisons for the Kruskal-Wallis test, as shown on the next page.

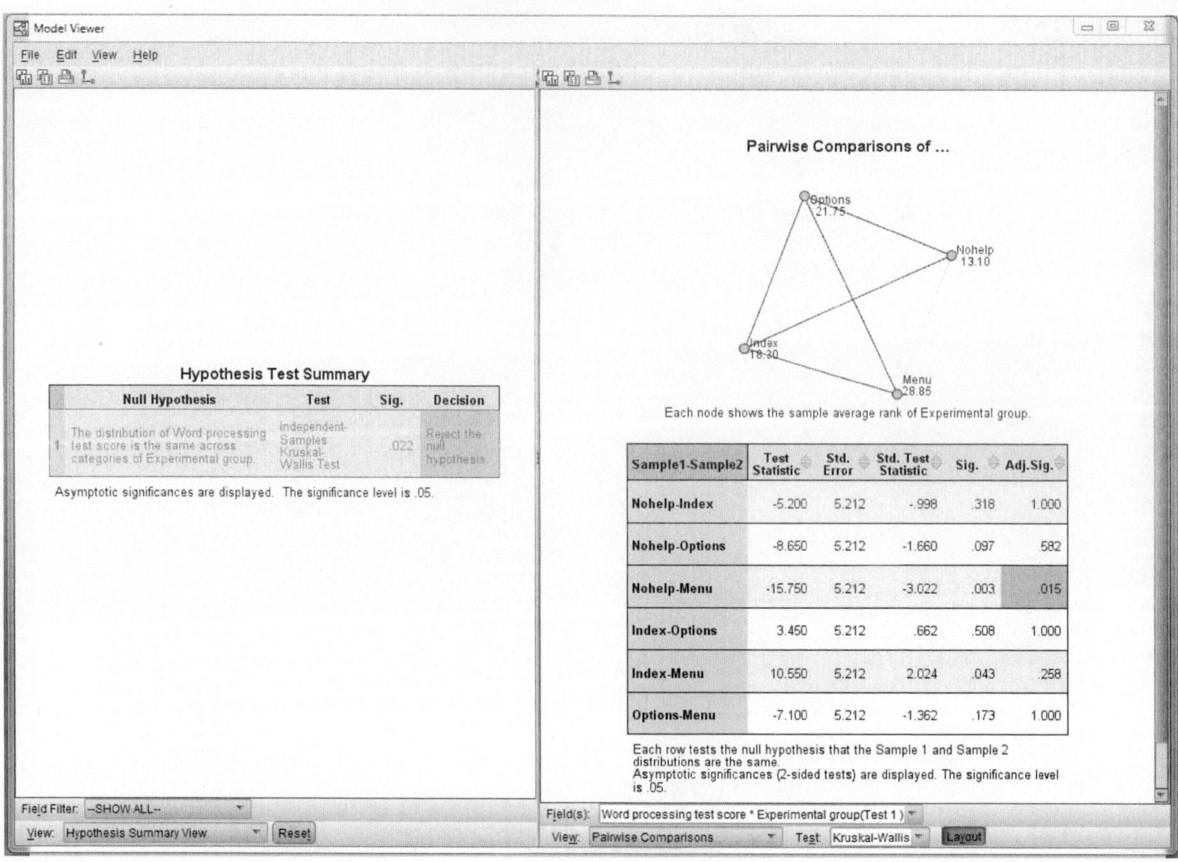

Hypothesis Test Summary

	Null Hypothesis	Test	Sig.	Decision
1	The distribution of Word processing test score is the same across categories of Experimental group.	Independent-Samples Kruskal-Wallis Test	.022	Reject the null hypothesis.

Asymptotic significances are displayed. The significance level is .05.

Sample1-Sample2	Test Statistic	Std. Error	Std. Test Statistic	Sig.	Adj.Sig.
Nohelp-Index	-5.200	5.212	-.998	.318	1.000
Nohelp-Options	-8.650	5.212	-1.660	.097	.582
Nohelp-Menu	-15.750	5.212	-3.022	.003	.015
Index-Options	3.450	5.212	.662	.508	1.000
Index-Menu	10.550	5.212	2.024	.043	.258
Options-Menu	-7.100	5.212	-1.362	.173	1.000

Each row tests the null hypothesis that the Sample 1 and Sample 2 distributions are the same.
Asymptotic significances (2-sided tests) are displayed. The significance level is .05.

Compare this output with that provided in Chapter 10. The post hoc comparisons are new. They automatically compute an adjusted significance level for maintaining familywise error at .05. This *adjusted* significance level is evaluated in comparison to the usual alpha level of .05, and significant differences are highlighted. What could be easier?

Appendix 2: New and old ways to create graphs

63 How to: Obtain a bar graph with error bars using Chart Builder

Chart Builder is the new way to create graphs in SPSS

Graphs
Chart Builder...
[A dialog box might now appear. Once the variables are correctly defined, click **OK** for the dialog box]

The Chart Builder window below then appears. It has already been set up for the required graph. Follow the numbers to see how this was done. Note that the **Legacy Dialogs** option in Graphs is much easier to use, but it is likely to be discontinued in future versions of SPSS

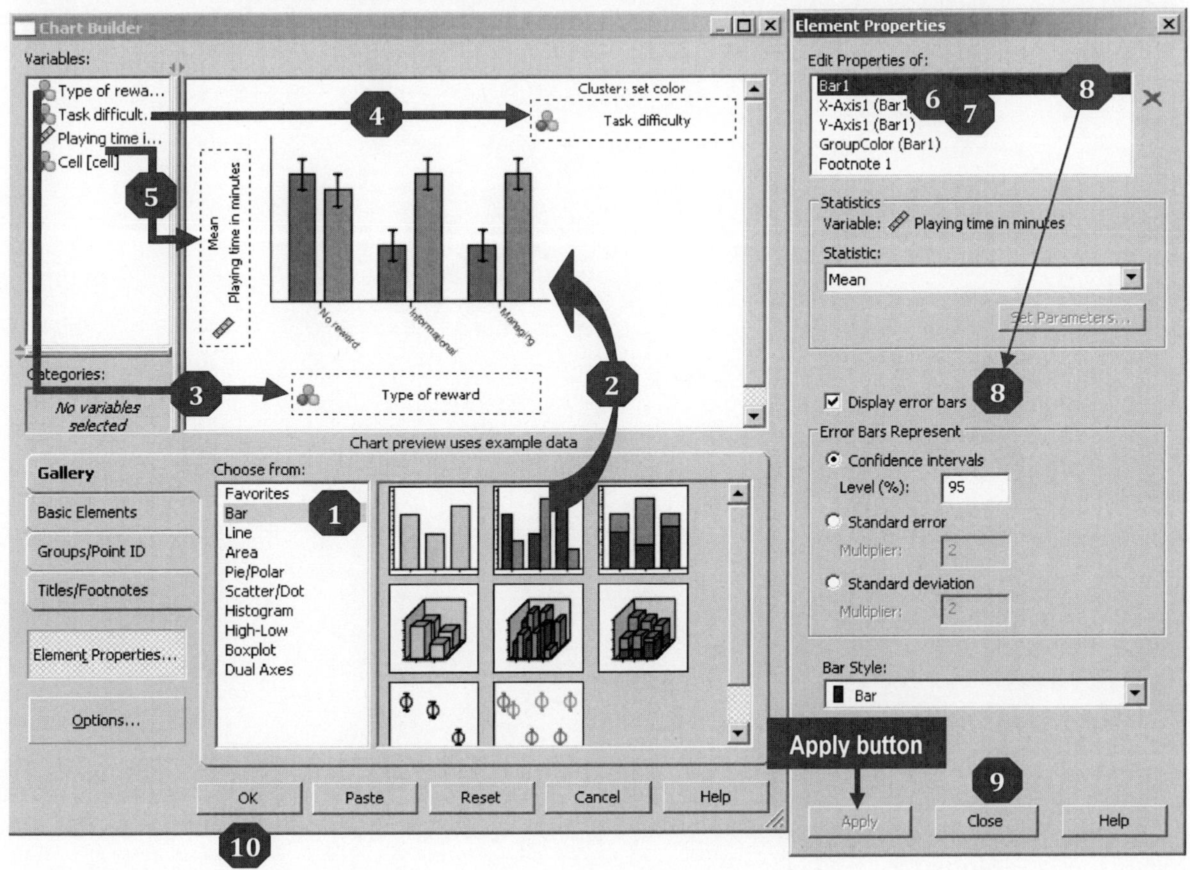

1. Click on **Bar**, if not already highlighted.
2. Click on the **clustered graph** and drag it into the **chart area** (the **Element Properties** window then appears to the right).
3. Click on **Type of reward** and drag it to the **X-axis?** box.
4. Click on **Task difficulty** and drag it to the **Cluster:Set color** box.
5. Click on **Playing time** and drag it into the **Count** box.

6. Move over to the **Element Properties Window**, and click on **X-Axis1 (Bar 1)**. You can then change the **Axis Label:** by typing **Type of Extrinsic Reward**. Then click the **Apply** button at the bottom of the window.
7. Click on **Y-Axis1 (Bar 1)**. You can then change the **Axis Label:** by typing **Mean Playing Time (Mins)**. Then click the **Apply** button at the bottom of the window.
8. Click on **Bar 1**, then click ☑ **Display error bars**, then select ⊙ **Confidence levels** 95. Click the **Apply** button at the bottom of the window.
9. Click the **Close** button.
10. Click the **OK** button. A basic graph then appears in the SPSS output window.

At each of the above steps there were more options available, but arguably, we are reaching a stage where SPSS complexity is becoming a source of sheer terror! There are all sorts of things that can be explored on a rainy day.

OLD WAY

64 How to: Obtain a repeated measures bar graph using Legacy Dialogs

Graphs
  Legacy Dialogs ▸
    Bar...
      [Click mouse on box for **Clustered**]
      [From **Data in Chart Are** select]
        ⊙ **Summaries of separate variables**
      Define
        [Select variable(s) you want]
        Juvenile pretest [Jpre] }
        Staff pretest [Spre] }
        Juvenile postttest [Jpost] }
        Staff postttest [Spost] }
        Juvenile follow-up [Jfollow] }
        Staff follow-up [Sfollow] } ▸ **Bars Represent:**

        Age group [Agegrp] ▸ **Category Axis**

        Options...
          ☑ **Display error bars**
          [From Error **Bars Represent** select]
            ⊙ **Confidence intervals**
              [For **Level (%):** type] **95.0** [assuming it is not already entered]
          Continue
      OK

References[1]

Anastasi, A. (1982). *Psychological testing* (5th ed.). New York: Macmillan. [Or any later edition.]

American Psychological Association. (2010). *Publication manual of the American Psychological Association* (6th ed.). Washington, DC: Author.

Baron, R. M., & Kenny, D. A. (1986). The moderator-mediator variable distinction in social psychological research: Conceptual, strategic, and statistical considerations. *Journal of Personality and Social Psychology, 51*, 1173-1182.

Cohen, J. (1988). *Statistical power analysis for the behavioral sciences* (2nd ed.). New York: Academic Press.

Cozby, P. C. (2007). *Methods in behavioural research* (9th ed.). New York, NY: McGraw-Hill.

de Vaus, D. (2002). *Surveys in social research* (5th ed.). Sydney: Allen & Unwin.

Field, A. (2005). *Discovering statistics using SPSS* (2nd ed.). London: Sage Publications.

Green, S. B., & Salkind, N. J. (2005). *Using SPSS for Windows and Macintosh: Analyzing and understanding data* (4th ed.). Upper Saddle River, NJ: Pearson Prentice Hall.

Hewson, K. (1999). *Relationships among negative life events, coping humour, positive and negative affect, and depression.* Unpublished Honours thesis. University of Western Sydney, Sydney, Australia.

Howell, D. C. (2002). *Statistical methods for psychology* (5th ed.). Pacific Grove, CA: Wadsworth Group.

Huberty, C. J., & Morris, J. D. (1989). Multivariate analysis versus multiple univariate analyses. *Psychological Bulletin*, 105, 302-308.

Keppel. G., & Wickens, T. D. (2004). *Design and analysis: A researcher's handbook* (4th ed.). Upper Saddle River, NJ: Pearson Education, Inc.

Loftus, G. R., & Masson, M. E. J. (1994). Using confidence intervals in within-subject designs. Psychonomic *Bulletin and Review, 1*, 476-490.

Martin, D.W. (2008). *Doing psychology experiments* (7th ed). Belmont, CA: Thomson/Wadsworth.

Pallant, J. (2007). *SPSS survival manual* (3rd ed.). Crows Nest, NSW: Allen &Unwin.

Payton, M. E., Greenstone, M. H., & Schenker, N. (2003). Overlapping confidence intervals or standard error intervals: What do they mean in terms of statistical significance? *Journal of Insect Science, 3*(34), 1-6. doi: 10.1672/1536-2442(2003)003[0001:OCIOSE]2.0.CO;2

Shaughnessy, J. J., Zechmeister, E. B., & Zechmeister, J. S. (2000). *Research methods in psychology* (5th ed.). Boston: McGraw-Hill.

Schenker, N., & Gentleman, J. F. (2001). On judging the significance of differences by examining the overlap between confidence intervals. *The American Statistician, 55*, 182-186

[1] Text books are updating all the time, so always check to see if a later edition is available.

Siegel, S., & Castellan, N. J. (1988). *Nonparametric statistics for the behavioral sciences* (2nd ed.). New York: McGraw-Hill.

Stevens, J. (2002). *Applied multivariate statistics for the social sciences* (4th ed.). Mahwah, NJ: Lawrence Erlbaum Associates.

Tabachnick, B. G., & Fidell, L. S. (2007). *Using multivariate statistics* (5th ed.). Boston: Pearson Education, Allyn & Bacon.

INDEX

The index to SPSS "How to" procedure boxes is provided under the Index entry: SPSS How to box for. The index to sample Results sections is provided under the Index entry: Results section examples

List of Main Statistical Symbols

Also see Table 4.5 (pp. 119-123) of the *APA Manual*.

α (Alpha) Probability level for determining statistical significance (also Type I error rate).

β (Beta) Type II error rate.

β (Beta) Standardised regression coefficient.

d Cohens' *d*, measure of effect size.

df Degrees of freedom.

η^2 Eta squared, measure of effect size (same as R^2).

η_p^2 Partial eta squared

F ANOVA test statistic.

f Frequency.

f_e Expected frequency.

h^2 Communality in factor analysis.

H_o Null hypothesis.

H_1 Alternative hypothesis.

H_a Alternative hypothesis.

H_o Null hypothesis.

κ Cohen's kappa, measure of inter-rater reliability.

Λ (Capital Lambda) Wilks' Lambda multivariate statistic.

MS Mean square (i.e., Variance).

μ (Mu) Population mean.

N	Total number of scores (participants) in a sample.
n	Number of scores in a subset of a sample.
ω^2	Omega squared, measure of effect size in ANOVA, variation on eta squared (η^2).
ϕ	Phi coefficient, measure of effect size in chi-square.
p	Probability (or significance).
r	Pearson product-moment parametric correlation coefficient for a sample (Squared Pearson correlation coefficient is r^2).
R	Multiple correlation coefficient (Squared multiple correlation coefficient is R^2).
ρ	(Rho) Population correlation parameter.
SS	**Sum of squares** (sum of squared deviations from mean).
s	Sample standard deviation symbol used in calculations, however when writing research reports use *SD* for the standard deviation.
s^2	Sample variance (standard deviation squared) symbol used in calculations.
Σ	(Capital Sigma) meaning "**the sum of**".
σ	(Sigma) Population standard deviation.
$s_{\overline{X}}$	Standard error of the mean.
t	*t* test symbol.
V	Pillai's Trace multivariate statistic.
X	Any score (used in calculations).
$\overline{X}$	"X-bar", the symbol used in **calculations** for the sample mean, however when writing research reports use *M* for the mean.
z	*z*-score or standard score (i.e., units of standard deviation).